Fundamentals of Biomechanics

Duane Knudson

Fundamentals of Biomechanics

Third Edition

 Springer

Duane Knudson
Department of Health and Human Performance
Texas State University
San Marcos, TX
USA

ISBN 978-3-030-51837-0   ISBN 978-3-030-51838-7   (eBook)
https://doi.org/10.1007/978-3-030-51838-7

This Springer imprint is published by the registered company Springer Nature Switzerland AG
The registered company address is: Gewerbestrasse 11, 6330 Cham, Switzerland

Preface

For Students:

Fundamentals of Biomechanics is written for Kinesiology/Exercise and Sports Science majors taking the required introductory biomechanics course. Welcome to the exciting world of biomechanics where anatomy comes alive using physics to document human motion and how that motion was created. As you read, you will discover that this is not just a physics book with lots of sport/exercise examples and problems, but a unique blend qualitative biomechanics concepts and nine simple principles for their application, all illustrated with real-world human movement examples and citations to research for further study. Whatever kinesiology profession you pursue, biomechanics will be your essential knowledge base for making recommendations to your clients on the most appropriate technique, equipment, and training exercises. Be sure to try out the Activities in each chapter and take advantage of any Lab Activities you may do with your instructor or fellow students. Doing the movement activities and physically experiencing biomechanical variables and principles will help you master this important professional knowledge. The review questions (odd questions have suggested answers in the back) will also help you learn and apply the fundamentals of biomechanics.

I hope that you master the fundamentals of biomechanics, integrate biomechanics into your professional practice, and are challenged to continuously update your biomechanical knowledge. Some of you will find advanced study and a career in biomechanics exciting opportunities to learn more about human movement and how to help people move more effectively and with less pain.

For Instructors:

Fundamentals of Biomechanics is written for students taking the introductory biomechanics course in Kinesiology/Exercise and Sports Science/HPERD departments. The third edition is designed for majors preparing for all kinds of human movement professions and therefore uses a variety of everyday exercise and sport movement examples to illustrate the application of biomechanics. The book targets the national guidelines for introductory biomechanics (Abraham et al. 2018) using narrative and active learning features that instructional research indicate improves mastery of mechanics and biomechanics concepts (Hake 1998; Freeman et al. 2014; Knudson and Wallace 2019).

The book is truly unique in its integration of biological and mechanical factors, as well as its qualitative approach to understanding biomechanics concepts and their application in human movement. There are many bioMECHANICS books that emphasize quantitative models and problem solving, essentially sport/human physics tomes. There are also several BIOmechanics books that are essentially functional anatomy sprinkled with a little mechanics. *Fundamentals of Biomechanics* has always forged the holistic path, integrating biological factors and the classical mechanics that are needed to measure and understand causes of human movement. This uniqueness of integrating biological and mechanical bases in analyzing and improving human movement has been expanded with more examples, color figures, and citations pointing students to important biomechanics discoveries.

While this approach to the application of biomechanics is critical, it is also essential that students be introduced to the scientific support or lack of support for these qualitative judgments. Throughout the text extensive citations are provided to support the principles developed and give students references for further study. This models the evidence-based practice we want from all kinesiology professionals. Algebraic level mathematics is used to teach mechanical concepts. The focus of the mathematical examples is to understand the mechanical variables and to highlight the relationship between various biomechanical variables, rather than to solve quantitative biomechanical word problems. It is obvious from research in physics instruction that merely solving numerous quantitative word problems does not increase the conceptual understanding of important mechanical laws (Elby 2001; Lawson and McDermott 1987; Kim and Pak 2002).

The book is also unique in the market because it presents a clear conceptual understanding of biomechanics with nine principles for the application of biomechanics. These nine principles form the **applied biomechanics tools** kinesiology professionals need. The application of these biomechanical principles is illustrated in qualitative diagnosis of a variety of human movements in several contexts for the kinesiology professional: teaching/coaching and allied health/sports medicine. The book gradually builds from anatomical knowledge familiar to students to new biomechanical principles and their application.

This book has several features that are designed to help students link personal experience to biomechanical concepts and that illustrate the application of biomechanics principles. First, nine general principles of biomechanics are proposed and developed throughout the text. Second, there are *activity boxes* that provide opportunities for students to see and feel the biomechanical variables discussed. Third, there are *practical application boxes* that highlight the applications of biomechanics in improving movement and in treating and preventing injury. Fourth, the *interdisciplinary issues boxes* show how biomechanics is integrated with other sport sciences in addressing human movement problems. Fifth, chapters have associated *lab activities* (located at the end of the book, after the index) that use simple movements and measurements to explore concepts and principles with active learning experiences. These lab activities do not require expensive lab equipment, large blocks of time, or dedicated lab space. Finally, real-life case studies of how the biomechanical principles can be qualitatively applied to improve human movement in a variety of professions are presented in Part III. No other text provides as many or as thorough guided examples of applying biomechanical principles in actual human movements in professional practice situations.

There are also features that make it easy for students to follow the material and study for examinations (*what and why introductions* & *review questions*). Graphs, photographs, and illustrations have been incorporated throughout. Aside from the visual appeal and improved communication of the updated color illustrations, these figures illustrate important points and relationships between biomechanical variables and performance. The book provides an extensive *glossary* of key terms and biomechanics research terminology so that students can read original biomechanical research. Each chapter provides a *summary,* extensive citations of important biomechanical research, and *suggested readings.*

Duane Knudson
San Marcos, TX, USA

Acknowledgments

The author would like to thank the numerous scholars and publishing professionals who have contributed so much to all editions this book. I am indebted to many colleagues who have shared their expertise with me, given permission to share their work, and contributed so much to students, biomechanics, and numerous professions.

To the ones I truly love—Lois, Josh (and Katie), and Mandy (Alex and Victor)—thanks for being such great people. Finally, I would like to thank God for knitting all of us so "fearfully and wonderfully made."

Contents

III   Applications of Biomechanics in Qualitative Diagnosis of Movement

Introduction

Kinesiology is the scholarly study of human movement and biomechanics is a core subdiscipline of kinesiology. Biomechanics in kinesiology involves the precise documentation of human movement and the study of the causes of that movement. The study of biomechanics is relevant to professional practice in many kinesiology professions. The physical educator, coach, or strength coach who is teaching movement technique, as well as the athletic trainer or physical therapist treating an injury use biomechanics to qualitatively diagnose movement technique. The chapters in Part I define biomechanics and kinesiology and illustrate that knowledge of gross anatomy alone is inadequate to understand the mechanics of human movement and implement professional qualitative movement diagnosis.

Introduction to Biomechanics of Human Movement

Key Questions
- ➤ **What do you need to know to begin?**
  Kinesiology majors should know the requirements of their degree and how they can access original research through their university library or professional organizations.
- ➤ **Why do you need to know this material?**
  Kinesiology professionals know biomechanical concepts and where to access this knowledge in peer-reviewed publications.
- ➤ **What is the key idea?**
  Biomechanics is an essential science of human movement technique and exercise used in kinesiology and many kinesiology professional occupations.

Most people are quite skilled in many everyday movements like standing, walking, catching, grasping, or climbing stairs. By the time children are two, they are skilled walkers with little instruction from parents aside from emotional encouragement. Unfortunately, modern living does not require enough movement to prevent several chronic diseases, disability, and premature death associated with low physical activity (USDHHS 1996, 2018) and many people suffer disability from aging, injury, or other factors.

Fortunately, many human movement professions help people to participate in beneficial physical activities. Physical educators, coaches, athletic trainers, strength and conditioning coaches, and occupational/physical therapists all help people reap the benefits of physical activity. These human movement professions rely on undergraduate training in kinesiology and require at least one course in biomechanics.

Kinesiology is the term referring to the whole scholarly area of human movement study, while **biomechanics** *is the study of motion and its causes in living things*. Biomechanics provides essential knowledge on the most effective and safest movement patterns, equipment, and relevant exercises to improve human movement. In a sense, kinesiology professionals solve human movement problems every day, and one of their most important tools is biomechanics. This chapter outlines the field of biomechanics, why biomechanics is such a foundational area to kinesiology and numerous kinesiology professions, and where biomechanics knowledge can be found.

1.1 What Is Biomechanics?

Biomechanics has been defined by one scholar as *the study of the movement of living things using the science of mechanics* (Hatze 1974). Mechanics is a branch of physics that is concerned with the description of motion and how forces create motion. Forces acting on living things can create motion, be a healthy stimulus for growth and development, but can also overload tissues, causing injury. Biomechanics provides conceptual and mathematical tools that are necessary for understanding how living things move and how kinesiology professionals might improve movement or make movement safer.

Most readers of this book will be majors in departments of Kinesiology, Exercise and Sport Science, Human Performance, or HPERD (Health, Physical Education, Recreation, and Dance). Kinesiology comes from two Greek verbs that translated literally means "the study of movement." Most American higher education programs now use "kinesiology" in the title of their department because this term has come to be known as the academic area for the study of human movement/physical activity (Corbin and Eckert 1990; Hoffman and Knudson 2018; American Kinesiology Association 2020). This change in terminology can be confusing because "kinesiology" is also the title of a foundational course on applied anatomy that was commonly required for a physical education degree in the first half of the twentieth century. This older meaning of kinesiology persists even today, because biomechanics has only recently (since

© Springer Nature Switzerland AG 2021
D. Knudson, *Fundamentals of Biomechanics*, https://doi.org/10.1007/978-3-030-51838-7_1

1

late1960) become a recognized term and specialization of scientific study (Atwater 1980; Wilkerson 1997).

This book will use the term kinesiology in the modern sense of the whole academic area of the study of human movement. Since kinesiology majors are pursuing careers focused on improving human movement, you and almost all kinesiology students world-wide are required to take at least one course on the biomechanics of human movement. It is a good thing that you are studying biomechanics. Once your friends and family know you are a kinesiology major, you will invariably be asked questions like: should I get one of those new rackets, why does my elbow hurt when I do this, or how can I stop my drive from slicing? Does it sometimes seem as if your friends and family have regressed to that preschool age when every other word out of their mouth is "why" when they are talking about movement or aches and pains? What is truly important about this kinesiology major common experience is that it is a metaphor for the life of a human movement professional. Professions require formal study of theoretical and specialized knowledge that allows for the reliable solution to problems. This is the traditional meaning of the word "professional," and it is different than the words common use today. Today people refer to pro-

fessional athletes or painters because people earn a living with these jobs, but I believe that kinesiology careers should strive to be more like true professions such as medicine or law. True professions are based on formal, specialized, and often scientific knowledge, along with legal requirements for licensure and ethical service.

People need help in improving human movement, and this help requires knowledge of "why" and "how" the human body moves. Since biomechanics gives the kinesiology professional much of the knowledge and many of the skills necessary to answer these "what works?" and "why?" questions, biomechanics is an essential science for solving human movement technique and exercise problems. However, biomechanics is but one of many sport and human movement science tools in a kinesiology professional's toolbox. This text is also based on the philosophy that your biomechanical knowledge should be combined with tools from other kinesiology sciences to most effectively deal with human movement problems. ◘ Figure 1.1 illustrates some of the scientific subdisciplines in kinesiology. These typically are the core sciences all kinesiology majors take in their undergraduate preparations. This overview should not be interpreted to diminish the other academic sub-

◘ **Fig. 1.1** Seven of the major academic subdisciplines of kinesiology. The darker shaded center illustrates the integration of all the subdisciplines in an interdisciplinary approach to solving human movement problems in kinesiology. Using all subdisciplinary knowledge together avoids conflicting recommendations and creates holistic knowledge that matches the holistic nature of our clients

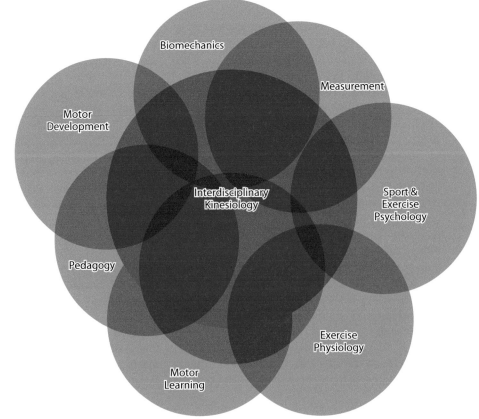

disciplines common in kinesiology departments like sport history, sport philosophy, dance, public health, analytics/measurement, and sport management, just to name a few.

The important point is that knowledge from all the subdisciplines must be *integrated* in professional practice since problems in human movement are multifaceted, with many interrelated factors. For the most part, the human movement problems you face as a kinesiology professional will be like those "trick" questions professors ask on exams: they are complicated by many factors (individual, task, and environment) and tend to defy simple, dualistic (black/white) answers. While the application examples discussed in this text will emphasize biomechanical principles, readers should bear in mind that this biomechanical knowledge should be *integrated* with professional experience and knowledge from the other subdisciplines of kinesiology. It is this **interdisciplinary** approach (◘ Fig. 1.1) that is essential to finding the best interventions to help people more effectively and safely. Dotson (1980) suggests that true kinesiology professionals can integrate the many factors that interact to affect movement, while the layman typically looks at things one factor at time. Unfortunately, this interdisciplinary approach to kinesiology instruction in higher education has been elusive (Harris 1993; Ives and Knudson 2007; Knudson 2019). Let's look at some examples of human movement problems where it is particularly important to integrate biomechanical knowledge into professional practice.

1.2 · Why Study Biomechanics?

Scientists from many different areas (e.g., kinesiology, engineering, physics, biology, zoology) are interested in biomechanics of plants and animals. Why are scholars from so many different academic backgrounds interested in animal movement? Biomechanics is interesting because many people marvel at the ability and beauty in animal movement. Some scholars have purely theoretical or academic interests in discovering the laws and principles that govern animal movement. The kind of research that answers the how/why questions of the ultimate nature and mechanisms of movement is basic biomechanical science. Within kinesiology, many biomechanists have been interested in the *application* of biomechanics to sport and exercise. Applied biomechanical science strives to answer specific, real-world movement problems. The applications of biomechanics to human movement have traditionally classified into two main areas: the improvement of performance and the reduction or treatment of injury (◘ Fig. 1.2).

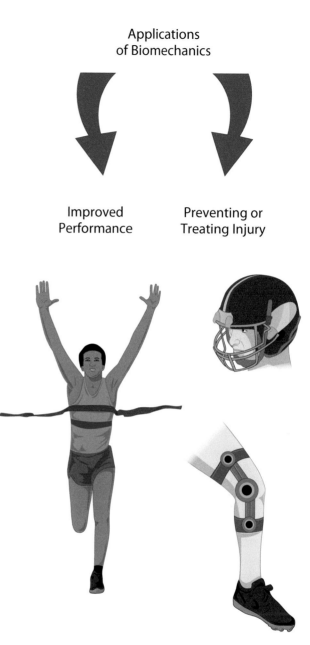

◘ Fig. 1.2 The two major applications of biomechanics are to improve human movement and the treatment or prevention of injury. Reducing injury with biomechanical knowledge can be based on exercise selection (not pictured) or improvements in protective equipment (braces and helmets)

Biomechanical research and knowledge actually extend beyond these two big areas of application. Other examples of application of biomechanics are the improvement of tools/equipment used by humans (shoes, rackets, orthotics, prosthetics, etc.), sport rules, and movement/sport instruction. Note these many applications of biomechanics knowledge can really be classified into one of the two main areas of performance improvement or prevention/treatment of injury/disability.

1

1.2.1 Improving Performance

Human movement performance can be enhanced many ways. Effective movement involves anatomical factors, neuromuscular skills, physiological capacities, and psychological/cognitive abilities. Most kinesiology professionals prescribe technique changes, exercises, and give instructions that allow a person to improve performance. Biomechanics is most useful in improving performance in sports or activities where technique is the dominant factor rather than physical structure or physiological capacity. Technique-centric sports like golf, gymnastics, or tennis would have greater performance benefits from biomechanics knowledge than a sport like marathon running. While an improved shoe might make small improvement in running (�‌ Fig. 1.2), it is likely that a customized racket or golf club might even be more effective these sports. Since biomechanics is essentially the science of movement technique, biomechanics is the main contributor to one of the most important skills of kinesiology professionals: the qualitative diagnosis of human movement (Knudson 2013). Teachers, coaches, and therapists all make qualitative judgments about the movement technique used by their clients. A **qualitative** evaluation would be classifying a joint range of motion as inadequate, normal, or excessive. The limited time some professionals have with individual clients sometimes means a qualitative assessment provides enough accuracy for treatment. **Quantitative** evaluation involves the measurement of a particular variable of interest and the subsequent calculations using those numerical values. A quantitative assessment of range of motion would be the measurement of joint angle with a goniometer or from video images with a low-cost app (e.g., Kinovea.org). ▸ Chapter 3 will introduce you to qualitative diagnosis of human movement, and ▸ Chaps. 9 and 10 illustrate its application in several kinesiology professions.

Imagine a coach is working with a gymnast who is having problems with her back handspring (◌ Fig. 1.3). The coach observes several attempts and judges that the angle of takeoff from the round off and body arch are performed poorly. The coach's experience tells him that this athlete is strong enough to perform this skill, but they must decide if the gymnast should concentrate on her takeoff angle or more back hyperextension in the block. The coach uses his knowledge of biomechanics to help in the qualitative diagnosis of this situation. Since the coach knows that a better arch affects the force the gymnast creates against the mat and affects the angle of takeoff of the gymnast, he decides to help the gymnast work on her "arch" following the round off. The qualitative diagnosis of movement goes beyond mere analysis/evaluation to evidence-based prescription.

Biomechanics research on sports techniques sometimes tends to lag behind the changes that are spontaneously occurring in athletes, sport instruction, or equipment design. Athletes and coaches experiment with new techniques all the time. Students of biomechanics may be surprised to find that there are often lim-

◌ **Fig. 1.3** Biomechanics principles must be integrated with other kinesiology sciences to solve human movement problems, like in the qualitative diagnosis of technique of a round off and back handspring ited biomechanical studies on many techniques, equipment, and instruction issues in many popular sports. The vast number of techniques, their variations, and their high rates of change and innovation tend to outdistance biomechanics research resources. Sport biomechanics research also lags behind the coaches and athletes because scientific research takes considerable time to conduct and report, and there is a lack of funding for this important research. There is less funding for biomechanical studies aimed at improving performance compared to studies focused on preventing and treating injuries that have obvious costs of pain and disability. Students looking for biomechanical research on improving sports technique often will have fewer sources than students researching the biomechanics of injury mechanism, prevention, or treatment. Ironically, research that would keep people active through safe, lifetime sports would logically improve productivity and reduce premature mortality but is rarely funded by the government, industry, or foundations.

While technique is always relevant in human movement, in some activities the psychological, anatomical, or physiological factors are more strongly related to success. Distance running is a good example of this kind of movement. There is a considerable amount of research on the biomechanics of running so coaches can fine tune a runner's technique to match the profile of elite runners (Cavanagh et al. 1985; Buckalew et al. 1985; Williams et al. 1987). While these technique adjustments make small improvements in performance, most of running performance is related to physiological abilities and their training. Studies that provide technique changes in running based on biomechanical measurements have found minimal effects on running economy (Cavanagh 1990; Lake and Cavanagh 1996; Messier and Cirillo 1989); however changing forces from external exoskeletons (Selinger et al. 2019) or shoes/surfaces with different properties will change muscle activation and function (Willwacher et al. 2020). This suggests that track coaches can use biomechanics to refine running technique, but they should only expect small changes in performance from these modifications.

Human performance can also be enhanced by improvements in the design of equipment. Many of these improvements are related to new materials and engineering designs. When these changes are integrated with information about the human performer, we can say the improvements in equipment were based on biomechanics. Engineers interested in sports equipment often belong to the International Sports Engineering Association and publish research in ISEA proceedings (Subic and Haake 2000) or the *Sports Engineering* journal. Research on all kinds of equipment is conducted in biomechanics labs at most major sporting goods manufacturers. Unfortunately, much of the results of these studies are closely guarded trade secrets, so it is difficult for the layperson to determine if marketing claims for "improvements" in equipment design are real engineering and biomechanical innovations or just creative marketing exaggeration.

There are numerous examples of how applying biomechanics in changing equipment designs has improved sports performance. When improved javelin designs in the early 1980s resulted in longer throws that endangered other athletes and spectators, redesigns in the weight distribution of the "new rules" javelin again shortened throws to safer distances (Hubbard and Alaways 1987). Biomechanics researchers (Ward and Groppel 1980; Elliott 1981) were some of the first to call for smaller tennis rackets that more closely matched the muscular strength of young players (◘ Fig. 1.4). Use of age appropriate equipment is important since large/adult equipment may interfere with the development of skilled coordination (Buszard et al. 2020). This is one example where the researchers were out ahead of the

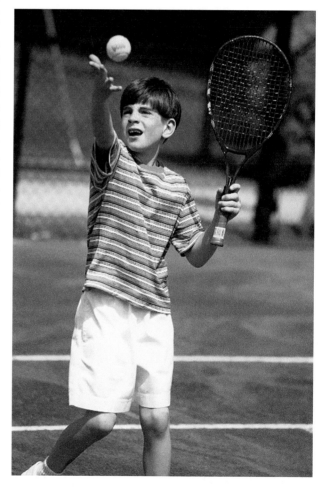

◘ **Fig. 1.4** The design of sports equipment must be appropriate for an athlete, so rackets for children are shorter and lighter than adult rackets. (Photo used with permission from Getty Images)

manufacturers and coaches recommending specialized equipment for children. ▶ Chapter 8 will discuss how changes in sports equipment are used to change fluid forces and improve performance.

While breaking world records using new equipment is exciting, not all changes in equipment are welcomed with open arms by sport governing bodies. Some equipment changes are so drastic they change the very nature of the game and are quickly outlawed by the rules committee of the sport. One biomechanist developed a way to measure the stiffness of basketball goals, hoping to improve the consistency of their response but found considerable resistance from basketball community who liked their unique home court advantages. Another biomechanist developed a new speed skate that increased the time and range of motion of each push off the ice, dramatically improving times and breaking world records (de Koning et al. 2000). They are called klap skates because the pivoting blade made a sound springing back after the longer, more powerful push offs athletes made with this skate. This gave quite an advantage to the country where these skates were developed, and there was controversy over the amount of time other skaters were able to practice with the new skates before competition. These dramatic equipment improvements in many sports have some people worried that winning Olympic medals may be more in the hands of the engineers than the athletes and coaches (Bjerklie 1993).

Another way biomechanics research improves performance is advances in exercise and conditioning programs. Biomechanical studies of exercise movements and training devices serve to determine the most effective training to improve performance (■ Fig. 1.5).

■ **Fig. 1.5** A computerized testing and exercise dynamometer by Biodex. The speed, muscle actions (isometric, concentric, eccentric), and pattern of loading (isokinetic, isotonic) can be selected. These isokinetic dynamometers provide the gold standard muscular strength measurement for most all major muscle groups. (Image courtesy of Biodex Medical Systems)

Biomechanical research on exercises is often compared to research on the sport or activity that is the focus of training. Strength and conditioning professionals can better apply the principle of specificity when biomechanical research is used in the development of exercise programs. Computer-controlled exercise and testing machines are another example of how biomechanics contributes to strength and conditioning (Ariel 1983). In the next section, the application of biomechanics in the medical areas of **orthotics** and **prosthetics** will be mentioned in relation to preventing injury, but many prosthetics are now being designed to improve the performance of disabled athletes.

1.2.2 Preventing and Treating Injury/ Disability

Movement safety, or injury prevention/ treatment, is another primary area where biomechanical knowledge can be applied. Sports medicine professionals have traditionally studied retrospective injury data to try to determine the potential causes of disease or injury (epidemiology). Biomechanical research is a powerful ally in the sports medicine quest to prevent and treat injury. Biomechanical studies help prevent injuries by providing information on the mechanical properties of tissues, mechanical loadings during movement, and preventative or rehabilitative therapies. Biomechanical studies provide important data to confirm potential injury mechanisms hypothesized by sports medicine physicians and epidemiological studies. The increased participation of girls and women in sports has made it clear that females are at a higher risk for anterior cruciate ligament (ACL) injuries than males due to several biomechanical factors (Boden et al. 2000). Biomechanical and sports medicine studies have begun unravel the mystery of this high risk and have developed effective prevention strategies.

Ergonomic/human factors engineers use biomechanics to design work tasks and assistive equipment to prevent overuse injuries related to specific jobs. When workers are injured occupational therapists provide rehabilitation exercise and training on assistive equipment. Combining biomechanics with other sport sciences has aided in the design of shoes for specific sports (Segesser and Pforringer 1989), especially running shoes (Frederick 1986; Nigg 1986). Since the 1980s, the design and engineering of most sports shoes has included research in company biomechanics labs. The biomechanical study of auto accidents has resulted in measures of the severity of head injuries, which has been applied in biomechanical testing, and in design of many kinds of helmets to prevent head injury in a variety of sports/activities (Calvano and

Berger 1979; Norman 1983; Torg 1992). When accidents result in amputation, **prosthetics** or artificial limbs can be designed to match the mechanical properties of the missing limb (Klute et al. 2001). Preventing acute injuries is also another area of biomechanics research. Forensic biomechanics involves reconstructing the likely causes of injury from accident measurements and witness testimony.

Biomechanics helps the physical therapist prescribe rehabilitative exercises, assistive devices, or orthotics. **Orthotics** are support objects/braces that correct deformities or joint positioning, while assistive devices are large tools to help patient function like canes or walkers. Recent work in external braces and exoskeletons by bioengineers may provide more than just support but through feedback from the person, additional force to assist with greater external loads (lifting, etc.) or help accommodate disability from a stroke or other injury (Shi et al. 2019; Yang et al. 2019). Qualitative diagnosis of gait (walking) also helps the therapist decide whether sufficient muscular strength and control have been regained in order to permit safe or cosmetically normal walking (◘ Fig. 1.6). An athletic trainer might observe the running and landing technique of an athlete for signs of pain and/or limited range of motion in undergoing long-term conditioning for future return to the field. An athletic or strength coach might use a similar qualitative diagnosis of warm-up, exercise, or sport activities of the same athlete several weeks later to judge their readiness for practice or competition. Many biomechanists work in hospitals providing quantitative assessments of gait function to document the effectiveness of surgery and therapy. The North American group interested in these quantitative assessments for medical purposes is the Gait and Clinical Movement Analysis Society (GCMAS). Good general sources for the clinical and biomechanical aspects of gait are Perry (1992), Whittle (1999), and Kirtley (2006).

Dramatic increases in computing memory and power have opened up new areas of application for biomecha-

nists. Many of these areas are related to treating and preventing human injury. Biomechanical studies are able to evaluate strategies for preventing falls and fractures in the elderly (Robinovitch et al. 2000). Biomechanical computer models can be used to simulate the effect of various orthopedic surgeries (Delp et al. 1990) or to educate with computer animation. Some biomechanists have developed software used to adapt human movement kinematic data so that computer game animations have the look of truly human movement, but with the superhuman speed that makes games exciting (◘ Fig. 1.7). Some people use biomechanics to perform forensic examinations of accidents. This reconstruction of events from physical measurements at the scene is combined with medical and other

◘ **Fig. 1.6** Qualitative diagnosis of gait (walking) is of importance in physical therapy and the treatment of many musculoskeletal conditions

◘ **Fig. 1.7** Biomechanics typically measure three-dimensional motion using reflective markers, synchronized infrared cameras, and specialized software to track the markers **a**. Wearable inertial measurement unit (IMU) sensors **b** can be also used to measure human motion. Modeling and calculation software are then used to obtain the final kinematic and kinetic data. Images permission of Vicon Motion Systems

1

evidence to determine the likely cause or factors influencing many kinds of accidents. This knowledge is valuable to manufacturers and parties in lawsuits.

Application

A variety of professions are interested in using biomechanics to modify human movement. A person that fabricates prosthetics (artificial limbs) would use biomechanics to understand the normal functioning of joints, the loadings the prosthetic must withstand, and how the prosthetic can be safely attached to the person. List possible questions biomechanics could answer for a(n):

➤ Athletic coach?
➤ Orthopedic surgeon?
➤ Physical educator?
➤ Physical therapist?
➤ Athletic trainer?
➤ Strength and conditioning coach? Occupational
➤ Fitness consultant?

You? What question about human movement technique are you curious or passionate about?

1.2.3 Qualitative and Quantitative Analysis

Biomechanics provides knowledge for a variety of kinesiology professions to analyze, evaluate, and diagnose human movement technique to improve effectiveness or decrease the risk of injury. How the movement is analyzed falls on a continuum between a qualitative analysis and a quantitative analysis. **Quantitative analysis** involves the *measurement of biomechanical variables* and usually requires a computer to do the voluminous numerical calculations necessary often with a biomechanical model. Even short movements will have thousands of samples of data to be collected over time, scaled, and numerically processed. In contrast, **qualitative analysis** is a subjective evaluation/ judgment about movement quality in achieving a goal. Analysis in both quantitative and qualitative contexts means identification of the factors that affect human movement performance, which is then interpreted using other higher levels of thinking (synthesis, evaluation) in applying the information to the movement of interest. Solving problems in human movement involves high levels of critical thinking and an interdisciplinary approach, integrating the many subdisciplines of kinesiology sciences.

The advantages of numerical measurements of quantitative approaches over those of qualitative analysis are greater accuracy, consistency, and precision. Most quantitative biomechanical analysis is performed in research settings; however, more and more devices are commercially available that inexpensively measure some biomechanical variables (e.g., wearable sensors, radar, timing lights, timing mats, quantitative videography systems). Unfortunately, the greater accuracy of quantitative measures comes at the cost of technical skills, calibration, computational and processing time. There are even dangers of increasing errors with the combination of variables and computations involved. Even with very fast modern computers, quantitative biomechanics is a labor-intensive task requiring considerable graduate training and experience. For these reasons and others, qualitative analysis of human movement remains the main approach kinesiology professionals use in solving many day-to-day human movement problems. Qualitative analysis will be the main focus of the applications of biomechanics presented in this book. ▶ Chapter 3 will introduce a specific model of qualitative diagnosis for the application of biomechanics in most kinesiology professions. Whether your future jobs use qualitative or quantitative biomechanical analysis, you will need to be able to access biomechanical knowledge. The next section will show you many sources of biomechanical knowledge, which will require the kinesiology professional to also use judgment to weight this evidence.

Activity: Video Replay

Record a sporting event from a television broadcast or using your own video camera. Find a sequence in the video where there is a movement of interest to you and where there is a good close-up shot of the action. Watch the replay at real-time speed and try to estimate the percentage of time taken up by the major temporal phases of the movement. Most skills can be broken down into three phases—preparation, action, and follow-through—but you can have as many phases as you think apply to the movement of interest. Replay the recording and use any "pause," "frame advance," or "slow motion" functions to count the number of video images in the skill and calculate the times and percentages for each phase of the skill. Videos can be captures at a variety of sampling rates (pictures per second). What could you video to check the frame rate of your camera and replay system in frame advance mode? How close was your qualitative judgment to the more accurate quantitative measure of time?

Application

Even though qualitative and quantitative analyses are not mutually exclusive, assume that qualitative-versus-quantitative biomechanical analysis is an either/or proposition in the following exercise. For the sports medicine and athletics career areas, discuss with other students what kind of analysis is most appropriate for the questions listed. Come to a consensus and be prepared to give your reasons (cost, time, accuracy, need, etc.) for believing that one approach might be better than another.

Sport Medicine

1. Is the patient doing the lunge exercise correctly?
2. Is athlete "A" ready to play following rehab for their injured ACL?

Athletics

1. Should pole vaulter "B" change to a longer pole?
2. Is athlete "A" ready to play following rehab for their injured ACL?

Work

1. Does office worker "A" need a different chair and desk to relieve their back pain?
2. What height does this conveyor belt need to be for workers to off-load and package product?

1.3 Where Can I Find Biomechanical Knowledge?

This text provides a general introduction to the biomechanics of human movement in kinesiology. Many students take advanced courses in biomechanics and do library research for term projects. This text will provide quite a few references on many topics that will help students find original sources of biomechanical data. The relative youth of the science of biomechanics and the many different academic areas interested in biomechanics (among others, biology, engineering, medicine, kinesiology, physics) makes the search for biomechanical knowledge challenging for many students. This section will give you a brief tour of some of the major areas where biomechanics research is of interest.

Where you find biomechanics research depends on the kind of data you are interested in. Many people are curious about human movement, but there are also many scholars who are interested in the biomechanics of a wide variety of animals. An excellent way to study the theoretical aspects of biomechanics is to study animals that have made adaptations to be good at certain kinds of movements: like fish, kangaroos, or frogs. Much of this biomechanical research on animals is relevant to the study of human movement.

Professionals from many fields are interested in human movement, so there is considerable interest and research in human biomechanics. As a science biomechanics is young and is more like the middle child within the subdisciplines of kinesiology. Biomechanics is not as mature as exercise physiology or motor learning but is a bit older than other subdisciplines like sport psychology or sport analytics. Biomechanics research on many sport techniques and basic human movements (walking, running, etc.) will have been conducted in the early to mid-twentieth century even though the term "biomechanics" was not in use. Biomechanics research in kinesiology since the 1970s has tended to become more narrowly focused and specialized and has branched into areas far beyond sport and education. As a result, students with basic sport technique interests now have to integrate biomechanics research over a 100-year period and a variety of scientific journals.

Depending on the depth of analysis and the human movement of interest, a student of biomechanics may find themselves reading original research reports in biomechanical, medical, physiological, engineering, or other specialized journals. The smaller and more narrow the area of biomechanical interest (e.g., specific fibers, myofibrils, ligaments, tendons), the more likely there will be recent research on the topic. Research on the effect of computerized retail check-out scanners would likely be found in recent journals related to engineering, human factors, and ergonomics. A student interested in a strength and conditioning career might find biomechanical studies on exercises in medical, physical education, physiology, and specialized strength and conditioning journals. Students with clinical career interests who want to know exactly what muscles do during movement may put together data from studies dealing with a variety of animals. Clues can come from classic research on the muscles of the frog (Hill 1970), the cat (Gregor and Abelew 1994) and many animals (Roberts et al. 1997; Dickinson et al. 2000), as well as human muscle (Ito et al. 1998; Finni et al. 2000; Finni 2006). While muscle force-measuring devices have been implanted in humans, the majority of the invasive research to determine the actions of muscles in movement is done on animals (◘ Fig. 1.8). Noninvasive imaging techniques have begun to expand our understanding of movement of soft tissues like muscles, tendons, and ligaments in both animals and humans. Biomechanics research is published in a wide variety of journals. A few select examples are listed in ◘ Table 1.1.

1

a

Tendon

b

1 mm   7-5 mm

3 mm

☐ **Fig. 1.8** Schematic of a buckle transducer for in vivo measurement of muscle forces in animal locomotion. (Adapted with permission from Biewener and Blickhan (1988)). Force is measured by a strain gage (in color)

☐ **Table 1.1**   Some Journals Publishing Biomechanics Research

Biomechanics

  Clinical Biomechanics

  Gait & Posture

  International Biomechanics

  Journal of Applied Biomechanics

  Journal of Biomechanics

  Journal of Electromyography and Kinesiology

  Sports Biomechanics

  Sports Engineering

Multi-Disciplinary

  Ergonomics

  European Journal of Sport Science

  Human Factors

  Medicine & Science in Sports & Exercise

  Journal of Science and Medicine in Sport

  Journal of Sports Sciences

  Journal of Strength and Conditioning Research

  Research Quarterly for Exercise and Sport

  Scandinavian Journal of Medicine & Science in Sport

1.3.1 **Scholarly Societies**

There are scholarly organizations exclusively dedicated to biomechanics. Scholarly societies typically sponsor meetings and publications to promote the development of their fields. Students of sport biomechanics should know that the International Society of Biomechanics in Sports (ISBS) is devoted to promotion of sport biomechanics research and to helping coaches apply biomechanical knowledge in instruction, training, and conditioning for sports. The ISBS publishes scholarly proceedings papers on sports biomechanics that are accepted from papers presented at their annual meetings and the journal *Sports Biomechanics*. Their website ▸ (http://isbs.org/) provides links to a variety of information on sport biomechanics. The websites for the societies discussed in this section are listed at the end of this chapter.

The International Society of Biomechanics (ISB) is the international society of scholars interested in biomechanics from all kinds of academic fields. The ISB hosts international meetings and sponsors journals. There are numerous national and regional biomechanics societies. Some examples include the American Society of Biomechanics (ASB), the Canadian Society of Biomechanics, and the European Society of Biomechanics. The ASB website has several links, including a list of graduate programs and papers accepted for presentation at ABS annual meetings. Another related scholarly society is the International Society for Electrophysiology and Kinesiology (ISEK), which promotes the study of **electromyography** (EMG) in animals and humans. Engineers interested in equipment design, sport, and human movement have founded the ISEA mentioned earlier. There are other scholarly organizations that have biomechanics interest groups related to the parent disciplines of medicine, biology, or physics.

Aside from the many specialized biomechanics societies, there are biomechanics interest groups in various scholarly/professional organizations that have an interest in human movement. The American College of Sports Medicine was founded in 1954 by physicians and exercise scientists to be a scholarly society interested in promotion of the study and application of exercise, sports medicine, and sports science. The ACSM substructure interested in biomechanics is the biomechanics interest group (BIG). Other professional organizations in medicine, physical therapy, athletic training, and/or strength and conditioning sponsor biomechanics programs related to their unique interests. Whatever career path you select, it is important that you join and participate in the related scholarly and professional organizations. Membership gives access to scholarly presentations at conferences and journals where biomechanical knowledge can be updated.

1.3.2 **Computer Searches**

One of the best ways to find knowledge on human biomechanics is to use electronic bibliographic databases that index scholarly (peer-reviewed) publications (books, chapters, and journal/proceedings articles). University libraries subscribe to many databases that will be relevant to finding original research in biomechanics. Ask your instructor and librarian what are the best databases to search for your specific research interests. It is important to remember that with tens of thousands of journals, no one database can index everything. Databases are run by academic publishers and other organizations that can only collect and manage limited publication data.

Some large databases relevant to kinesiology students are *MEDLINE, Physical Education Index, Scopus, SportDiscus, and Web of Science. Google Scholar* can also be searched; however there are fewer controls on what is indexed by this system. For example, searching *Google Scholar* will return well viewed/cited books and chapters, while most other academic databases focus on indexing original research articles. It is important that students search multiple databases to ensure they find all the most relevant biomechanics publications to get a good sense of the current knowledge of their topic.

One example of the need for multiple searches can be found in rehabilitation medicine. The medical literature has been well cataloged by *Index Medicus* (1879–2004), the print precursor to *Medline* that is available for free from your tax dollars at the National Library of Medicine (*Pubmed*). Physical rehabilitation researchers searched two medical databases (*MEDLINE* and *EMBASE)* and reported very low percentages of relevance (there are many kinds of rehabilitation) and overlap (coverage of same journals) in the returned records (Minozzi et al. 2000). Biomechanics is not as large as medicine, but like physical rehabilitation it is informed by many kinds of studies published in a wide variety of journals. Students should search multiple databases, with careful use of keywords and Boolean logic operators (and, or, not). Remember that much of the power of indexing is the cross-referencing as well as the direct listings for your search items.

Most all journals publish keywords with articles to facilitate the searching for the articles with similar terms. The search request for "biomechanics" in some databases will return all items (probably too many) beginning with these letters in the title, abstract, or keywords including biomechanics or biomechanical. Searching for "kinematic and ankle" will find sources documenting the motion at the ankle joint. Even better would be "kinematic or ankle or subtalar," because any one of the three search terms matching would select a resource. You miss very little with this search, but it is necessary to go through quite a few sources to find the most relevant ones. Be persistent in your search and let your readings refine your search strategy. A student interested in occupational overuse injuries (sports medicine term) will find that the human factors field may refer to this topic as "cumulative trauma disorder," "work-related musculoskeletal disorders," or "occupational overuse syndrome" just to name a few (Grieco et al. 1998).

1.3.3 **Biomechanics Textbooks**

Good sources for general background knowledge in the field and references to sources sometimes missed by students are biomechanics textbooks. Biomechanics students can refer to scholarly biomechanics books and review their coverage of a research topic (See some examples in the Suggested Reading at the end of the chapter). Scholars often write specialized biomechanics books that provide extensive reference lists for students. Remember that writing books takes considerable time, so references in a particular text may not be totally up-to-date, but they do give students leads and clues on many good sources. The quality of a biomechanical source will be difficult for many introductory biomechanics students to judge, so the next section will coach you in evaluating biomechanical sources you find in your searches.

> **Interdisciplinary Issue: Collaborative Biomechanics**
> Finding biomechanics information is like a scavenger hunt that will lead students all over a library. We have seen that biomechanics research can be found in biology, engineering, medical, and other specialized journals. "Interdisciplinary" means using several different disciplines simultaneously to solve a problem. Do some preliminary research for sources (journals and edited proceedings/books) on a human movement of interest to you. Do the titles and abstracts of the sources you found suggest scholars from different disciplines are working together to solve problems or are scholars working on a problem primarily from their own area or discipline? What have other students found in their research?

1.4 **Biomechanical Knowledge Versus Information**

Knowledge is different from information. **Knowledge** is contextual, theory-based, and data-supported ideas that make the best current explanation for reality. Scientific knowledge is a theoretical structure of laws and principles that is built on the consensus of experimental/

1

research evidence by scientists in that field. Students often fail to realize that knowledge is a structure that is constantly being constructed and remodeled as new theories and evidence are examined, and research instruments improve Transitions in the knowledge structure are often controversial since so much previous experimental data must be critically weighted/integrated with the new evidence.

Biomechanical knowledge is built by a consensus of scientists from a variety of disciplines interested in human movement (e.g., biology, engineering, kinesiology, medicine). Most real-world human movement problems have only partial answers because of limited biomechanical research or knowledge that is specifically related to the context of the person, movement task, and environment. Although the current structure of biomechanical knowledge (or any science) is not perfect, a critical review of this will be the best guide and closest to the truth.

The modification of human movement based on biomechanical knowledge is difficult because movement is a multifaceted problem, with many factors related to the performer, environment, and activity all interacting to affect the outcome. The next chapter will present nine general principles of biomechanical knowledge that are useful in applying biomechanics in general to improve human movement. There will be a few bits of the knowledge puzzle that are well-known and rise to the level of scientific law. While most biomechanical knowledge is not perfect and can only be organized into some general principles, it is much better at guiding professional practice than merely using information or trial and error.

Living in an information age, it is easy for people to become insensitive to the important distinction between information and knowledge. The most important difference is that information has a much higher chance of being incorrect than knowledge. **Information** is merely access to opinions or data, with no implied degree of accuracy. Information is also much easier to access in the age of the Internet and wireless communications. Do not confuse ease of access with accuracy or value. This distinction is clearer as you look at the hierarchy of the kinds of sources used for scholarly research and a simple strategy for the evaluation of the quality of a source.

1.4.1 Kinds of Sources

When searching for specific biomechanical knowledge, it is important to keep in mind the kind of source you are reading. There is a hierarchy of the research design and scholarly rigor of published research and writing. ▣ Figure 1.9 illustrates typical examples of the hierarchy for kind of publication. Although there are exceptions to most rules, it is generally true that the higher up

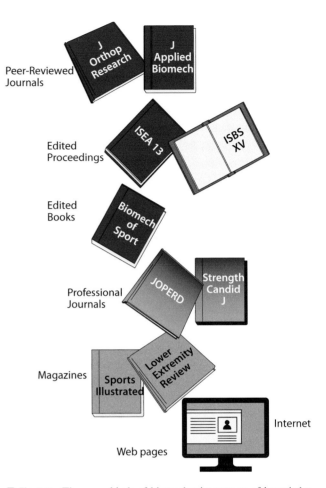

▣ **Fig. 1.9** The many kinds of biomechanics sources of knowledge and the hierarchy of their academic rigor. Peer-reviewed original research and review sources (in color) have greater potential for accuracy and less potential bias

a source on the hierarchy the better the chance that the data and knowledge presented is closer to the current state of knowledge and the truth. For this reason, professionals and scholars focus their attention on original research articles published peer-reviewed journals to maintain a knowledge base for practice. Most journals are now "publishing" electronic versions on the internet for subscribers instead of printed versions. Major publishers give each article a digital object identifier (doi) number that provide a unique electronic location for the article. While most journal articles require some subscription cost to access, some journals make papers available for free after a certain waiting period, or immediately provide "open access." Over the past several years, the open access movement has allowed journals to collect publication costs directly from authors/universities/grants so peer-reviewed articles can be made immediately available for free.

Most scholarly journals publish original research that extends the body of knowledge or review papers that attempt to summarize a body of knowledge. Some systematic reviews (meta-analysis) can create numerical estimates of effects based on all the research studies on a topic or intervention. Many journals also publish supplements that contain abstracts (short 100–200 word summaries of a research study) of papers that have been accepted for presentation at a scholarly meeting or were published in another journal. The review of these abstracts is not as rigorous as a full journal article, so they are not usually considered archival, peer-reviewed research reports. Abstracts and presentations at scientific conferences do provide scholars and students with clues about what the most recent research is focusing on.

Reading original biomechanics research will be challenging for many undergraduates because of the specialized language and mechanical variables. Appendix A provides a comprehensive glossary of biomechanics terms that will help you when reading the biomechanics literature related to your professional interests. Typically, masters and doctoral training in kinesiology or professional fields (DAT, DPT, MD, DO, etc.) teach research design and interpretation principles needed to fully understand research and apply it in evidence-based practice (EBP). EBP involves several systems to formally evaluate/grade and synthesize all the research on a topic to inform clinical decisions. These methods began in the late 1990s at the Center for Evidence-Based Medicine at Oxford. Since then many professions have updated graduate training programs to include (EBP) to foster more effective professional treatment (Hadorn et al. 1996; Sackett et al. 1996) and expanded clinical practice guidelines (Kredo et al. 2016). ◘ Figure 1.10 shows a typical hierarchy of kinds of research study designs by

their rigor (Knudson 2005). Basic science related to biomechanics might include measures of the mechanical strength of tissue samples from animals with varying levels of training or detraining/rest. The many clinical research designs allow for quick presentation of medical treatments (case studies and case series) and increase in rigor/cost to large clinical interventions with randomization and control groups. Research that critically reviews all the research on a topic helps professionals understand the consensus of research results and theory on a topic. Meta-analyses utilize statistical procedures to integrate all the research on a topic to provide an estimate of the likely size of differences or associations between variables.

Authors have called the sports medicine and kinesiology professions to more consistently focus on using critical review of evidence to support practice (Knudson 2005; Faulkner et al. 2006; Shrier 2006; Coutts 2017). Sports performance professionals from the United States working with top athletes rate research a preferred source of knowledge for their work (Fullagar et al. 2019). Kinesiology professionals should strive to weigh biomechanics evidence more heavily when the research uses a design higher on the hierarchy that has a lower risk of bias/error. This book cites biomechanics research to support key points to both illustrate the use of evidence and to give interested students places to access influential research.

In the middle of academic rigor are edited proceedings, edited books, and professional journals. These publications have more varied degrees of peer review before publication, as well as varying rules on what constitutes acceptable evidence. Unfortunately, an unintended consequence of the open access journal movement is the development of deceptive/predatory publishers. These phony open access journals use numerous deceptive strategies (similar names to prestigious journals) to trick scholars and readers into thinking the articles have undergone traditional, rigorous peer review. The majority of articles published in these "journals" have not been reviewed by top scholars in the field (Eriksson and Helgesson 2018). Publication science scholars have exposed articles published by these journals with joke authors like Margaret Simpson (Maggie Simpson of The Simpsons TV show) and titles like "Get me off your #$%@ mailing list."

At the bottom of the credibility chain with the deceptive publishers are popular press publications (magazines/newspapers) and hypertext on the internet. While these sources are appropriate for more subjective observations of laypersons, there are serious threats to the validity of the observations from these sources. The major problems with webpages are their impermanence (unlike archival research literature) and the lack of review (anyone can post a webpage). It is possible that

Review

Meta-analysis

Systematic Reviews

Randomized control trial

Clinical

Cohort studies

Case-control studies

Case studies

Experience

Basic

Animal/cellular research

◘ **Fig. 1.10** Typical hierarchy of research design/evidence for professional practice. (Adapted from Knudson (2005))

some of these opinions are correct and useful, but there is little evidence used to verify the advice, so kinesiology professionals should verify with other primary sources before using the advice. The next section will summarize a quick method for checking the credibility of various sources for biomechanical knowledge.

1.4.2 Evaluating Sources

The previous section clearly indicates that certain sources and kinds of evidence are more likely to be accurate. Until you receive additional training in research design or a system of EBP, consider evaluating the credibility of sources that fall at similar levels of publication rigor with the "me" test. The "m" stands for motivation. What is the motivation for the person or source providing the evidence or purported knowledge? Sources with little financial interest in to making the observations/claims and who are dedicated to advancing a body of knowledge or human potential (scholarly journals) are much more likely to provide accurate observations and knowledge. The motivation of the popular press (TV, newspapers, magazines) and the Internet involves profit and self-promotion based on numbers of viewers (hits, likes, shares, etc.) and, therefore, is more prone to sensationalize and to not weigh all the evidence.

The "e" in the acronym stands for the key element of all science: evidence. Science is based on logical analysis and the balance of data from many controlled studies. This sober, qualified weighing of all the evidence stands in stark contrast to the more emotional claims of the popular press. The more emotional and sensational the language, even if it talks about "the latest study," the more likely you are reading only a recent part of a larger picture. Remember that the structure of knowledge is a complicated structure built over time using many small pieces. The "latest" piece of the knowledge puzzle may be in error (see the next section) or may be rejected by most scholars as having flaws that make it less valuable as other research.

One formidable barrier to a kinesiology professional's ability to weigh biomechanical evidence is the technical and specialized terminology employed in each subdiscipline. Students need to read carefully because sometimes terms differ across subdiscpline confusing communication in reports (Knudson 2009, 2019). Throughout this text many of these measurement systems and mechanical terms are covered. Appendix A provides an extensive glossary of biomechanical terms and quantitative measurement systems. Two papers that provide good summaries of biomechanical and exercise science terms are available (Knuttgen and Kraemer 1987; Rogers and Cavanagh 1984). Students reviewing biomechanical studies should ask their instructor for assistance when the text or these sources do not clear up their understanding.

> **Application**
> On your next trip to a physician or other allied health professional's waiting room, evaluate the nature of the articles and advertisements in the magazines and displays you encounter. Do advertisements related to claims in the articles appear near the article? Do the articles talk about several studies, their relative merits, as well as the percentage of subjects with various responses? Does the professional you are visiting sell supplements or products to patients? If so, what does this tell you about motivation and potential conflicts of interest between practice and profits? The biomechanics of most health and human performance problems in human movement are classic examples of complicated problems, with many interrelated factors and variability in the response of individuals to treatment.

1.4.3 A Word About Right and Wrong Answers

The increasing amount and complexity of research and technology tends to give many people a false sense of the correctness of numbers. Few people will question a measurement if some machine output numbers on a printout, unless they are very familiar with the measurement. Like our knowledge-versus-information discussion, it is important for kinesiology professionals to understand that the process of reviewing and weighing the evidence is often more important than finding the perfect or "right" answer. Such absolutes in a complicated world are quite rare, usually only occurring when a technique change would run against a law of physics or one of our principles of biomechanics. These principles (theories and laws) of mechanics are the application tools developed throughout this book.

So the good news is that biomechanics helps kinesiology professionals solve problems, while the bad news is that most of these everyday questions/problems do not have easy, dichotomous (right/wrong) answers. The human body is complex, and in most cases, several biomechanical principles will be relevant, so some trade-off must be considered between them in prescribing an intervention. There are many factors that affect most phenomena and there is variation in nearly all phenomena. In fact, all true science is written using statistics to account for this variation. Statistics use estimates of data variation to attach a probability to any yes/no decision about the data. If you read a study that says an observation was significant at the 0.05 level, this only means that the result is not likely a fluke or observation due to chance variation alone. It is possible that chance alone created this "difference," and $p < 0.05$ means that in the long run there is about a 1-in-20 chance that the observation or decision from the data is wrong. Since most studies use this error standard ($p < 0.05$), this means that, out of 20 studies on a particular topic, one likely reports an incorrect observation from chance variation alone.

A common misconception among laypersons is that statistics in a scientific study "proves" things. Statistics only provide tools that allow scientists to place probability values on the interpretation of the numbers/data observed in their research. Proof is a long-term process requiring critical review of the whole body of research on the issue. Remember this when television news broadcasts sensationalize the results of the "latest" study on some health issue or you are tempted to believe that one biomechanical study settles a particular issue.

Interdisciplinary Issue: Too Much Performance?
Recent controversies about sport performance enhancement through steroids and genetics parallel the issues related to biomechanics and improvements in equipment. Engineers and biomechanists have used advances in technology to improve the materials and design of sports equipment, although the use of tools in sport has a long history (Minetti 2004). Jenkins (2002) presents a nice review of how improvements in equipment materials has dramatically affected performance in several sports. These are truly interdisciplinary controversies because there are ethical, safety, athlete, coaching, and sport/historical perspectives on performance. One example of technology correcting too much performance is the new rules for the javelin in the mid-1980s. The center of gravity of the javelin was moved forward to decrease throwing distances because many athletes were throwing the old javelin over 100 m (into positions where officials and other athletes would be injured). Advances in biomechanics and computer technologies have also been used to modify technique, training, and equipment for the Olympics (Legwold 1984; Sheppard 2006). Sport philosophers and sport governing bodies will all be interested in the topic of when is performance getting too good, changing the nature of competition, the sport, and potentially compromising player and spectator safety.

Biomechanical knowledge is constantly changing and usually cannot be easily classified into always right or wrong answers, so there are two important professional tools you must not forget to use. These tools will work quite well with the biomechanical tools (nine principles) developed in this text. These two tools are the Swiss Army Knives™ or Leathermen™ of your professional toolbox because of they are so flexible and important. One is your *ability to access biomechanical knowledge*, and the other is the *critical thinking* necessary to evaluate and integrate knowledge so it can be applied in solv-

ing human movement problems. You are not likely going to remember everything in this book (though you would be wise to), but you should have the knowledge to access, and critical thinking tools that allow you to find, evaluate, and integrate biomechanics knowledge with other kinesiology subdisciplinary knowledge to help your clients. The rest of this text will illustrate and explicate the nine principles of biomechanics, which are tools you would do well to never forget when helping people improve their movement or deal with injury.

1.5 Summary

Kinesiology is the scholarly study of human movement/physical activity. A core science in the academic discipline of kinesiology is biomechanics. Biomechanics in kinesiology is the study of motion and its causes in human movement. The field of biomechanics is relatively new and only has a few principles and laws that can be used to inform professional practice. Kinesiology professionals often use biomechanical knowledge in the qualitative diagnosis of human movement to decide on how to intervene to improve movement and prevent or remediate injury, while biomechanical research is based on measurement (quantitative analysis). Applying biomechanics in qualitative diagnosis is most effective when a professional integrates biomechanical knowledge with professional experience and the other subdisciplines of kinesiology. Biomechanical knowledge is found in a wide variety of journals because there are many academic and professional areas interested in the movement of living things. Students studying human biomechanics might find relevant biomechanical knowledge in peer-reviewed journals in applied physics, biology, engineering, ergonomics, medicine, physiology, and biomechanics.

? **Review Questions**
1. What is biomechanics and how is it different from the two common meanings of kinesiology?
2. Biomechanical knowledge is useful for solving what kinds of problems?
3. What are the advantages and disadvantages of a qualitative biomechanical analysis?
4. What are the advantages and disadvantages of a quantitative biomechanical analysis?
5. What kinds of journals publish biomechanics research and how can articles in these be found?
6. What is the difference between knowledge and information?
7. Why should biomechanical knowledge be integrated with other sport and exercise sciences in solving most human movement problems?
8. Biomechanics will be most relevant to helping solve what kinds of human movement problems?

1

In what movements or problems is biomechanics knowledge likely secondary to other kinesiology subdisciplinary sciences?

🛈 Key Terms

Biomechanics · Electromyography (EMG) · Information · Interdisciplinary · Kinesiology · Knowledge · Orthotics · Prosthetics · Qualitative analysis · Quantitative analysis · Science

★ Suggested Reading

Abraham, L. et al. (2018). *Guidelines for undergraduate biomechanics* (4th ed.). [Guidance Document] Retrieved from SHAPE America website: *https://www.shapeamerica.org/uploads/pdfs/2018/guidelines/Guidelines-for-UG-Biomechanics.pdf*

American Kinesiology Association (2020). About AKA. Retrieved from *http://www.americankinesiology.org/SubPages/Pages/About*.

Chaffin, D. B., Andersson, G. B. J., & Martin, B. J. (2006). *Occupational biomechanics* (4th ed.). New York: Wiley.

Dvir, Z. (2000). *Clinical biomechanics*. New York: Churchill Livingstone.

Hoffman, S. J., & Knudson, D. V. (Eds.). (2018). *Introduction to kinesiology* (5th ed.). Champaign: Human Kinetics.

Hughes, M. (2019). *Essentials of performance analysis in sport* (3rd ed.). New York: Routledge.

Knudson, D. V. (2013). *Qualitative diagnosis of human movement: Improving performance in sport and exercise* (3rd ed.). Champaign: Human Kinetics.

Kumar, S. (1999). *Biomechanics in ergonomics*. London: Taylor & Francis.

Nigg, B. M., & Herzog, W. (2007). *Biomechanics of the musculoskeletal system* (3rd ed.). West Sussex: Wiley.

Ozkaya, N., et al. (2017). *Fundamentals of biomechanics: Equilibrium, motion, and deformation* (4th ed.). New York: Springer.

Richards, J. (2018). *The comprehensive textbook of clinical biomechanics* (2nd ed.). London: Elsevier.

Robertson, et al. (2014). *Research methods in biomechanics* (2nd ed.). Champaign: Human Kinetics.

Rogers, M. M., & Cavanagh, P. R. (1984). Glossary of biomechanical terms, concepts, and units. *Physical Therapy, 64*, 82–98.

Sheppard, L. M. (2006). Visual effects and video analysis lead to Olympics victories. *IEEE Computer Graphics and Applications, 26*(2), 6–11.

Whiting, W. C., & Zernicke, R. F. (2008). *Biomechanics of musculoskeletal injury* (2nd ed.). Champaign: Human Kinetics.

📡 Internet Resources

- ACSM–American College of Sports Medicine is a leader in the clinical and scientific aspects of sports medicine and exercise. ACSM provides the leading professional certi-fications in sports medicine. *acsm.org*
- AKA–American Kinesiology Association is the organization of academic units in higher education promoting the field of kinesiology. *americankinesiology.org*
- ASB–American Society of Biomechanics posts meeting abstracts from a variety of biomechanical scholars. *asbweb.org*
- BIOMCH-L–Web-based biomechanics listserve/internet forum hosted by the ISB. *https://biomch-l.isbweb.org/*
- Deluca Foundation–What is biomechanics? *youtube.com/watch?v=OhN1d0xxUjE*
- ISB–International Society of Biomechanics was the first biomechanics scholarly society. *isbweb.org*
- ISBS–International Society of Biomechanics in Sports hosts annual conferences and indexes papers published in their proceedings and journal (*Sports Biomechanics*). *isbs.org*
- ISEA–International Sports Engineering Association hosts international meetings and publishes the journal *Sports Engineering*. *sportsengineering.org*
- NAK–National Academy of Kinesiology is the premier, honorary society in kinesiology. *nationalacademyofkinesiology.org*
- Ritual Calling of an Engineer: Iron Ring Ceremony, University of Waterloo, Canada. *https://uwaterloo.ca/engineering/current-undergraduate-students/iron-ring*

Anatomical Description and Its Limitations

Key Questions

➤ **What do you need to know to begin?**
Kinesiology majors need the qualitative descriptions/terminology of the gross musculoskeletal anatomy to study biomechanics and communicate with other kinesiology and allied health professionals.

➤ **Why do you need to know this material?**
Kinesiology professionals know the limitations of qualitative anatomical description (form/structure) and the need for biomechanics to understand human movement (form and function).

➤ **What is the key idea?**
Gross anatomy is an essential language for kinesiology professionals but must be integrated with biomechanics knowledge to truly document and understand the causes of human movement.

In order to understand any complex system like the human body, one must establish basic terms describing structure. **Anatomy** is the study of the structure of the body. Human anatomy provides essential labels for musculoskeletal structures and qualitative descriptions of joint motions relevant to human movement. Knowledge of anatomy also provides a common professional "language" of the human body and motions for kinesiology and medical professionals. Anatomy is an important prerequisite for kinesiology professionals trying to improve movement and prevent or treat injury. Anatomy is primarily a descriptive field of study and is not, by itself, enough to *explain the function* of the musculoskeletal system in creation of movement. Knowledge of anatomy must be combined with biomechanics to accurately determine the musculoskeletal and external force causes or the "how" human movement is created. This chapter reviews key anatomical concepts, shows how functional anatomy traditionally classifies muscle concentric actions, and shows how biomechanics is needed to determine the actual, more numerous muscle actions creating movement.

2.1 Review of Key Anatomical Concepts

This section reviews several key concepts from human anatomy. A course in gross anatomy (macroscopic structures) is a typical prerequisite for the introductory biomechanics course. This section does not review all the bones, muscle, joints, and terms of anatomy. Students and kinesiology professionals must continuously review and refresh their knowledge of anatomy given the great volume/complexity of the body and continuous discoveries of anatomists. Anatomy qualitatively describes the human body relative to the *anatomical position*. The anatomical position is approximated in ◼ Fig. 2.1. The three spatial dimensions of the body correspond to the three anatomical planes: frontal, sagittal, and transverse. Recall that a plane of motion is a particular spatial direction or dimension of motion, and an axis is an imaginary line about which a body rotates. The anatomical axes associated with motion in each of these planes are the anteroposterior, mediolateral, and longitudinal axes. Knowing these planes and axes is important to understanding medical descriptions of motion or movements. Even more important may be the functional implications of the orientation of these axes to the planes of motion they create. Note that motion in a particular plane (e.g., sagittal) occurs by rotation about an axis oriented 90° (mediolateral axis) to that plane. A person supinating their forearm to illustrate the anatomical position is creating motion in a transverse plane about a longitudinal axis roughly along the forearm. Functional anatomy applies knowledge of joint axes of rotation and muscle positions to hypothesize which muscles contribute to motion in an anatomical plane.

D. Knudson, *Fundamentals of Biomechanics*, https://doi.org/10.1007/978-3-030-51838-7_2

2

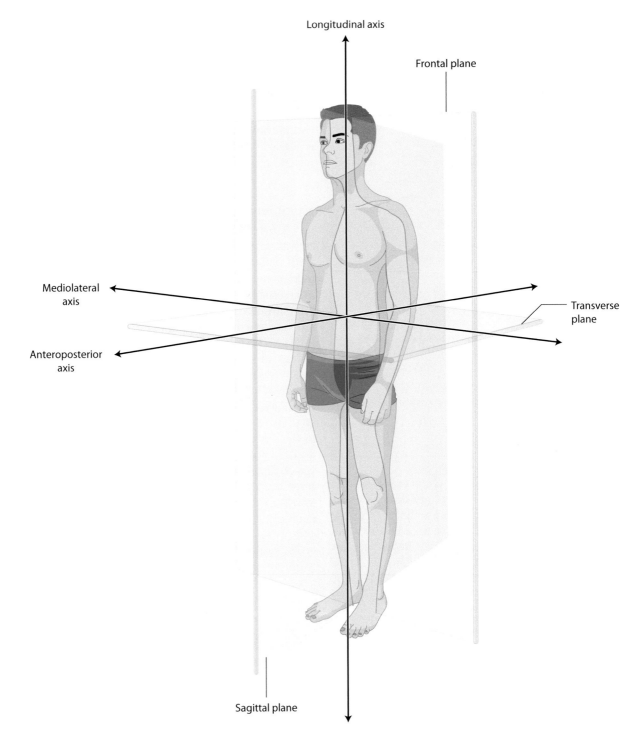

□ **Fig. 2.1**   The major anatomical planes of motion and axes of rotation

2.1.1   **Directional Terms**

In addition to planes and axes, anatomy uses several directional terms to help describe the position of structures relative to the anatomical position. Toward the head is called superior, while toward the feet is inferior. Body parts toward the front of the body are anterior, and objects to the back are in the posterior direction. Parts or motion toward the midline of the body is said to be medial, while motion or position toward the sides of the body is lateral. There are many other anatomical terms that have similar meanings as these but retain the original Latin or Greek form of classical anatomy. For example, superior is synonymous with cephalic, while inferior is the same a caudal. Anatomy is also difficult for many because of the variation in terms, and the Latin or Greek terms (that were descriptive of the structure and made sense) are often not obvious to us in other languages. Despite great efforts by the International Federation of Associations of Anatomists (IFAA 2002) to standardize terminology in the field, there remain constant updates and inconsistencies in use of anatomical terms. This book will use the more familiar English anatomical terms whenever possible.

Students with interests in sports medicine careers would do well to keep a medical dictionary handy and become familiar with the variety of classical anatomical terms used in medicine. Careful use of terminology is important in science and professions to prevent confusion. One example of the confusion that can occur with using unfamiliar Greek or Latin terms is the debate over the directional terms *valgus* (*valgum*) and *varus* (*varum*). The original Greek meanings of these terms (valgus [bowlegged] and varus [knock-kneed]) can be at odds with their typical use in orthopedic medicine. Medicine usually defines genu (knee) valgus as an inward deviation of the knee joint, resulting in a knock-kneed appearance (◘ Fig. 2.2). Genu varus or varum usually corresponds to an outward deviating knee, which results in a bowlegged appearance. This leads to considerable confusion in describing anatomical abnormalities at the knee and other locations in the body, and some have suggested that these terms be dropped or at least defined every time they are used (Houston and Swischuk 1980). Some would look at ◘ Fig. 2.2a and say the knee deviates medially, while others would say the lower leg deviates laterally. We will see that this little problem of qualitative anatomical description is similar to the multiple kinematic frames of reference (▶ Chap. 5) that are all correct descriptions of a single motion and the different units of measurement that can be used. Mechanics and anatomy both share the minor problem that there are several standards that have grown up with these sciences since people all over the world have been working on these same problems. Students should strive to read and write with special attention to the meaning of professional/scholarly terminology. Clarity of expression in qualitative terms of anatomy and quantitative terms of science is necessary for accurate professional and scientific communication.

2.1.2 Joint Motions

Anatomy also has specific terminology describing the major rotations of bones at joints. "Flexion" refers to a decrease in joint angle in the sagittal plane, while "extension" is motion increasing joint angle (◘ Fig. 2.3a). Motion into the extremes of the range of motion is often noted as "hyper," as in hyperextension. Motion of a segment away from the midline in the frontal plane is "abduction," while movement back toward the midline is called "adduction" (◘ Fig. 2.3b). Joint motions in the transverse plane are usually called inward rotation (rotation of the anterior aspect of the segment toward the midline) and outward rotation (◘ Fig. 2.4). Some examples of special joint motion terms are "pronation," which refers to internal rotation of the forearm at the radioulnar joint, or "horizontal adduction," which is drawing the shoulder (glenohumeral joint) toward the midline in a transverse plane. Like the directional terms, anatomical terminology related to the rotations of joints are also sometimes used incorrectly. It is incorrect to say "a person is flexing a muscle" because flexion is a joint movement. It is important for kinesiology majors to use anatomical terms correctly since these terms provide the professional language of numerous health-related professions. Refer to your anatomy book frequently to keep all the joint motion terminology (this section does not review them all) fresh in your mind.

While there are attempts to standardize anatomical description throughout the world (IFAA 2002; Greathouse et al. 2004), there remain regional inconsistencies in terminology. For example, some refer to the frontal plane as the "coronal" plane. Applied sciences such as medicine often develop specialized terms that are borrowed from anatomy, but that go against anatomical convention. A good example is related to how the foot acts during the stance phase of running. Medical and biomechanical studies have adopted the terms "pronation" and "supination" to refer to the complex tri-planar actions of the subtalar joint. In normal running the foot strikes the ground on the lateral aspect of the foot; the combined anatomical action of eversion, plantar flexion, and abduction in the first part of stance is called pronation. This pronation serves to absorb the shock of the collision of the foot with the ground (◘ Fig. 2.5). The opposite motion (supination) stiffens the foot for the push-off phase of stance. Here is another example of how anatomical terms are not always used in a consistent way. In your studies of biomechanics and other kinesiology disciplines, remember that adaptations and variations in anatomical terminology make it important to read carefully and often check background information. Modern biomechanical studies often assume quite a bit about reader expertise in the area and may not cite sources giving necessary terminology and background sources/citations. This saves journal space but places a burden on the kinesiology professional to be knowledgeable about variations in descriptive terminology.

2

■ **Fig. 2.2**   Orthopedic and sports medicine often refer to the lower extremity deviation illustrated by **a**. as genu (knee) valgus/valgum because of the knock-kneed medial deviation of the knee. Normal leg orientation in the frontal plane is illustrated in **b**. Notice that this wording does not explain how this position was created: varus femur alignment, valgus lower leg alignment (from stretched ligaments or knee joint deformity), or all of these. The use of valgus and varus terminology is often inconsistent in the literature and should be clearly defined when used (Houston and Swischuk 1980)

■ **Fig. 2.3** **a** Flexion and extension movements occur in a sagittal plane about a mediolateral axis; **b** adduction/abduction of the hip joint occurs in a frontal plane about an anteroposterior axis

■ **Fig. 2.4** Inward and outward rotation of the shoulder joint occurs in a transverse plane about a longitudinal axis

2.1.3 Review of Muscle Structure

The anatomical structure and microstructure of skeletal muscle have considerable functional importance. We will see later that the function of the complex structures of skeletal muscle can be easily modeled as coming from active and passive sources. This section will review a few of the structural components of skeletal muscle that are believed to be important in these active and passive tissue properties.

Careful dissection of skeletal muscle shows that muscles are composed of many distinct bundles of muscle fibers called *fascicles*. In cutting across a piece of beef or chicken, you may have noticed the tissue is in small bundles. The connective tissue sheath that surrounds the whole muscle, bundling the fascicles together, is called *epimysium* (meaning over/above the muscle). Each fascicle is covered by connective tissue called *perimysium*, meaning "around the muscle." There are hundreds of muscle fibers within a fascicle, and an individual fiber is essentially a muscle cell. Muscle fibers are also covered with connective tissue called *endomysium* (within the muscle). The gradual blending of these connective tissue components of muscle forms a distinct tendon or fuses with the calcified connective tissue, the periosteum of bones. A schematic of the macrostructure of skeletal muscle is shown in ◘ Fig. 2.6.

The specific arrangement of fascicles has a dramatic effect on the force and range of motion capability of the muscle (Lieber and Friden 2000). Anatomically, this fiber arrangement has been classified as either **parallel** or **pennate**. A **parallel** arrangement means that the muscle

Supination   Neutral   Pronation

◘ **Fig. 2.5** Frontal plane view of rear-foot motion in the first half of the stance phase of running. The foot lands in a supinated position. The motion of the foot and ankle to accommodate to the surface and absorb shock is called *pronation*

◘ **Fig. 2.6** The macroscopic structure of muscle includes several layers of connective tissue and bundles of muscle fibers called fascicles. Muscle fibers (cells) are multinucleated and composed of many myofibrils

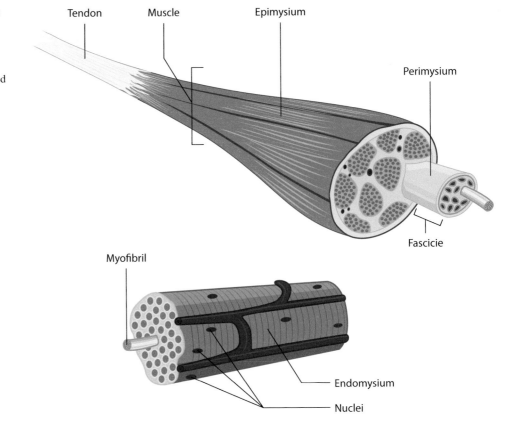

Fig. 2.7 **a** Parallel arrangement of muscle fibers with the tendon favors range of motion over force. **b** Pennate arrangements of fibers are angled into the tendon and create greater force but less range of motion

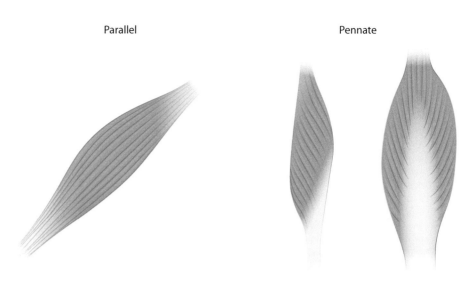

Parallel

Pennate

Unipennate   Bipennate

fascicles are aligned parallel to the long axis or line of pull of the muscle. Muscles like the rectus abdominis, sartorius, and biceps brachii have predominantly a parallel architecture (■ Fig. 2.7a). **Pennate** muscles have fibers aligned at a small angle (usually less than 15°) to a tendon or **aponeurosis** running along the long axis of the muscle. An **aponeurosis** is a distinct connective tissue band within a muscle. This arrangement is called pennate because of the feathered appearance. The tibialis posterior and semi-membranosus are primarily unipennate, while rectus femoris and gastrocnemius are bipennate (■ Fig. 2.7b). An example of a multipennate muscle is the deltoid.

Muscles with parallel architecture favor range of motion over force development. The greater muscle excursion and velocity of parallel muscles comes from the greater number of sarcomeres aligned in series. The rectus abdominis can shorten from 1/3 to 1/2 of its length because of the parallel arrangement of fibers and fascicles. Small muscles may have a simple parallel design with fibers that run the length of the muscle, while larger parallel muscles have fibers aligned in series or end to end. These end-to-end connections and transverse connections within muscles make force transmission in muscle quite complex (Patel and Lieber 1997; Sheard 2000). Fiber architecture also interacts with the connective tissue within muscle to affect force or fiber shortening. The fibers in the center of the biceps do not shorten uniformly due to differences in the distal and proximal aponeurosis (Pappas et al. 2002). The amount of tendon a muscle has and the ratio of tendon to fibers also affect the force and range of motion potential of a muscle.

In essence, pennate muscles can create a greater tension because of a greater physiological cross-sectional area per anatomical cross-sectional area, but have less range of shortening than a muscle with a parallel architecture. Physiological cross-sectional area is the total area of the muscle at right angles to the muscle fibers.

Muscle fibers are some of the largest cells in the body and are long cylindrical structures with multiple nuclei. A typical muscle cell is between 10 and 100 μm in diameter. The lengths of muscle fibers vary widely from a few centimeters to 30 cm long. Besides many nuclei there are hundreds to thousands of smaller protein filaments called **myofibrils** in every muscle fiber. If a muscle cell were to be imagined as a cylindrical straw dispenser, the myofibrils would be like the straws packed in this dispenser. ■ Figure 2.8 illustrates the microstructure of a muscle fiber.

The microstructure of a muscle becomes even more fascinating and complex as you pull out a straw (myofibril), only to notice that there are even smaller threads or cylindrical structures within a myofibril. These many smaller fibers within each myofibril are all well organized and aligned with other adjacent myofibrils in a fiber. This is why looking at skeletal muscle under a light microscope gives the appearance of a consistent pattern of dark and light bands. This is how skeletal muscle came to be called *striated muscle* (■ Fig. 2.8). These small sections of a myofibril between two Z lines (thin dark band) are called **sarcomeres**. Sarcomeres are the basic contractile structures of muscle.

Biomechanists model the active tension of whole muscles based on the behavior of the interaction of two contractile proteins in sarcomeres: **actin** and **myosin**. Actin is the thin protein filaments within the sarcomeres of a myofibril, and myosin the thicker protein filaments. Cross-bridges between myosin and actin are attached and detached with the chemical energy stored in adenosine triphosphate (ATP). You may be familiar with the

2

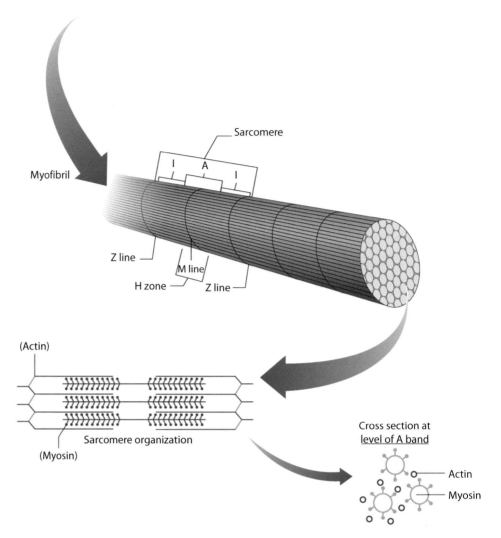

◻ **Fig. 2.8**   The microscopic structure of myofibril components of muscle fibers. Schematics of the sarcomere as well as of the actin and myosin filaments are illustrated. The light and dark bands resulting in the adjective striated are not illustrated

names of the various zones (Z line, A band, and I band) and other substructures of a sarcomere.

While most biomechanists use simple models of the active tension of whole muscles, some biomechanists are interested in researching the mechanical behavior of the microstructures of myofibrils to increase our understanding of where active and passive forces originate. Considerable research is being done to understand muscle actions at this microscopic level from variations in myosin isoforms (Lutz and Lieber 1999) to force transmission throughout the muscle fiber and muscle (Patel and Lieber 1997; Sheard 2000). Some muscle injuries could be due to this complex force production behavior and to nonuniform stresses in the sarcomeres of fibers (Talbot and Morgan 1996; Morgan et al. 2000). Muscle force production is also complicated by passive (connective tissue) forces transmitted laterally (Mass and Finni 2018) as well as longitudinally in the sarcomere through the molecular spring titin (Herzog 2019). Muscle force transmission is a fascinating mix of active and passive tension.

Many kinesiology students are familiar with muscular **hypertrophy** (increased muscle fiber diameter as a result of training), but they are unaware that chronic elongation of muscles (like in stretching) increases the number of sarcomeres in series within muscle fibers to increase their functional range of motion (Williams and Goldspink 1978; Cox et al. 2000). The number of sarcomeres and muscle fiber length are adaptable and strongly related to muscle performance (Burkholder et al. 1994).

It is clear that biomechanics plays a role in understanding the functional significance of the gross and microstructural factors of the muscle–tendon unit. Most general concepts related to human movement, like muscular strength or range of motion, have many biomechanical factors and levels of structure that interact to determine how the concept actually affects movement. This is our first example of the paradox of learning: the more you know, the more you know what you don't know. Now that we have reviewed some of the major structural factors that affect muscle force and range of motion, let's define the kinds of actions muscles have.

In order to do this, we need quick definitions of two mechanical variables, force and moment. **Force** is a straight line push or pull acting between two objects. When a force acts off-center of an object, it creates a rotation (or bending effect in a non-rigid material). This rotation/bending effect is called a **moment of force** or **torque**. We will need to use this understanding of moment or torque to study what rotation effects are created by the linear forces of the active and passive tension created by muscles.

2.2 Muscle Actions

Muscle forces are the main internal motors, springs, struts, and brakes for animal and human movement (Dickinson et al. 2000; Kuhman and Hurt 2019). While gravity and other external forces can be used to help us move, it is the torques created by skeletal muscles that are coordinated with the torques from external forces to obtain the human joint rotations and overall motion of interest. While some biomechanists are interested in the forces and motions created by smooth (visceral) or cardiac (heart) muscle, this text will focus on the actions of skeletal muscle that create human movement.

The activation of skeletal muscle has traditionally been called *contraction*. I will avoid this term because there are several good reasons why it is often inappro-

priate for describing what muscles actually do during movement (Cavanagh 1988; Faulkner 2003). Contraction implies whole muscle shortening, which may only consistently be accurate in describing the general interaction of actin and myosin in activated muscle. Contraction also conflicts with the many actions of muscles beyond shortening to overcome a resistance. Saying "eccentric contraction" is essentially saying "lengthening shortening"! Cavanagh suggests that the term "action" is most appropriate, and this book adopts this terminology. **Muscle action** is the neuromuscular activation of muscles that contributes to movement (shortening or lengthening) and stabilization of the musculoskeletal system. We will see that muscles have three major actions (eccentric, isometric, concentric) resulting from both active and passive components of muscle tension. It could also be said that a fourth action of muscle is inaction, not being activated because their use at that time would be inefficient or counterproductive to the task at hand.

Mechanically, the three kinds of actions are based on the balance of the forces and torques present at any given instant (◘ Fig. 2.9). If the torque the activated muscles creates is exactly equal to the torque of the re-

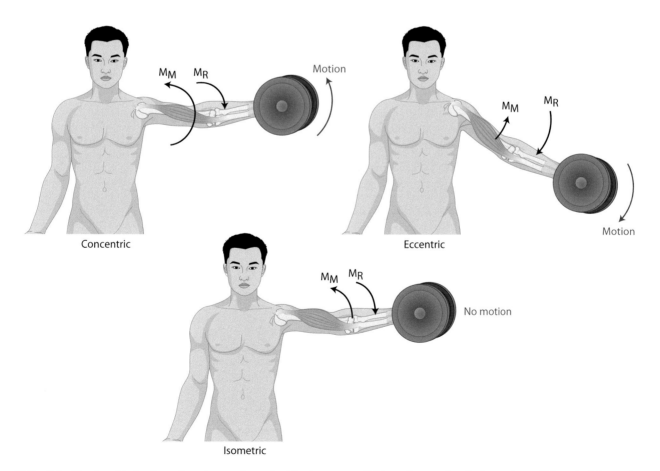

Concentric

Eccentric

Isometric

◘ **Fig. 2.9** The three kinds of muscle action are determined by the balance of torques (moments of force: **M**). In concentric action the torque of the abductors ($\mathbf{M}_M$) is greater than the torque of the resistance ($\mathbf{M}_R$), so the arm rises. In isometric conditions the joint angle does not change because $\mathbf{M}_M$ and $\mathbf{M}_R$ are equal. In eccentric action $\mathbf{M}_M$ is less than $\mathbf{M}_R$, so the arm is lowered

2

sistance, an **isometric** action results. A bodybuilder's pose is a good example of isometric muscle actions of opposing muscle groups. Recall that isometric literally means "same length."

A **concentric** action occurs when the torque the muscle group makes is larger than the torque of a resistance, resulting in muscle shortening. The upward lift of a dumbbell in an arm curl is the concentric phase of the exercise. In essence a concentric action occurs when a muscle activation results in shortening of the whole muscle–tendon unit. When the lifter gradually lowers the weight in an arm curl, the torque the muscle group makes is less than the torque of the resistance. This lowering of the dumbbell is an **eccentric** muscle action or the lengthening of an activated muscle. In eccentric actions muscles are used as brakes on external forces or motion like the brakes of your car.

The importance of these different muscle actions cannot be overemphasized. Functional anatomical analysis and most people tend to focus primarily on the concentric actions of muscles. This overemphasis of what is usually in the minority of muscle actions and inaction for most movements gives a false impression of how muscles create human movement. Kinesiology professionals should not remain in the cult of concentric contractions when trying to understand muscle function! The following section on the limits of functional anatomy will expand on this idea by showing that muscles create movement in a variety of ways using all three muscle actions, not just concentric action.

> **Application: Eccentric Actions and Muscle Injury**
> Eccentric actions are common to all muscles and virtually every human movement. Eccentric actions of high intensity, repetitive nature, or during fatigue are associated with muscle injury. When eccentrically active muscles are rapidly overcome by external forces, a *muscle strain* injury can occur. When people perform physical activity beyond typical levels, especially eccentric muscle actions, the result is usually *delayed-onset muscle soreness*. This is why it is important in conditioning to include both eccentric and concentric phases of exercises. Some athletic events would benefit from emphasis on eccentric training. For example, long jumpers and javelin throwers need strong eccentric strength in the takeoff and plant leg.

2.2.1 Active and Passive Tension of Muscle

Activated muscles create forces by pulling about equally on all their attachments. This tensile force really has two sources: active and passive tension.

Active tension refers to the forces created between actin and myosin filaments in the sarcomere of activated motor units. So active tension is the force created by the contractile proteins (actin and myosin) using chemical energy stored in ATP. This ability of muscles to create active tensile forces is unique compared to the connective tissue components (ligaments, tendons, bone) of the musculoskeletal system. The shape of this active tension potential of skeletal muscle is called the force–velocity relationship of muscle and is described in ► Chap. 4.

Passive tension is the force that comes from an elongation of the connective tissue components of the muscle–tendon unit. When a person does a stretching exercise, the tension she feels in the muscles is the internal resistance of the muscle–tendon unit to the elongation of the stretch. This passive tension in stretching exercises can be quite large and may be responsible for the muscular weakness seen in muscles following stretching (Knudson et al. 2000; Simic et al. 2013). In the midranges of joint motion, passive tension does not initially contribute to muscle forces in normal movement (Siegler and Moskowitz 1984); however, it is more a factor in low-force movements (Muraoka et al. 2005), in various neuromuscular disorders (Lamontagne et al. 2000), and near the end of the range of motion. Muscle passive tension is a significant factor affecting movement at the extremes of joint range of motion. The increase in passive tension limiting range of joint motion is quite apparent in multiarticular muscles and is clinically referred to as **passive insufficiency**. We will see in the following chapter that passive tension is an important component of the force–length relationship of muscle. The passive insufficiency of poor hamstring flexibility could lead to poor performance or risk of injury in activities that require combined hip flexion and knee extension, such as in a karate front kick (◘ Fig. 2.10). The passive tension in the hamstring muscles is high in ◘ Fig. 2.10 because these muscles are simultaneously stretched across the hip and knee joint. We will learn later on in this chapter that the concept of range of motion is a complicated phenomenon that involves several mechanical variables.

2.2.2 Hill Muscle Model

One of the most widely used mechanical models of muscle that takes into account both the active and passive components of muscle tension is the three-component model developed by A. V. Hill in 1938 (Hill 1970). Hill was an English scientist who made substantial contributions to the understanding of the energetics (heat and force production) of isolated muscle actions. Hill was also interested in muscular work in athletics and some of his experimental techniques represent ingenious early work in biomechanics (Hill 1926, 1927). The **Hill muscle**

model has two elements in series and one element in parallel (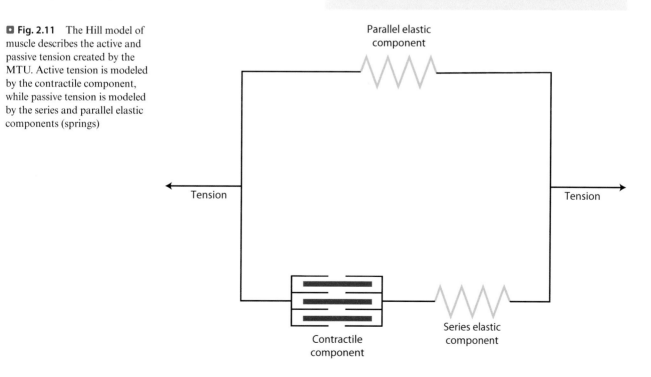 Fig. 2.11). The contractile component (CC) represents the active tension of skeletal muscle, while the parallel elastic component (PEC) and series elastic component (SEC) represent two key sources of passive tension in muscle. The Hill muscle model has been the dominant theoretical model for understanding muscle mechanics and is usually used in biomechanical computer models employed to simulate human movement. In addition to being an interdisciplinary scientific giant with a Nobel Prize, Hill was an outstanding scientific mentor and a hero saving many scientists from Nazi Germany through the Society for the Protection of Science and Learning (Rall 2017).

> **Activity: Passive Tension**
>
> The effect of passive tension on joint motions can be felt easily in multi-joint muscles when the muscles are stretched across multiple joints. This phenomenon is called *passive insufficiency*. Lie down in a supine (face upward) position, and note the difference in hip flexion range of motion when the knee is flexed and extended. The hamstring muscle group limits hip flexion when the knee is extended because these muscles cross both the hip and the knee joints. Clinical tests like the straight-leg raise (Eksstrand et al. 1982), active knee extension (Gajdosik and Lusin 1983), and the sit-and-reach (Wells and Dillon 1952) all use passive insufficiency to evaluate hamstring static flexibility. Careful

☐ Fig. 2.10 The combined hip flexion and knee extension of a karate front kick may be limited by the passive insufficiency of the hamstring muscles. This technique requires excellent static and dynamic hamstring flexibility. (Image courtesy of Master Steven J. Frey, 4th- Degree Black Belt)

☐ Fig. 2.11 The Hill model of muscle describes the active and passive tension created by the MTU. Active tension is modeled by the contractile component, while passive tension is modeled by the series and parallel elastic components (springs)

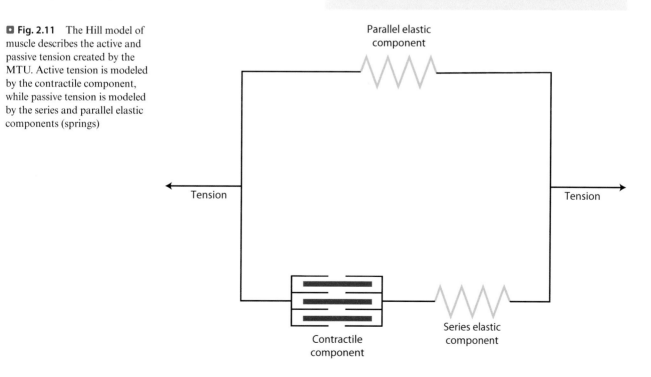

Parallel elastic component

Tension

Tension

Contractile component

Series elastic component

2

body positioning is required in flexibility tests because of passive insufficiency and other mechanical factors across several joints. Some aspects of this issue are explored in Lab Activity 3.

We can make several functional generalizations about the mechanical behavior of muscle based on ◘ Fig. 2.11. First, there is elasticity (connective tissue) in the production of active muscle tension modeled by the series elastic component. The source of this series elasticity is likely a mixture of the actin/myosin filaments, cross-bridge stiffness, sarcomere nonuniformity, and other sarcomere connective tissue components like titin. Second, the passive tension of relaxed muscle that is easily felt in stretching exercises or in passive insufficiency affects motion at the extremes of joint range of motion. The "p" in the parallel elastic component is a key for students to remember this as the primary source of passive tension in the Hill muscle model. Third, muscle tension results from a complex interaction of active and passive sources of tension. This third point can be generalized beyond the simple Hill muscle model as a result of recent research that has focused on the complex transmission of force within the connective tissue components of muscle (Patel and Lieber 1997). Muscles may not create equal forces at their attachments because of force transmitted to extramuscular connective tissues (Huijing 1999; Huijing and Baan 2001; Maas et al. 2004; Maas and Finni 2018).

Biomechanical modeling of the separation of the passive tension into series and parallel components in the Hill model and the exact equations used to represent the elastic and contractile components are, however, still active research issues. The molecular spring titin that forms part of the series elasticity of sarcomeres (returning shortened contractile components to resting length) may be somewhat active as it is sensitive to activation conditions (Herzog 2019). Whatever the eventual source and complexity of elastic tension, it is important to remember that the stretch and recoil of elastic structures are an integral part of all muscle actions. It is likely that future research will increase our understanding of the interaction of active and passive components of muscle tension in creating human movement. There are many other complexities in how muscles create movement. The next section will briefly review the logic of functional anatomical analysis and how biomechanics must be combined with anatomy to understand how muscles actually create movement.

2.3 The Limitations of Functional Anatomical Analysis

Anatomy classifies muscles into functional groups (flexors/extensors, abductors/adductors, etc.) based on hypothesized actions. These muscle groups are useful general classifications and are commonly used in fitness education, weight training, and rehabilitation. These hypothesized muscle actions by anatomy and so-called Anatomical Kinesiology books for movements and exercises are sometimes used to judge the relevance of various exercise training or rehabilitation programs. This section will show that such qualitative estimations of muscle actions are often incorrect in many movements and exercises, so they should be used with caution. Similarly, many of the muscle actions hypothesized by coaches and therapists from subjective observation of movement are not correct (Herbert et al. 1993; Bartlett 2012). Kinesiology professionals can only determine the true actions of muscle by examining and integrating several kinds of biomechanics research that build on anatomical hypotheses.

2.3.1 Mechanical Method of Muscle Action Analysis

Functional anatomy, while not an oxymoron, is certainly a phrase that often stretches the truth. Functional anatomy classifies muscle actions based on the mechanical method of muscle action analysis. This method essentially examines one muscle's line of action relative to one joint axis of rotation, inferring a joint action based on assumed attachment stability, orientation, and pulls of the muscle in the anatomical position (◘ Fig. 2.12). In the sagittal plane, the biceps brachii is classified as an elbow flexor because it is assumed that (1) the origins are at the shoulder joint, (2) the insertion is on the radial tuberosity, and (3) the anterior orientation and superior pull, as well as the superior orientation and posterior pull, would create elbow flexion. When a muscle is activated, however, it pulls both attachments approximately equally so that which end moves (if one does at all) depends on many biomechanical factors. Recall that there are three kinds of muscle actions, so that what the biceps brachii muscle does at the elbow in a particular situation depends on many biomechanical factors this book will explore. Notice that the tension at both ends of a muscle often might not be the same because of the force transmitted to nearby muscles and extramuscular connective tissue (Huijing 1999; Huijing and Baan 2001; Maas et al. 2004; Maas and Finni 2018).

While the biceps is clearly an elbow flexor, this analysis *assumes quite a bit* and does not take into consideration other muscles or external forces. The anatomical analysis presented so far is only part of the mechanical reality. The mechanical system in ◘ Fig. 2.12 is a linked segment system of two rigid bodies (upper arm and forearm) connected at a common axis of rotation (elbow joint). We will explore various mechanical models in ► Chap. 3 and the rest of this book. What biceps does to move the forearm about the elbow joint axis cannot

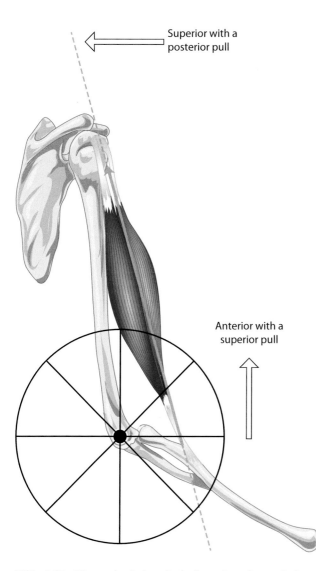

Superior with a posterior pull

Anterior with a superior pull

Fig. 2.12 The mechanical method of muscle action analysis applied to biceps and elbow flexion in the sagittal plane. It is assumed that in the anatomical position the biceps pulls upward toward its anatomical origin from its anatomical insertion (radial tuberosity). The motion can be visualized using a bicycle wheel with the axle aligned on the joint axis. If the muscle were pulling on the wheel from its illustrated direction and orientations relative to that joint axis (visualize where the line of action crosses the medial–lateral and superior–inferior axes of the sagittal plane), the wheel would rotate to the left, corresponding to elbow flexion. Unfortunately, the actions of other muscles, external forces, or other body positions are not accounted for in these analyses. More thorough and mathematical biomechanical analyses of the whole body are required to determine the true actions of muscles

Fig. 2.13 In the eccentric phase of the pull-over exercise, the motion primarily occurs at the shoulder joint, with the elbow angle remaining unchanged. The isolated mechanical method of muscle action does not help in this situation to determine if the long head biceps (crossing both the elbow and shoulder joints) is isometrically active, concentrically active, or inactive. Do you think a biarticular muscle like the biceps can be doing two kinds of muscle actions at once? We will see later that extensive kinetic biomechanical models and EMG research must be combined to determine the actual action of muscles in many movements. Image courtesy of VHI Kits, Tacoma, WA

joint, muscle, and external forces that influence actual human movement.

Another complication in determining muscle action is the biarticular nature of the biceps brachii. The long head of the biceps brachii crosses the shoulder joint. What if the movement of interest was the lowering or "eccentric phase" of the pull-over exercise (■ Fig. 2.13), where the shoulder was the origin because the elbow angle essentially did not change while shoulder flexion and extension were occurring? It is not entirely clear if the long head of the biceps is in isometric or concentric action in this pull-over exercise example. Biomechanical data and analysis are necessary to determine the actual actions of muscles in movement. There are even cases where muscles accelerate a joint in the opposite direction to that inferred by functional anatomy because of the interaction of a series of linked body segments (Zajac and Gordon 1989; Zajac 1991; Hof and Otten 2005; Kulmala et al. 2016).

Rather invasive biomechanical measurements are usually required to determine exactly what muscle actions are occurring in normal movement. Many studies conducted on animals have shown that muscles often have surprising and complex actions (Herzog 1996a, b; Biewener 1998). One such study of the turkey (Roberts et al. 1997) gastrocnemius (plantar flexor) found the muscle acted in essentially an isometric fashion in the stance phase of level running (■ Fig. 2.14a), while some concentric action was used running uphill (■ Fig. 2.14b). The invasive nature of these kinds of measurements and

be determined just looking at the segments and this one muscle force in isolation. Some readers will have had a physics class where levers are used to illustrate the rotation of a segment and how its geometry can either amplify force or motion. A **lever** is a simple machine consisting of a rigid bar (body segment) and an axis of rotation. What a particular lever configuration does depends on knowing the geometry and all the forces acting on the lever (see ▶ Chap. 7). ■ Figure 2.12 only illustrates the biceps force pulling on the forearm, not other

2

the interesting variations in the musculoskeletal structure of animals (fish, kangaroo rats, wallabies; Griffiths 1989; Shadwick et al. 1998; Biewener 1998) make animal studies a major area of interest for the biomechanics of muscle function.

Similar complex behavior of muscle actions has been observed in humans using recent improvements in ultrasound imaging with video (Finni et al. 2000; Finni et al. 2001: Fukunaga et al. 1997, 2002; Kubo et al. 1999) and implantable fiber-optic force sensors (Komi et al. 1996). Recent studies of the human tibialis anterior have also documented nonlinear and nonisometric behavior of the muscle (lengthening of tendon and aponeurosis while fibers shorten) in isometric actions (Ito et al. 1998; Maganaris and Paul 2000). This is an area of intense research in biomechanics because the lengthening and shortening of muscle fibers, aponeurosis, and tendon from several different muscles can all be documented *in vivo* during human movements (Finni 2006; Fukashiro et al. 2006; Kawakami and Fukunaga 2006). It is clear now that muscle actions in all animal move-

ments are more complicated than can be predicted by the concentric, single-joint analysis of functional anatomy.

Given these many examples of the complexity of muscle actions at the macro- and microscopic levels, the hypothesized muscle actions from functional anatomy in many human movements should be interpreted with caution. Seemingly simple questions of what muscles contribute most to walking, jumping, or any movement represent surprisingly complex biomechanical issues. For example, should the word "eccentric" be used as an adjective to describe phases in weight training exercise (eccentric phase), when all the active muscles are clearly not in eccentric actions in the movement? If the active muscle group, body position, and resistance are well defined, this terminology is likely accurate. When the lifter "cheats" with other muscles in the exercise, modifies exercise technique, or performs a similar sporting movement, the eccentric adjective may not be accurate. The actions of other muscles, external forces like gravity, and the complexity of the musculoskeletal system can make

◨ **Fig. 2.14** Simultaneous muscle force, length, and activation (EMG) measurements of the gastrocnemius of running turkeys. **a** Note that in level running the muscle creates considerable force but the fibers do not shorten, so the muscle is in isometric action and length changes are in the stretching and recoiling of the tendon (muscle acts like a spring). **b** In the stance phase of uphill running, the muscle fibers shorten (concentric action), doing mechanical work to lift the turkey's body (muscle acts as a motor). (Image reprinted with permission from Roberts et al. (1997). Copyright © 1997 American Association for the Advancement of Science)

the isolated analyses of functional anatomy in the anatomical position inaccurate for dynamic movement. Several biomechanical issues that illustrate this point are summarized here and developed throughout the book.

Interdisciplinary Issue: Anthropometry

Anthropometry is the science concerned with measurement of the physical properties (length, mass, density, moment of inertia, etc.) of a human body. Kinanthropometry is an area within kinesiology that studies how differences in anthropometry affect sport performance (see Chaps. 5 and 7 in Bloomfield et al. 1994). *Sports Illustrated* has popularized the large differences across sports with photo spreads of the body types of Olympic athletes. The main organization in this area is the International Society for the Advancement of Kinanthropometry (ISAK). Since humans move in a wide variety of activities, many professionals use anthropometric data. Engineers use these measurements to design tools and workstations that fit most people and decrease risk of overuse injuries. Prosthetic and orthotic manufacturers often make anthropometric measurements on individuals to customize the device to the individual. Motor development scholars track the changes in anthropometric characteristics with growth and development. While people seem to have a wide variety of shapes and sizes, the relative (scaled to size) size of many anthropometric variables is more consistent. Biomechanists use many of these average physical measurements to make quite accurate kinetic or center-of-gravity calculations and models.

2.3.2 The Need for Biomechanics to Understand Muscle Actions

The traditional "kinesiological" analysis of movements of the early twentieth century essentially hypothesized how muscles contributed to motion in each phase of the skill by noting anatomical joint rotations and assuming muscles that create that joint rotation are active. Muscle actions in human movements, however, are not as simple as functional anatomy assumes (Herbert et al. 1993; Bartlett 2012). Several kinds of biomechanical research bear this out and show that the combination of several kinds of quantitative biomechanical research is necessary to understand the functions of muscles in movements.

First, **electromyographic (EMG)** studies have documented general trends in activation of muscles in a particular muscle group, but with considerable potential variation in that trend or in activation between people

(Basmajian and De Luca 1985). The primary source of this variation may be the considerable redundancy (muscles with the same joint actions) of the muscular system. Nearly identical movements can be created by widely varying muscular forces or joint torques (Winter 1984; Patla 1987; Hatze 2000).

EMG studies generally show that the activation patterns of individual muscles are not representative of all muscles in the same functional group (Bouisset 1973; Arndt et al. 1998) and there are differences in how muscles within a muscle group respond to training (Rabita et al. 2000). Even individual muscles are quite sophisticated, with different motor unit activation depending on the task or muscle action (Enoka 1996; Gandevia 1999; Gielen 1999; Babault et al. 2001). Muscles within a muscle group can alternate periods of activity in low-level activities to minimize fatigue (Kouzaki et al. 2002). Muscle activation can vary because of differences in joint angle, muscle action (Nakazawa et al. 1993; Kasprisin and Grabiner 2000), or the degree of stabilization required in the task (Kornecki et al. 2001). For example, a manual muscle test for the biceps used by physical therapists typically evaluates isometric elbow flexion with the forearm in supination to minimize brachioradialis activity and maximize biceps activity (Basmajian and De Luca 1985). EMG studies, however, have also demonstrated that some of these procedures used to isolate specific muscles in physical therapy do not always isolate the muscle hypothesized as being tested (Rowlands et al. 1995; Kelly et al. 1996).

The activation of many muscles to create a specific force or action is called a muscle **synergy**. A muscle **synergy** is a combination of muscle actions that serves to optimally achieve a motor task. There is considerable recognition of the importance of muscle synergies and force sharing of muscles in biomechanical research (Herzog 1996b, 2000; Arndt et al. 1998) and in current rehabilitation and conditioning trends (see Interdisciplinary Issue on training muscles versus movements). How individual muscles share the load is complicated, depending on fiber type, contractile properties, cross-sectional area, moment arm, and antagonism (Ait-Haddou et al. 2000). Motor control uses the term *synergy* to refer to underlying rules of the neuromuscular system for using muscles to coordinate or create movements (Bernstein 1967; Aruin 2001).

Activity: Muscle Synergy

Make a tight fist in your dominant hand as forcefully and quickly as you can. Observe the actions of the superficial muscles of your arm. Why do you think biceps and triceps are isometrically coactivated in a power grip muscle synergy?

2

Recent EMG research has in addition begun to focus on different activation of intramuscular sections within a muscle beyond the traditional gross segmentation in classical anatomy (Paton and Brown 1994; Mirka et al. 1997; Wickham and Brown 1998; Wickham et al. 2004; Brown, et al. 2007; Szucs and Borstad 2013). Wickham and Brown (1998) have confirmed different activation of seven distinct segments of the deltoid muscle, rather than the typical three sections (anterior, intermediate, posterior) of muscle fibers usually identified in anatomy. Activation variability within regions of a muscle may also contribute to risk of injury (Falla and Gallina 2020). These lines of research support the EMG studies mentioned earlier which indicate that activation of muscles is much more complex than had been previously thought or predicted by functional anatomy. Further microanatomy and micro-electrode EMG research on muscles, particularly those with large attachments, will most likely increase our understanding of how parts of the muscles are activated differently to create movement (Vieira and Botter 2021).

Second, the descriptions of musculoskeletal anatomy often do not account for variations in muscle attachment sites across individuals. The numbers and sites of attachments for the rhomboid and scalene muscles vary (Kamibayashi and Richmond 1998). A person born with missing middle and lower fibers of trapezius on one side of their body must primarily rely on rhomboids for scapular retraction. Variations in skeletal structure are also hypothesized to contribute to risk of injury. For example, the shape of the acromion process of the scapula is believed to be related to a risk of impingement syndrome (Whiting and Zernicke 2008). The role of anatomical variation in gross anatomy (attachments, angles of long axes, etc.) or in muscle architecture (Richmond 1998) and their biomechanical effects of muscles actions and injury risk remain an important area of study.

Third, the linked nature of the human body makes the isolated functional anatomical analysis incomplete. This linking of body segments means that muscle actions have dramatic effects on adjacent and other joints quite distant from the ones the muscles cross (Zajac and Gordon 1989; Zajac 1991; Hof and Otten 2005; Kulmala et al. 2016). This redistribution of mechanical energy at distant joints may be more important to some movements than the traditional joint action hypothesized by functional anatomy (Zajac et al. 2002). Zajac and Gordon (1989), for example, showed how soleus activity in a sit-to-stand movement tends to extend the knee joint more than it plantar flexes the ankle joint. Physical therapists know that the pectoralis major muscle can be used to extend the elbow in closed kinetic chain (see ▶ Chap. 6) situations for patients with triceps paralysis (Smith et al. 1996). Functional anatomy does not analyze how forces and torques created by a muscle are distributed throughout all the joints of the skeletal system or how

these loads interact between segments. Zajac and Gordon (1989) have provided a convincing argument that the classification of muscles as **agonists** or **antagonists** should be based on biomechanical models and joint accelerations, rather than just torques the muscles create.

Dramatic examples of this wide variety of effects of muscles can be seen in multiarticular muscles (van Ingen Schenau et al. 1989; Zajac 1991). There is considerable interest in the topic of biarticular or multiarticular muscles, and it is known that they have different roles compared to similar monoarticular muscles (Prilutsky and Zatsiorsky 1994; van Ingen Schenau et al. 1995; Hof 2001; Cleather et al. 2015; Flaxman et al. 2017). Another example of the complexity of movement is how small differences in foot placement (angle of ankle plantar/dorsiflexion) dramatically affect which joint torques are used to cushion the shock in landing (DeVita and Skelly 1992; Kovacs et al. 1999). A flat-footed landing minimizes a plantar flexor's ability to absorb shock, increasing the torque output of the hip and knee extensors. Small differences in foot angle in walking also affect the flexor or extensor dominance of the knee torque (Simonsen et al. 1997) and the frontal plane knee torques that may be related to knee injury (Gregersen et al. 2006; Teichtahl et al. 2006). The kinematics and kinetics Chapters (▶ 5, 6, and 7, respectively) will expand on the effects of joint and segment interactions in human movement.

Interdisciplinary Issue: Training Muscles Versus Movements

In the strength and conditioning field, a long-running philosophical debate is related to a greater emphasis on training functional movements rather than training specific muscle groups (Gambetta 1995, 1997). The most recent reincarnation of this debate is a purported emphasis on "functional training." This debate is quite similar to the debate about the relative benefits of training with free weights or with machines. Training with free weights can more easily simulate the balance and stabilizing muscle actions in normal and sport movements. The advantage of machines is that they provide more muscle group-specific training with greater resistance throughout the range of motion that is not as dependent on position relative to gravity as free weights are. How might rehabilitation and conditioning professionals use biomechanics and EMG research to help match training to the demands of normal movement? How might this philosophical debate be resolved by a clear definition of function and focused EMG and biomechanics research?

The fourth line of biomechanical research documenting the complexity of muscular actions creating movement are **modeling** and **simulation**. **Modeling** involves the development of a mathematical representation of the biomechanical system, while **simulation** uses biomechanical models to examine how changes in various techniques and parameters affect the movement or body. Biomechanical models of the human body can be used to simulate the effects of changes in any of the parameters of the model. The more simple the model, the easier the interpretation and application of results. For example, models of the motion of body segments in airborne skills in gymnastics and diving are quite effective in determining their effect on flight and rotation (Yeadon 1998; Hiley et al. 2015; Yeadon and Hiley 2017). As biomechanics models get more complicated and include more elements of the musculoskeletal system, the more difficult it is to validate the model. Interpretation is even complicated because of the many interrelated factors and variations in model parameters across people (Chow et al. 1999).

Despite the many controversial issues in biomechanical modeling, these kinds of studies show that the actions of muscles in movements are quite complex and are related to segment and muscle geometry (Bobbert and van Ingen Schenau 1988; Doorenbosch et al. 1997), muscle elasticity (Anderson and Pandy 1993), coordination (Hatze 1974; Bobbert and van Soest 1994; Nagano and Gerritsen 2001), and accuracy or injury (Fujii and Hubbard 2002; Thelen et al. 2006). One simulation found that non-extensor muscles of the legs could be used to improve jumping performance (Nagano et al. 2005), and it is also possible that coordination in a movement even varies slightly across people because of differences in muscle mechanics (Chowdhary and Challis 2001). Here we have the paradox of learning again. What muscles do to create movement is quite complex, so scholars and professionals must decide what level of biomechanical system to study to best understand movement. The strength and conditioning field commonly groups muscles into functional groups like the knee extensors (quadriceps) or knee flexors (hamstrings). Whatever movements or level of analysis a kinesiology professional chooses, biomechanics needs to be added to anatomical knowledge to make valid inferences about human movement. The next section briefly shows how the sports medicine professions have integrated more biomechanical information into their professional practice.

> **Application: Muscle Groups**
>
> If muscles create movement in complex synergies that are adaptable, should kinesiology professionals abandon the common practice of naming muscle groups according to anatomical function (quads [knee extensors] or calf [ankle plantar flexors])? Such an extreme reaction to the complexity of biomechanics is not necessary. This common terminology is likely appropriate for prescribing general strength and conditioning exercises. It may even be an appropriate way to communicate anatomical areas and movements in working with athletes knowledgeable and interested in performance. Kinesiology professionals, however, need to qualitatively diagnose movements at a higher level than their clients and remember that this simplified terminology does not always give an accurate picture of how muscles really act in human movement. Biomechanical and other kinesiology research must be integrated with professional experience in analyzing muscles and muscle group contributions to movement. Access to biomechanics research is critical to this deeper understanding of actual muscle function.

2.3.3 Sports Medicine and Rehabilitation Applications

Musculoskeletal anatomy and its motion terminology are important in kinesiology and sports medicine, but it cannot be the sole basis for determining the function of muscles in human movement. Medical doctors specializing in sports medicine found that their extensive training in anatomy was not enough to understand injuries and musculoskeletal function in the athletes they treated (McGregor and Devereux 1982). This recognition by MDs that their strong knowledge of anatomy was incomplete to understand function and that they needed the sciences of kinesiology was a factor in the fusion of medical and kinesiology professionals that formed the American College of Sports Medicine (ACSM) in 1954.

Today, many kinesiology students prepare for careers in medicine and sports medicine/allied health careers (athletic training, physical therapy, orthotics, prosthetics, strength and conditioning). These professions are concerned with analyzing the actions of muscles in movement. Where can sports medicine professionals (athletic trainers, physical therapists, physical medicine, strength and conditioning) get the most accurate knowledge on the biomechanical function of specific areas of the human body? Fortunately, there are several sources that strive to weigh the anatomical/clinical observations with biomechanical research. These sources focus on both normal and pathomechanical function of the human body. The following sources are recommended since they represent this balanced treatment of the subject, not relying solely on experience or research (Basmajian and Wolf 1990; Smith et al. 1996; Kendall et al. 2005; Nordin & Frankel, 2021).

It is important to remember that biomechanics is an indispensable tool for all kinesiology professionals trying to understand how muscles create movement, how to improve movement, and how problems in the musculoskeletal system can be compensated for.

2

> **Interdisciplinary Issue: Anatomy and Biomechanics**
> Many women active in sports suffer breast discomfort from the jarring movements of running, jumping, and landing. Biomechanics, anatomical/medical, and engineers in sports apparel industries work closely together try and understand this problem and design supportive and protective garments. Success in this endeavor depends on interdisciplinary, collaborative work (McGhee and Steele 2020). The mechanics of deformable bodies like human breast and other soft tissues is introduced in ▶ Chap. 4.

2.4 Summary

Anatomy is the descriptive study of the structure of the human body. This structural knowledge is an important prerequisite, qualitative language for the study of human movement, but must be combined with biomechanical knowledge to determine how muscle actions create human movement. Kinesiology professionals and students of biomechanics need to continually review their knowledge of musculoskeletal anatomy. Muscles tend to be activated in synergies to cooperate or coordinate with other forces to achieve movement goals. Muscle tension is created from active or passive components, and the action muscles create are either eccentric, concentric, or isometric. Many kinds of biomechanical research have shown that the actions of muscles in normal movement are more complicated than what is hypothesized by functional anatomy.

? Review Questions

1. What are the major anatomical terms used in kinesiology and medicine to describe the position and motion of the body?
2. What structural and functional properties of muscle cells are different from other body cells?
3. How do fiber properties and arrangement affect force and range of motion potential of a muscle?
4. Name and define the three kinds of muscle actions.
5. What are the two major sources of muscle tension, and where in the range of motion are they most influential?
6. Explain the Hill three-component model of muscle and how the components relate to the sources of muscle tension.
7. Why is the mechanical method of muscle action analysis used in functional anatomy inadequate to determine the actions of muscles in human movement?
8. How does biomechanics help kinesiology professionals understand the causes and potential improvement of human movement?
9. What is the relationship between a plane of motion and the axis of rotation creating motion in that plane? Given an example using an anatomical plane of joint motion.

! Key Terms

Active tension · Agonist · Antagonist · Anatomy · Anthropometry · Actin · Concentric · Contractile component · Eccentric · Fascicle · Force · Hill muscle model · Hypertrophy · Isometric · Lever · Modeling · Moment of force (torque) · Muscle action · Myofibril · Myosin · Parallel elastic component · Passive insufficiency · Passive tension · Pennation · Sarcomere · Series elastic component · Simulation · Synergy

★ Suggested Reading

Basmajian, J. V., & De Luca, C. J. (1985). *Muscles alive: Their functions revealed by elec- tromyography* (5th ed.). Baltimore: Williams & Wilkins.

Dickinson, M. H., et al. (2000). How animals move: A integrated view. *Nature, 288,* 100–106.

Faulkner, J. A. (2003). Terminology for contractions of muscles during shortening, while isometric, and during lengthening. *Journal of Applied Physiology, 95,* 455–459.

Fitts, R. H., & Widrick, J. J. (1996). Muscle mechanics: Adaptations in muscle resulting from exercise training. *Exercise and Sport Sciences Reviews, 24,* 427–473.

Hellebrandt, F. A. (1963). Living anatomy. *Quest, 1,* 43–58.

Herbert, R., et al. (1993). Making inferences about muscles forces from clinical observations. *Australian Journal of Physiotherapy, 39,* 195–202.

Herzog, W. (2000). Muscle properties and coordination during voluntary movement. *Journal of Sports Sciences, 18,* 141–152.

IFAA. (2002). *Terminologia anatomica: International anatomical terminology* (2nd ed.). Stuttgart: Theme.

Kendall, F. P., et al. (2005). *Muscles: Testing and function* (5th ed.). Baltimore: Lippincott Williams & Wilkins.

Kleissen, R. F. M., et al. (1998). Electromyography in the biomechanical analysis of human movement and its clinical application. *Gait and Posture, 8,* 143–158.

Lieber, R. L., & Friden, J. (2000). Functional and clinical significance of skeletal muscle architecture. *Muscle and Nerve, 23,* 1647–1666.

Merlletti, R., & Mucceli, S. (2019). Tutorial. Surface EMG detection in space and time: Best practices. *Journal of Electromyography and Kinesioloyg, 49.* https://doi.org/10.1016/j.jelekin.2019.102363.

Roberts, T. J., et al. (1997). Muscular force in running turkeys: The economy of minimizing work. *Science, 275,* 1113–1115.

Zajac, F. E. (2002). Understanding muscle coordination of the human leg with dynamical simulations. *Journal of Biomechanics, 35,* 1011–1018.

⟳ Internet Resources

▶ Anatomage Virtual Human Dissection. *anatomage.com/anatomage-medical/*
▶ Body Worlds–von Hagens' plastic-preserved human bodies. *koerperwelten.de/en/pages/home.asp*
▶ Hypermuscle: Review of anatomical joint motion terminology from the University of Michigan. *med.umich.edu/lrc/Hypermuscle/Hyper.html*
▶ International Federation of Associations of Anatomists *ifaa.net.*
▶ Tour of Visible Human Project–Simple review of anatomical planes and structures using images from the NIH visible human project. *madsci.org/~lynn/VH/*

Fundamentals of Biomechanics and Qualitative Movement Diagnosis

Key Questions

➤ **What do you need to know to begin?**
The definitions of kinesiology and biomechanics, as well as their application of human movement professions

➤ **Why do you need to know this material?**
Specialized biomechanical terms and principles are needed to observe, evaluate, and diagnosis human movement technique.

➤ **What is key the idea?**
Biomechanics principles are essential tools in the qualitative diagnosis of human movement technique in numerous professions.

In ▶ Chaps. 1 and 2, we found that biomechanics provides tools that are needed to understand the causes of human motion and use that knowledge to improve performance and reduce the risk of injury. In order to facilitate the use of these biomechanical tools, this text will emphasize the qualitative understanding of mechanical laws, concepts, and principles. Many chapters, however, will include several quantitative examples using the algebraic definitions of the mechanical variables being discussed. Mathematical formulas are the precise language of science (like biomechanics) needed to show the importance, interactions, and relationships between biomechanical variables. While more rigorous calculus forms of these equations provide the most accurate answers commonly used by scientists in biomechanics (Beer et al. 2018; Hamill Knutzen, and Derrick 2014; Ozkaya et al. 2017; Zatsiorsky 1998, 2002), the majority of kinesiology majors will benefit most from a qualitative understanding of these mechanical concepts. This chapter begins with key mechanical variables and terminology essential for introducing other biomechanical concepts. This chapter will emphasize the conceptual understanding of these mechanical variables and leave more detailed development and some quantitative examples for later in the text. Next, nine general principles of biomechanics are introduced that will be developed throughout the rest of the text. These principles use less technical language and are the theoretical tools for applying biomechanical knowledge in the qualitative diagnosis of human movement. The chapter concludes by summarizing a model of qualitative diagnosis (Knudson 2013) that is illustrated in the application chapters of the book.

3.1 Key Mechanical Concepts

3.1.1 Mechanics

Before we can begin to understand how humans move, there are several mechanical terms and concepts that must be clarified. **Mechanics** is the branch of physics that studies the motion of objects and the forces that cause that motion. The science of mechanics is divided into three main areas identified in ◘ Fig. 3.1. The three main areas most relevant to sport biomechanics are **rigid body**, **deformable body**, and **fluids**.

Mechanics studies the motion of objects using models. Models are theoretical simplifications (physical and mathematical) of a system because objects interact with other objects in motion. Study of linear motion can be considerably simplified by assuming an object is a **point mass**. An object represented by a point mass model does not have dimensions, and motion is assumed to be represented by a dimensionless point in space with all the mass located at that point.

In **rigid body mechanics**, the object being analyzed is assumed to be rigid and the deformations in its shape so small they can be ignored. While this almost never happens in any material, this assumption is quite reasonable for most biomechanical studies of the motion and causes of motion

D. Knudson, *Fundamentals of Biomechanics*, https://doi.org/10.1007/978-3-030-51838-7_3

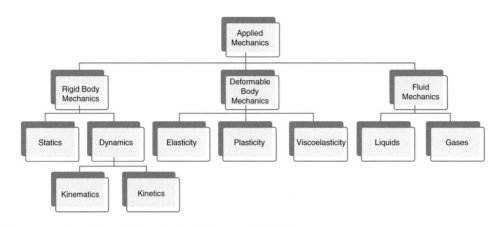

■ **Fig. 3.1**   Primary areas of study in applied mechanics. (Ozkaya et al. (2017))

■ **Fig. 3.2**   The major branches of mechanics used in most biomechanical studies

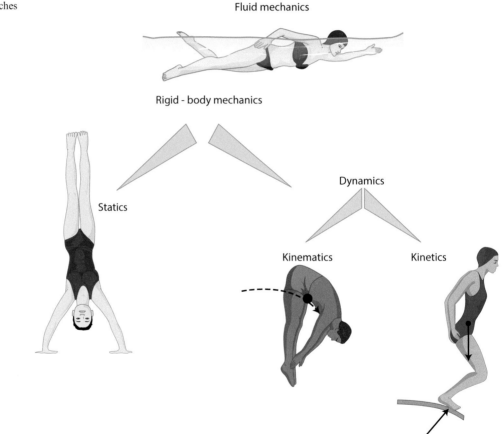

of the major segments of the body. The rigid body assumption in studies saves considerable mathematical and modeling work without great loss of accuracy. Some biomechanists, however, use deformable body mechanics to study how biological materials respond to external forces that are applied to them and the internal force distributions in them. **Deformable body** mechanics studies how forces are distributed within a material and can be focused at many levels (cellular to tissues/organs/system) to examine how forces stimulate growth or cause damage.

▶ Chapter 4 will cover some of these concepts in how body tissues are deformed by forces. **Fluid mechanics** is concerned with the forces in fluids (liquids and gasses). A biomechanist would use fluid mechanics to study heart valves, swimming, or adapting sports clothing and equipment to minimize air resistance.

Most sports biomechanics studies are based on rigid body models of the skeletal system. Rigid body mechanics is divided into **statics** and **dynamics** (■ Fig. 3.2). **Statics** is the study of objects at rest or in uniform (constant) motion. **Dynamics** is the study of objects being accelerated by the actions of forces. Most importantly, dynamics is divided into two branches: **kinematics** and **kinetics**. **Kinematics** is motion description. In kinematics the motions of objects are usually measured in linear (meters, feet, etc.) or angular (radians, degrees, etc.) terms. Examples of the kinematics of running could be the speed of the athlete, the length of the stride, or the angular velocity of hip extension. Most angular mechanical variables have the adjective "angular" before them. **Kinetics** is concerned with determining the causes of motion. Examples of kinetic variables in running are the forces between the feet and the ground or the forces of air resistance. Understanding these variables gives the track coach knowledge of the causes of running performance. Kinetic information is often more powerful in improving human motion because the causes of poor performance have been identified. For example, knowing that the timing and size of hip extensor action are weak in the takeoff phase for a long jumper may be more useful in improving performance than knowing that the jump was shorter than expected.

3.1.2 Basic Units

The language of science is mathematics. Biomechanics often uses some of the most complex kinds of mathematical calculations, especially in deformable body mechanics. Fortunately, most of the concepts and laws in classical (Newtonian) rigid body mechanics can be understood in qualitative terms. A conceptual understanding of biomechanics is the focus of this book, but algebraic definitions of mechanical variables will be presented and will make your understanding of mechanical variables and their relationships deeper and more powerful.

First, let's look at how even concepts seemingly as simple as numbers can differ in their complexity. **Scalars** are variables that can be completely represented by a number and the units of measurement. The number and units of measurement (10 kg, 100 m) must be reported to completely identify a scalar quantity. It makes no sense for a track athlete to call home and say, "Hey mom, I did 16 and 0"; they need to say, "I made 16 feet with 0 fouls." The positive number given a scalar quantity represents the magnitude or size of that variable.

Vectors are more complicated quantities, where size, units, and *direction* must be specified. ◘ Figure 3.3 shows several scalars and the associated vectors common in biomechanics. For example, **mass** is the scalar quantity that represents the quantity of matter for an object. That same object's **weight** is the gravitational force of attraction between the earth and the object. The difference between mass and weight is dramatically illustrated with pictures of astronauts in orbit about the earth. Their masses are essentially unchanged, but their weights are virtually zero because of the microgravity when far from and orbiting the mass of the earth.

Biomechanics commonly uses directions at right angles (horizontal/vertical, longitudinal/transverse) to mathematically handle vectors. Calculations of velocity vectors in a two-dimensional (2D) analysis of a long jump are usually done in one direction (e.g., horizontal) and then the other (vertical). The directions chosen depend on the needs of the analysis. Symbols representing vector quantities like velocity (**v**) in this text will be identified with **bold letters**. Physics and mechanics books also use underlining or an arrow over the symbol to identify vector quantities. These and other rules for vector calculations will be summarized in ▶ Chap. 6. These rules are important because when adding vectors, one plus one is often not two because the directions of the vectors were different. When adding scalars with the same units, one plus one is always equal to two. Another important point related to vectors is that the sign (+ or –) corresponds to vector directions. A –10 lb force is not less than a +10 lb force; they are the same size but in opposite directions. The addition of vectors to determine their net effect is called the **resultant** and requires right-angle trigonometry. In ▶ Chap. 6 we will also subtract or break apart a vector into right-angle **components** to take advantage of these trigonometry relationships to solve problems and to "see" other important pushes/pulls of a force.

There are two important vector quantities at the root of kinetics: **force** and **torque**. A **force** is a straight-line push or pull, usually expressed in pounds (lbs) or Newtons (N). The symbol for force is **F**. Note that this push or pull is an interactional effect between two bodies. Sometimes this "push" appears obvious as in a ball hitting a bat, while other times the objects are quite distant as with the "pull" of magnetic or gravitational forces. Forces are vectors, and vectors can be physically represented or drawn as arrows (◘ Fig. 3.4). The important characteristics of vectors (size and direction) are directly apparent on the figure. The length of the arrow represents the size or magnitude (500 N or 112 lbs), and the orientation in space represents its direction (15 degrees above horizontal). Forces tend to create linear motion acting on rigid bodies but can create motion and deformation in non-rigid bodies.

The corresponding angular variable to force is a **moment of force** or **torque**. A moment is the rotating effect of a force and will be symbolized by an **M** for moment of force or **T** for torque. This book will use the term "torque" synonymously with "moment of force." This is a common English meaning for torque, although there is a more specific mechanics-of-materials meaning (a torsion or twisting moment) that leads some scientists to prefer the term "moment" or "moment of force." When a force is applied to an object that is not on line with the

3

■ **Fig. 3.3** Comparison of various scalar and vector quantities in biomechanics. Vector quantities must specify magnitude and direction. Vectors are drawn/visualized by an arrow (length = magnitude and arrow head indicates direction)

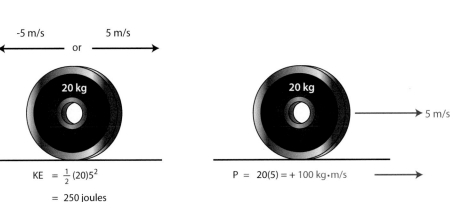

center of the object, the force will also create a torque that tends to rotate the object. In ■ Fig. 3.4 the impact force acts below the center of the ball and would create a torque that causes the soccer ball to acquire backspin. We will see later that the most common units of torque are pound–feet (lb ft) and Newton–meters (N•m).

Let's look at an example of how kinematic and kinetic variables are used in a typical biomechanical measurement of **isometric** muscular strength. Recall that isometric is a muscle research term referring to muscle actions performed in approximately constant (*iso*) length (*metric*) conditions, where the muscle forces and moment are equal and opposite to external forces and moments. The example of a spring is important for learning how mathematics and graphs can be used to understand the relationship between variables, as well as how a new variable (stiffness) can be created from the

combination of two variables. This example will also help to understand our future examples of how muscles, tendons, and ligaments can be said to have spring-like behavior when they stretch, storing energy in passive tension. ■ Figure 3.5 illustrates the force–displacement graphs for two springs in a handgrip **dynamometer**. A **dynamometer** is a force-measuring device. As a positive force (**F**) pulls on the spring, the spring is stretched a positive linear distance (displacement = **d**). Displacement is a kinematic variable; force is a kinetic variable.

Therapists often measure a person's grip strength in essentially isometric conditions because the springs in hand dynamometers are very stiff and only elongate very small distances. The force–displacement graph in ■ Fig. 3.5 shows a very simple (predictable) and linear relationship between the force in the spring (**F**) and the resulting elongation (**d**). In other words, there is a uni-

500 N

15°

◻ Fig. 3.4 Vector representation of the force applied by a foot to a soccer ball. The magnitude and direction properties of a vector are both apparent on the diagram: the length of the arrow represents 500 Newtons of force, while the orientation and tip of the arrow rep- re-sent the direction (15° above horizontal) of the force

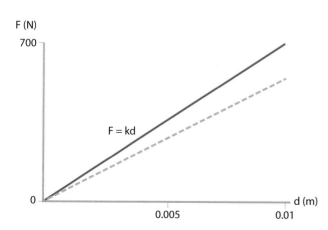

F (N)

700

F = kd

0

0.005   0.01   d (m)

◻ Fig. 3.5 A graph (solid line) of the relationship between the force (**F**) required to stretch a spring a given displacement (**d**) and a less stiff spring (dashed line). The elasticity of the springs is the slope of the lines. The slope is the constant (k) in Hooke's Law: (**F** = k • **d**)

form increase (constant slope of the line) in force with increasing spring stretch. We will see later on in ▶ Chap. 4 that biological tissues have much more complex (curved and timing dependent) mechanical behaviors when loaded by forces, but there will be mostly linear regions of their load–deformation graphs late in the range of motion that are representative of their elastic properties.

Let's extend our example and see how another mechanical variable can be derived from force and displacement. Many simple force measuring devices (e.g., bathroom and fishing scales) take advantage of the elastic behavior of stiff metal springs that are stretched or compressed short distances. This relationship is essentially the mathematical equation (**F** = k • **d**) of the calibration line illustrated in ◻ Fig. 3.5 and is called Hooke's Law. Hooke's Law is valid for small deformations of stiff materials like springs. The **stiffness** (potential elasticity) of the spring is symbolized as k, which represents the slope of the **F-d** line, so the spring with the solid line would store more energy than the spring with the dashed line. The energy stored in the spring can be visualized as the area under the **F-d** line. In ▶ Chap. 4 we will look at the stiffness of biological tissues as the slope of the linear region of a graph like this. If we plug in the largest force and displacement (700 = k • 0.01), we can solve for the stiffness of the spring and find it to be 70,000 N/m. This says that the spring force will increase 70,000 Newtons every meter it is stretched. This is about 15,730 pounds of tension if the spring was stretched to about 1.1 yards! Sounds pretty impressive, but remember that the springs are rarely elongated that much, and you might be surprised how stiff muscle-tendon units can get when strongly activated.

Engineers measure the stiffness or elasticity of a material with special machines that simultaneously record the force and deformation of the material. The slope of the load–deformation graph (force/length) in the linear region of loading is used to define stiffness. **Stiffness** is the measure of elasticity of the material, but this definition often conflicts with most people's common understanding of elasticity. People often incorrectly think elasticity means an object that is easily deformed with a low force, which is really compliance (length/force) which is the opposite of stiffness. An engineer would say that there was less stiffness or greater compliance in the spring illustrated as a dashed line.

Can you find the stiffness (spring constant, k) that corresponds to the dashed calibration line in ▶ Fig. 3.5? Remember that the stiffness, k, corresponds to the slope of the line illustrated in the figure and represents the change in force for a given change in length. The slope or rate of change of a variable or graph will be an important concept repeated again and again in biomechanics. Remember that forces and displacements are vectors, so directions are indicated by the sign (+ or −) attached to the number. What do you think the graph would look like if the force were reversed, i.e., to push and compress the spring rather than stretching it? What would happen to the sign of **F** and **d**?

Activity: Elasticity

Take a rubber band and loop it between the index fingers of your hands. Slowly stretch the rubber band by moving one hand away from the other. The tension in the rubber band creates a torque that tends to abduct the metacarpophalangeal joints of your index finger. Does the tension your fingers sense resisting the torque from the rubber band uniformly increase as the band is stretched? Does a slightly faster stretch feel different? According to Hooke's Law, elastic materials like springs (rubber bands are mostly linear) create forces directly proportional to the deformation of the material, but the timing of the stretch does not significantly affect the resistance. ▶ Chapter 4 will deal with the mechanical responses of biological tissues, which are not perfectly elastic, so the rate of stretch affects the mechanical response of the tissue.

There are many other mechanical variables that help us understand how human movement is created. These variables (e.g., impulse, angular momentum, kinetic energy) often have special units of measurement. What all these mechanical variables and units have in common is that they can be expressed as combinations of only a few base units (Martin-Delgado, 2020). These base units are length, mass, and time. In the International System (SI), these units are the second (s), kilogram (kg), meter (m), and radian (rad). Scientific research commonly uses SI units because they are base 10, are used throughout the world, and move smoothly between traditional sciences. A Joule of mechanical energy is the same as a Joule of chemical energy stored in food. When this book uses mathematics to reinforce a conceptual understanding of mechanics in human movement (like in ◘ Fig. 3.4), the SI system will usually be used along with the corresponding English units for a better intuitive feel for many students. The symbols used are based on the recommendations of the International Society of Biomechanics (ISB 1987).

These many mechanical variables are vitally important to the science of biomechanics and the integration of biomechanics with other kinesiological sciences. Accurate quantification is better than qualitative evaluation; however, application of biomechanics by kinesiology professionals most often is based on qualitative evaluation and diagnosis of human movement technique. The next section will outline biomechanical principles based on the science and specialized terminology

of biomechanics that facilitate the qualitative diagnosis of human movement.

3.2 Fundamentals of Biomechanics

Biomechanists measure all kinds of linear and angular mechanical variables to measure motion and find the causes of that motion. While these variables and studies are extremely interesting to biomechanists, some kinesiology students and professionals may not find them quite so inherently stimulating. Most kinesiology professionals want to know the basic theories or principles of biomechanics that they can apply in their jobs. This section proposes nine such principles of biomechanics and demonstrates how some come from Newton's laws of motion. These biomechanical principles must be combined with other tools from your kinesiology toolbox to most effectively solve movement problems. Because these principles are the application rules for kinesiology professionals, they have usually been given less scientific names so that we can communicate effectively with our clients.

3.2.1 Principles and Laws

The nine principles of biomechanics that follow take the form of general principles related to human movement. It is important to realize that principles for application are not the same as scientific laws. **Science** is a systematic method for testing hypotheses with experimental evidence for the purpose of improving our understanding of reality. Science uses a process, known as the scientific method, for testing a theory about a phenomenon with measurements, then reevaluating the theory based on the data. Ultimately, science is interested in finding the truth, facts, or laws of nature that provide the best understanding of reality. When experimentation shows data always consistent with a theory (given certain conditions), then the theory becomes a law. Scientists must always be open to new data and theories that may provide a more accurate description or improved understanding of a phenomenon. True scientific revolutions that throw out long-held and major theories are not as common as most people think. Though news reporters often herald scientific "breakthroughs," they are usually exaggerating the importance of a small step in what is a slow, thoughtful process of weighing a great deal of evidence by the scientists in a particular field like biomechanics.

Note that science is not defined as a method for making practical applications of knowledge. **Technology** is the term usually used to refer to the tools and methods of applying scientific knowledge to solve problems or perform tasks. Remember that in ▶ Chap. 1, we noted the belief of some scholars that studying academic disciplines and doing theoretical research are worthy enterprises without any need to show any practical application of knowledge. Even in more "applied" fields like kinesiology, there is a long history of a theory-to-practice or a science-to-profession gap (Harris 1993; Knudson 2005). Why does this gap exist and get in the way of evidence-based practice? It might exist because some scholars are hesitant to propose application based on what is often less-than-conclusive data or they might be concerned about receiving less recognition for applied scholarship. Practitioners contribute to this gap as well by refusing to recognize the theoretical nature of science, by not reading widely to compile the necessary evidence for practice or seeking clinical practice guidelines, and by demanding simple "how-to" and "one size fits all" rules of human movements when these simple answers often do not exist.

This text is based on the philosophy that the best use of the science of biomechanics is in its translation to principles for improving human movement. These principles are general rules for the application of biomechanics that are useful for most all human movements. Some of the principles are based on major laws of mechanics, many of which are hundreds of years old. For example, Newton's laws of motion are still used at NASA because they accurately model the motion of spacecraft. Recent advancements in theoretical physics are only an improvement in very extreme conditions (near the speed of light like satellite communications) that are not relevant to human motion. Unfortunately, the human body is a much more complicated system than simple mechanical structures like carts and bridges, and biomechanists have not had hundreds of years to make progress on theories of human movement. For these reasons, these nine principles of application should be viewed as general rules that currently fit what we currently know about the biomechanics of human movement.

3.2.2 Nine Principles for Application of Biomechanics

The nine principles of biomechanics proposed in this text were selected because they constitute the minimum number or core principles that can be applied to all human movements and because they provide a simple paradigm or structure to apply biomechanical knowledge. The names of the principles are put in the common language of application; however, each can be directly linked to the concepts and laws of biomechanics. Special attention has been paid to make application of these principles both friendly and consistent with the specialized terminology of mechanics. As kinesiology professionals, you will know the names of the biomechanical laws and theories behind these principles, but you will need to use more applied terminology when communicating with clients. This section will provide a description of each principle, and the application of these principles will be developed throughout the text. The principles can be organized (◘ Fig. 3.6) into ones dealing primarily with the creation of movement (process) and ones dealing with the outcome of various projectiles (product).

I want to point out that these principles are based primarily on work of several biomechanists (Norman 1975; Hudson 1995) who have developed generic biomechanical principles for all human movements. Past biomechanics textbooks have proposed general principles for all movements (Meinel and Schnabel 1998); various categories of human movements like throwing, catching, and running (Bunn 1972; Broer and Zernicke 1979; Luttgens and Wells 1982; Dyson 1986; Kreighbaum and Barthels 1996); or specific movements (Groves and Camaione 1975). Some biomechanists believe that general principles applicable to all sports are difficult to identify and have limited practical application due to unique goals and environmental contexts of skills (Hochmuch and Marhold 1978). This book is based on the opposite philosophy. Kinesiology professionals should keep in mind the specific goals and contextual factors affecting a movement, but the nine principles of biomechanics are important tools for improving all human movements.

The first principle in biomechanics is the **force–motion principle**. Force–motion says that unbalanced forces are acting on our bodies or objects when we either create or modify movement. In quiet standing the force of gravity is balanced by ground reaction forces under our feet (◘ Fig. 3.7), so to move from this position, a person creates larger horizontal and vertical forces with their legs. This simple illustration of the body is our first example of what in mechanics is called a **free-body diagram**. A **free-body diagram** is a simplified model of any system or object drawn with the significant forces acting on it. The complexity and detail of the free-body diagram depend on the purpose of the analysis. Inspection of ◘ Fig. 3.7 should make it qualitatively obvious that the addition of the two vertical forces illustrated would cancel each other out, keeping the person essentially motionless in the vertical direction (standing position). The force–motion principle here correctly predicts no change in motion, since there is no unbalanced force acting on the person. Later on in the text, we will use free-body diagrams to actually calculate the effect of forces and torques on the motion of the human body, and we will study the effects of forces acting over time to change the motion of the human body. We will also

◻ **Fig. 3.6** The nine principles of biomechanics can be classified into those related to movement of the body or a projectile. The human body can be a projectile, so all nine principles can be applied to the human body

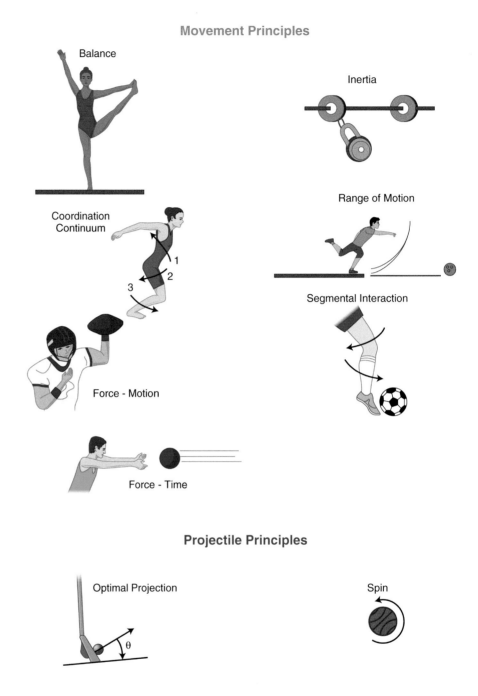

come to see later that this principle is based on Newton's three laws of motion. The application of the force–motion principle in qualitative diagnosis will be explored throughout the text.

An important thing to notice in this principle is the sequence of events. Forces must act first, before changes in motion can occur. Detailed study of kinematics will illustrate when the motion occurred relative to the acceleration and force causing it. Suppose a person is running on a sidewalk and a small child darts directly in the

runner's path to grab a bouncing ball. In order to avoid the child, the runner must change the state of motion. The force–motion principle tells the kinesiology professional that the runner's sideward movement (a change in direction and speed) had to be created by large forces applied by the leg to the ground. The force applied by the leg comes first, and the sideward motion to avoid the collision was the result.

Substantial changes in motion do not instantly occur but are created over time, which leads us to the next

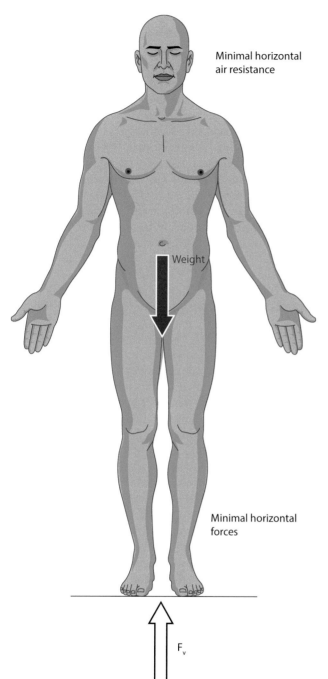

Minimal horizontal
air resistance

Weight

Minimal horizontal
forces

F_v

◼ Fig. 3.7 A free-body diagram of a person quietly standing. The major vertical forces acting on the person (weight and vertical ground reaction force F_v) are illustrated, while horizontal forces are small enough to ignore and are not illustrated

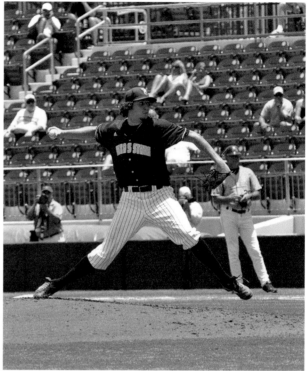

◼ Fig. 3.8 The forward stride of a pitcher increases the range of motion used to accelerate the body and eventually the baseball. Photo used by permission of Texas State University Department of Athletics

relationship, the original language of Newton's second law, is the mathematical explanation of this important principle.

Another important principle to understand in the modification of motion is **inertia**. **Inertia** can be defined as the property of all objects to resist changes in their state of motion. Newton's first law of motion outlines the principle of inertia. The Newtonian view of inertia as a fundamental property of objects and motion was a major conceptual leap, rejecting the old Aristotelian view that constant application of force was required for motion. The linear and angular measures of inertia are mass (m) and moment of inertia (I). We will see that inertia can be viewed as a resistance to state of motion in the traditional sense, but this property can also be used to an advantage when modifying motion or transferring force from one body segment to another.

The next principle involves the **range of motion** the body uses in movement. **Range of motion** is the overall motion used in a movement and can be specified by linear or angular motion of the body segments. The purpose of some movements might require that some body segments limit range of motion, while others requiring maximum speed or force might require larger ranges of motion. Increasing the range of motion in a movement can be an effective way to increase speed or to gradually slow down from a high speed. A baseball pitcher taking a longer stride (◼ Fig. 3.8) is increasing the range of

principle of **force–time**. It is not only the amount of force that can increase the motion of an object; the amount of time over which force can be applied also affects the resulting motion. A person using a longer approach in bowling has more time to apply forces to increase ball speed. Increasing the time to apply force is also an important technique in slowing down objects (catching) and landing safely. The impulse–momentum

3

motion of the weight shift. Since moving through a range of motion takes time, this principle is related to the force–time principle.

The next biomechanical principle is **balance**. **Balance** is a person's ability to control their body position relative to some base of support. Stability and mobility of body postures are inversely related, and several biomechanical factors are involved in manipulating a person's stability and mobility. A handstand is a difficult gymnastic skill not only because of the upper extremity muscular strength required but also because of the small base of support in the anterior and posterior directions. Athletes in the starting blocks for sprints choose body postures with less stability in favor of increased mobility in the direction of the race.

How the muscle actions and body segment motions are timed in a human movement is usually referred to as coordination. The **Coordination continuum** principle says that determining the optimal timing of muscle actions and segmental motions depends on the goal of the movement. If high forces are the goal of the movement, more simultaneous muscle actions and joint rotations are usually observed, while low-force and high-speed movements tend to have more sequential muscle and joint actions (Hudson 1995; Kreighbaum and Barthels 1996). These two strategies (simultaneous/sequential) can be viewed as a continuum, with the coordination of most motor skills falling somewhere between these two strategies.

The principle of **Segmental interaction** says that the forces acting in a system of linked rigid bodies can be transferred through the links and joints. Muscles normally act in short bursts to produce torques that are precisely coordinated to complement the effects of torques created by forces at the joints. A wide variety of terms have been used to describe this phenomenon (transfer,

summation, sequential) because there are many ways to study human movement. This variety of approaches has also created a confusing array of terminology classifying movements as either open or closed (kinematic or kinetic) chains. We will see that the exact mechanism of this principle of biomechanics is not entirely clear, and common classification of movements as open or closed chains is not clear or useful in analyzing movement (Dillman et al. 1994; Blackard et al. 1999; di Fabio 1999).

The biomechanical principle of **Optimal projection** says that for most human movements involving projectiles, there is an effective range of projection angles for a specific goal. Biomechanical research shows that optimal angles of projection provide the right compromise between vertical velocity (determines time of flight) and horizontal velocity (determines range given the time of flight) within the typical conditions encountered in many sports. For example, in throwing most sport projectiles for horizontal distance, the typical air resistance and heights of release combine to make it beneficial for an athlete to use projection angles below 45 degrees. ▶ Chapter 5 will give several examples of how biomechanical studies have determined desirable release angles for various activities. This research makes it easier for coaches to determine if athletes are optimizing their performance.

The last principle involves the **spin** or rotations imparted to projectiles and particularly sport balls. Spin is desirable on thrown and struck balls because it stabilizes flight and creates a fluid force called **lift**. This lift force is used to create a curve or to counter gravity, which affects the trajectory and bounce of the ball. A volleyball player performing a jump serve should strike above the center of the ball to impart topspin to the ball. The topspin creates a downward lift force, making the ball dive steeply and making it difficult for the opponent to pass. The spin put on a pass in American football (◻ Fig. 3.9) stabilizes

◻ **Fig. 3.9**   The spin imparted to a football during a forward pass serves to stabilize ball flight, to provide aerodynamically efficient flight the orientation of the ball, which ensures aerodynamically efficient flight. The natural application of these biomechanical principles is in qualitative diagnosis of human movement. Lets see how the range of motion principle can be applied to human movement.

> **Interdisciplinary Issue: The Vertical Jump**
> Now that the principles are out of the bag, let's use them to look at a common sport movement, the vertical jump. Imagine an athlete is doing a standing vertical jump test. Which principles of biomechanics would be of most interest to scholars from motor development, motor learning, exercise physiology, or sport psychology studying the vertical jump test? What combinations of the sport sciences are most relevant to the concept of *skill* in vertical jumping? What sports science provides the most relevant information to the physical determinants of jumping ability? How could someone determine if the success of elite jumpers is more strongly related to genetics (nature/physical) than coaching (nurture/training)? How could a strength coach integrate jump training studies with biomechanical studies of jumping techniques?

3.3 Range of Motion Principle

Human movement can be accurately described as combinations of joint angular motions. A person can modify the number of joints, specific anatomical joint rotations, and amount of those rotations to tailor range of motion. Range of motion in movement can be imagined on a continuum from negligible motion to 100% of the physically possible motion. The **range of motion principle** states that less range of motion is most effective for low-effort (force and speed) and high-accuracy movements, while greater range of motion favors maximum efforts related to speed and overall force production (Hudson 1989). A person playing darts "freezes" or stabilizes most of the joints of the body with isometric muscle actions and limits the dart throw to a small range of motion focused on elbow and wrist. The javelin thrower uses a long running approach and total body action to use considerable range of motion to maximize the speed of javelin release. The great accuracy required in golf putting favors limiting range of motion by using very few segments and limiting their motion to only what is needed to move the ball near the hole (Fig. 3.10). Specific joint motions can be of interest, but so too can the overall linear motions of the whole body or an extremity. Coaches can speak of the range of motion of a "stride" in running or an "approach" in the high jump. Therapists can talk about the range of motion for a joint in the transverse or other anatomical planes.

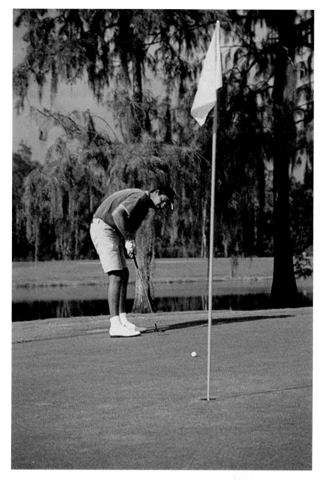

 Fig. 3.10 Very accurate movements like putting in golf limit range of motion by freezing most segments and using only a few segments. (Photo courtesy of Getty Images)

The application of the range of motion principle is more complicated when the effort of the movement is not maximal and when the load cannot be easily classified at the extremes of the continuum. A baseball or softball seems pretty light, but where on the range of motion continuum are these intermediate load activities? How much range of motion should you use when the load is a javelin, a shot, or your bodyweight (vertical jump)? Biomechanical studies can help kinesiology professionals decide how much range of motion is qualitatively "about right." In the qualitative diagnosis of movement, this approach of identifying a range of correctness (like in range of motion) is quite useful because the professional can either reinforce the performer's good performance or suggest less or more range of motion be used (Knudson 2013). The continuum of range of motion can also be qualitatively evaluated as a sliding scale (Knudson 1999c) or volume knob (Hudson 1995) where the performer can be told to fine tune range of motion by feedback (Fig. 3.11). Let's briefly look at how biomechanical research can help professionals evaluate the range of motion in a vertical jump.

3

Range of motion

◼ **Fig. 3.11**   Range of motion can be evaluated and pictured as an analog scale or a volume knob. If a change in range of motion is inappropriate, the performer can be instructed to "increase" or "decrease" the range of motion in their movement

The amount and speed of countermovement in a vertical jump is essential to a high jump. This range of motion variable can be expressed as a linear distance (lowering of center of mass as percentage of height) or as body configuration, like minimum knee angle. We use the knee angle in this example because it may be more easily visualized than a whole-body center of mass. One can hypothesize that maximizing the drop (range of motion) with a small knee angle in the countermovement would increase the height of the jump; however, this is not the case. Skilled jumpers tend to have minimum knee angles between 90 and 110° (Ross and Hudson 1997). The potential benefits of range of motion beyond this point seems to be lost because of poorer muscular leverage, change in coordination, or diminishing bene-

fits of extra time to apply force. The exact amount of countermovement will depend on the strength and skill of the jumper, but coaches can generally expect the knee angles in this range.

Another example of the complexity of applying the range of motion principle would be in maximal effort overarm throwing. In overarm throwing, the athlete uses range of motion from virtually the entire body to transfer force from the ground, through the body and to the ball. The range of motion (kinematics) of skilled overarm throwing has been extensively studied. Early motor development studies show that one range of motion variable (the length of the forward stride is usually greater than 50% of height) is important in a mature and forceful overarm throw (Roberton and Halverson

1984). Stride length in throwing is the horizontal distance from the rear (push-off) foot to the front foot. This linear range of motion from leg drive tends to contribute 10–20% of the ball speed in skilled throwers (Miller 1980). The skill of baseball pitching uses more stride range of motion, usually between 75% and 90% of standing height, and has been shown to significantly affect pitch speed (Montgomery and Knudson 2002). Recent research indicates this influence of stride length may be influenced by fatigue (Crotin et al. 2014) and trunk motion (Solomito et al. 2020).

The axial rotations of the hips and trunk are the range of motion links between stride and arm action. The differentiation of hip and trunk rotation is believed to be an important milestone in mature throwing (Roberton and Halverson 1984), and these movements contribute about 40–50% of ball speed in skilled throwers. In coaching high-speed throwing, coaches should look for hip and trunk opposition (turning the non-throwing side toward the target) in preparation for the throw. The optimal use of this range of motion is related to the coordination continuum and segmental interaction principles discussed in detail later. Students interested in the skilled pattern of hip and trunk range of motion should look at the extensive research on skilled pitchers (Hong and Roberts 1993; Fleisig et al. 1999, 2018; Stodden et al. 2005; Escamilla et al. 2007).

Arm action is the final contributor to the range of motion used in overarm throwing. The complex joint actions of throwing contribute significantly (30–50% of ball velocity) to skilled throwing (Miller 1980). To take advantage of the trunk rotation, the shoulder stays at roughly 90° of abduction to the spine (Atwater 1979) and has been called the strong throwing position (Plagenhoef 1971). With initiation of the stride, the elbow angle stays near 90° to minimize resistance to rotating the arm, so the major increase in ball speed is delayed until the last 75 ms (a millisecond [ms] is a thousandth of a second) before release (Roberts 1991). Contrary to most coaching cues to "extend the arm at release," the elbow is typically 20° short of complete extension at release to prevent injury (Fleisig et al. 1999). Inward rotation of the humerus, radioulnar pronation, and wrist flexion also contribute to the propulsion of the ball (Roberts 1991), but the fingers usually do not flex, but in applying large forces and increase the speed of the ball while they are forced into extension (Hore et al. 1996; Matsuo et al. 2018).

In overarm throwing, it appears that the range of motion principle can be easily applied in some motions like stride length using biomechanical research as benchmarks; however, it is much more difficult and perhaps impossible to define *optimal* amounts of joint motions or body actions in complex movements like overarm throwing. How range of motion might be changed to accommodate different level of effort throws, more specific

tasks/techniques (e.g., curveball, slider), or individual differences is not clear. Currently, professionals can only use biomechanical studies of elite and skilled performers as a guide for defining desirable ranges of motion for movements. More data on a variety of performers and advances in modeling or simulation of movement are needed to make better recommendations on how modifications of range of motion may optimize performance.

Interdisciplinary Issue: Variability

Scientists from a variety of disciplines have been interested in the variability of human movement performance for many years. Interest in movement variability has increased recently, particularly in biomechanics, motor control, and sports medicine. Biomechanical studies have often documented variability of kinematic and kinetic variables to determine the number of trials that must be analyzed to obtain reliable data (Bates et al. 1983; Winter 1984; Rodano and Squadrone 2002; Forrester 2015; Taylor et al. 2015). Motor learning studies have focused on variability as a measure of neuromuscular control (Slifkin and Newell 2000; Davids et al. 2003; Stergiou et al. 2006). The high variability of novice movement becomes more consistent at intermediate skill levels; however highly skilled performers have both great consistency but also flexibility/variability to meet complex movement task demands. The study of variability has also indicated that variability may play a role in potential injury (James et al. 2000) and pain (Falla and Gallina 2020). The study of coordination variability is currently a vigorously studied area within biomechanics (Bartlett et al. 2007). Multiple biomechanical measurements of kinematics, kinetics, and EMG may provide important contributions to interdisciplinary studies of human movement variability.

Application: Decline Squats

Rehabilitation and conditioning professionals often used incline and decline support surfaces to modify exercises for clients. People with limited ankle dorsiflexion range of motion often do squats with support under their heels. In rehabilitation, similar squat exercises emphasizing the eccentric phase on decline surfaces are used in treating patellar tendinopathy (Kongsgaard et al. 2006). How does the different orientation of body to gravity and the joint angles compare to a regular squat exercise? What other data or knowledge would help you in making this comparison or understanding the influence of variations in the squat exercise?

3.4 Qualitative Movement Diagnosis

The examples that illustrate the application of the principles of biomechanics in the solution of human movement problems in this book will be based on a qualitative diagnosis (Knudson 2013). Research has shown that general principles of biomechanics provide a useful structure for qualitative diagnosis of human movement (Johnson 1990; Matanin 1993; Nielsen and Beauchamp 1992; Williams and Tannehill 1999; Wilkinson 1996). Quantitative biomechanical analysis can also be used, but most kinesiology professionals will primarily be using qualitative diagnosis of human movement in their daily work.

There are several models of qualitative diagnosis of human movement. Traditionally, kinesiology professionals have used a simple error detection and correction approach to qualitative analysis. Here the analyst relies on a mental image of the correct technique to identify "errors" in the performance and provide a correction. This approach has several negative consequences and is too simplistic a model for professional judgments (Knudson 2013). The application of the principles of biomechanics is illustrated in this book using a more comprehensive, professional model of qualitative diagnosis than the simple error detection/correction of the past. This text uses the Knudson (2013) model of qualitative diagnosis (◘ Fig. 3.12). This model provides a simple four-task structure: preparation, observation, evaluation/diagnosis, and intervention. This model of qualitative diagnosis is equally relevant to athletic or clinical applications of biomechanics to improving human movement.

In the preparation task of qualitative diagnosis, the professional gathers relevant kinesiology knowledge about the activity, performer, and context/environment and then selects an observational strategy. Recall that biomechanics will be the primary science for planning observation for technique issues, although kinesiology professionals should remember to consider all subdisciplines in preparing for observation in qualitative diagnosis.

In the observation task, the professional executes the observational strategy to gather all relevant sensory information about the performance of the movement. Often video capture and replay are used to extend observational accuracy. Human vision of motion (dynamic visual acuity) is weakly associated at slow movement speeds and unrelated at fast speeds (Morris 1977). The numerous video replays reviewing referee's calls in professional sports clearly shows the imperfect ability of humans to visually evaluate fast movements. Just give the referees and coaches a break because they may not be able to see the movement live because it is too fast or their eyes were essentially off. The eyes/brain cannot gather visual information during blinks and when eyes saccade (jump from one fixation to another). The moderate ability to qualitatively evaluate human in real time argues for extending observation using video replay and slow motion (Knudson 2013). There are several free or low-cost video replay software programs (e.g., Dartfish, Hudl, Kinovea, SiliconCoach, Vstrator) designed for two-dimensional qualitative and quantitative assessments from video (Knudson 2013; Mills 2015). Real precision in documenting motion requires quantification (kinematic research in biomechanics) summarized in ▶ Chap. 5.

The third task has two difficult components: evaluation and then diagnosis of performance. In evaluation the analyst identifies strengths and weaknesses of performance. Diagnosis involves the prioritizing of the potential interventions to separate causes of poor performance from minor or symptomatic weaknesses.

Intervention is the last task of qualitative diagnosis. In this task the professional executes some action on behalf of the performer. Examples of intervention are feedback, changing practice, and exercise training. Often in live qualitative diagnosis, the analyst will return immediately to the observation task to monitor the intervention and the mover's progress.

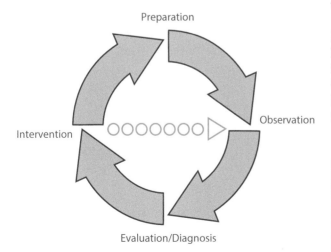

◘ **Fig. 3.12** The four-task model of qualitative diagnosis. (Adapted from Knudson and Morrison (2002))

Application: Quantitative Diagnosis
An athletic trainer is planning a qualitative diagnosis of the lower extremity muscular function of an athlete finishing up an anterior cruciate ligament (ACL) rehabilitation program. The trainer has run the athlete through the rehabilitation program but wants a more

functional evaluation of the athlete's ability and readiness for play. The athlete will be doing several drills, including multiple one-legged hops and squats, shuttle runs, landings, jumps, and lateral cutting movements. For the preparation task of qualitative diagnosis, give examples of research or biomechanical principles that you think would be relevant to analyzing the athlete's ability to prevent damage to the ACL in these dynamic locomotion tasks of sport. Is there a task of qualitative diagnosis that more heavily relies on biomechanics than other kinesiology subdisciplines?

3.5 Summary

Most human biomechanical research has been based on rigid body, linked segment models of the skeletal system. Kinematics involves the description/measurement of the motion, while kinetics focuses on the forces and moments that created the motion. There are many biomechanical variables, and they can be classified as either scalars or vectors. Despite the precision of quantitative biomechanics, most kinesiology professionals however apply biomechanics at a qualitative or conceptual level. The nine principles of biomechanics that can be used to apply biomechanics knowledge in professional practice are force–motion, force–time, inertia, range of motion, balance, Coordination continuum, Segmental interaction, Optimal projection, and spin. These nine principles can be applied using a comprehensive model (Knudson 2013) of qualitative diagnosis.

❓ Review Questions

1. What are the major branches of mechanics, and which are most commonly used in performing biomechanical analyses of human movement?
2. What are the specific foci of kinematic and kinetic analyses, and provide some examples?
3. How are vector variables different from scalar variables?
4. How is a scientific principle different from a law?
5. The nine principles of biomechanics can be classified into which two areas of interest?
6. What are the nine principles of biomechanics?
7. What are some other factors that affect human movement and the application of the principles of biomechanics?
8. List as many reasons as possible for the apparent theory-to-practice gap between scholars and practitioners.
9. What factors should a kinesiology professional consider when trying to apply the range of motion principle in a client's movement?

❗ Key Terms

Components · Deformable body · Dynamics · Dynamometer · Fluid · Free-body diagram · Isometric · Kinematics · Kinetics · Mass · Mechanics · Qualitative movement diagnosis · Range of motion principle · Resultant · Scalar · Strength (muscular) · Stiffness · Technology · Torque/moment of force · Vector · Weight

⭐ Suggested Reading

Amonette, W. E. et al. (2016). *Evidence-based practice in exercise science: The six-step approach*. Champaign, IL: Human Kinetics.

Challis, J. H. (2021). *Experimental methods in biomechanics*. Cham, Switzerland: Springer Nature.

Hudson, J. L. (1995). Core concepts in kinesiology. *JOPERD, 66*(5), 54–55, 59–60.

Knudson, D. (2000). What can professionals qualitatively analyze? *Journal of Physical Education, Recreation, and Dance, 71*(2), 19–23.

Knudson, D. (2005). Evidence-based practice in kinesiology: The theory to practice gap revisited. *The Physical Educator, 62*, 212–221.

Knudson, D. (2007). Qualitative biomechanical principles for coaches. *Sports Biomechanics, 6*, 108–117.

Knudson, D. (2013). *Qualitative dianosis of human movement: Improving performance in sport and exercise* (3rd ed.). Champaign: Human Kinetics.

Knudson, D., & Kluka, D. (1997). The impact of vision and vision training on sport performance. *Journal of Physical Education, Recreation, and Dance, 68*(4), 17–24.

Knuttgen, H. G., & Kraemer, W. J. (1987). Terminology and measurement in exercise performance. *Journal of Applied Sport Science Research, 1*, 1–10.

Norman, R. (1975). Biomechanics for the community coach. *JOPERD, 46*(3), 49–52.

Rogers, M. M., & Cavanagh, P. R. (1984). Glossary of biomechanical terms, concepts, and units. *Physical Therapy, 64*, 82–98.

🔊 Internet Resources

▶ International System of Units (SI)–Base unit summary by the National Institute of Standards and Technology (US Department of Commerce). *nist.gov/pml/weights-and-measures/metric-si/si-units*

▶ Kinovea–Two-dimensional system for qualitative and quantitative assessments from video images. *kinovea.org*

▶ Knudson. (2013). A link to the only book on the qualitative diagnosis of human movement. *humankinetics.com/qualitativediagnosisofhumanmovement*

▶ Physics and Mathematics Review provided by the physics department of the University of Guelph in Canada. *physics.uoguelph.ca/tutorials/tutorials.html*

Biomechanical Basis of Human Movement

The study of human biomechanics requires an understanding of anatomy. The structure of musculoskeletal systems and the mechanical properties of tissues inform training effects and risk of injury. This requires the study of biomechanics. Precise measurement of motion (kinematics) is needed to understand exactly what motions happen, like illustrated in the key segments of the lower extremity in cycling. The resultant force applied by the foot to the pedal represent the kinetics or causes of the movement. Knowledge of biomechanics allows kinesiology professionals to understand human movement technique, develop specific training exercises, and recommend interventions to change movement. Part II introduces you to key mechanical properties of biological tissues, kinematics, kinetics, and fluids. The related lab activities explore qualitative and quantitative analyses of key mechanical variables important in understanding human movement, as well as their application in qualitative movement diagnosis.

Mechanics of the Musculoskeletal System

Key Questions

➤ **What do you need to know to begin?**
The definition of the kinematic variables length and angle, and the kinetic variables force and moment/torque

➤ **Why do you need to know this material?**
Understanding of mechanical response of biological tissues informs growth, training, and injures.

➤ **What is the key idea?**
Biomechanics principles are essential tools in the qualitative diagnosis of human movement technique in numerous professions.

Many professionals interested in human movement function need knowledge on how forces act on and within the tissues of the body. The deformations of muscles, tendons, and bones created by external forces, as well as the internal forces created by these same structures, are relevant to understanding human movement, growth, and injury. This chapter will provide a brief overview of the mechanics of biological tissues, specifically muscles, tendons, ligaments, and bone. The neuromuscular control of muscle forces and the mechanical characteristics of the muscle will also be summarized. The application of these concepts is illustrated using the **force–motion principle** of biomechanics. An understanding of mechanics of musculoskeletal tissues is important in understanding the organization of movement, injury, and designing conditioning programs.

4.1 Tissue Loads

When forces are applied to a deformable material, like human musculoskeletal tissues, they create **loads**. A load is force acting on and within a body. Engineers use various names to describe how loads tend to change the shape of a material. These include the principal or axial loadings of compression, tension, and shear (◻ Fig. 4.1). **Compression** is when an external force tends to squeeze the molecules of a material together. **Tension** is when the load acts to stretch or pull apart the material. For example, the weight of a body tends to compress the foot against the ground in the stance phase of running, which is resisted by tensile loading of the plantar fascia and the longitudinal ligament in the foot. **Shear** is a right-angle loading acting in opposite directions. A trainer creates a shearing load across athletic tape with scissor blades or their fingers when they tear the tape. Note that loads are not vectors (individual forces) acting in one direction but are illustrated by two arrows (◻ Fig. 4.1) to show that the load results from forces from both directions.

When many forces are acting on a body, they can combine to create combined loads called **torsion** and **bending** (◻ Fig. 4.2). In bending one side of the material is loaded in compression, while the other side experiences tensile loading. When a person is in single support in walking (essentially a one-legged chair), the femur experiences bending loading. The medial aspect of the femur is in compression, while the lateral aspect is in tension.

4.2 Response of Tissues to Forces

The immediate response of tissues to loading depends on a variety of factors. The size and direction of forces, as well as the mechanical strength and shape of the tissue, affect how the material structure will change. We will see in this section that mechanical strength and muscular strength are different concepts. This text will strive to use "muscular" or "mechanical" modifiers with the term strength to help avoid

© Springer Nature Switzerland AG 2021
D. Knudson, *Fundamentals of Biomechanics*, https://doi.org/10.1007/978-3-030-51838-7_4

▫ Fig. 4.1   The principal axial loads of **a** compression, **b** tension, and **c** shear

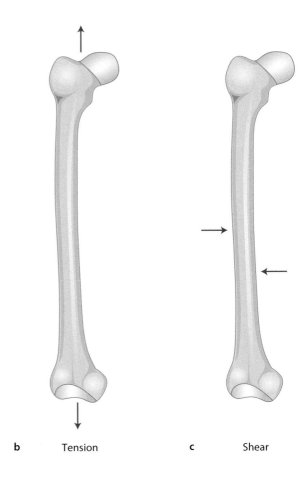

| a | Compression | b | Tension | c | Shear |

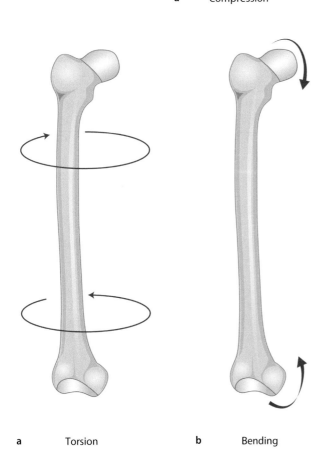

| a | Torsion | b | Bending |

▫ Fig. 4.2   Combined loads of **a** bending and **b** torsion. A bending load results in one side of the material experiencing tension and the other compression

confusion. There are several mechanical variables that explain how musculoskeletal tissues immediately respond to forces or loading. This chapter will not deal extensively with submaximal loading and their likely long-term training or overuse injury effects.

4.2.1 Stress

How hard a load works to change the shape of a material is measured by **mechanical stress**. **Mechanical stress** is symbolized with the Greek letter sigma (σ) and is defined as the force per unit area within a material (σ = F/A). Mechanical stress is similar to the concept of pressure and has the same units (N/m^2 and lbs/in^2). In the SI system, one Newton per meter squared is one Pascal (Pa) of stress or pressure. As you read this book, you are sitting in a sea of atmospheric gasses that typically exert a pressure of 1 atm, 101.3 KPa (kilopascals), or $14.7 \ lbs/in^2$ on your body. Note that mechanical stress is not vector quantity but an even more complex quantity called a tensor. Tensors are generalized vectors that have multiple directions that must be accounted for, much like resolving a force into anatomically relevant axes like along a longitudinal axis and at right angles (shear). The maximum force capacity of skeletal muscle is usually expressed as a maximum stress of about $25–40 \ N/cm^2$ or $36–57 \ lbs/in^2$ (Herzog 1996). This force potential per unit of cross-sectional area is the same across gender, with females tending to have about two-thirds of the muscular strength of males because they have about two-thirds as much muscle mass as males.

4.2.2 Strain

The measure of the deformation of a material created by a load is called **strain**. This deformation is usually expressed as a ratio of the normal or resting length (L_0) of the material. Strain (ε) can be calculated as a change in length divided by normal length: $(L - L_0)/L_0$. Imagine stretching a rubber band between two fingers. If the band is elongated to 1.5 times its original length, you could say the band experiences 0.5 or 50% tensile strain. This text will discuss the typical strains in musculoskeletal tissues in percentage units. Most engineers use much more rigid materials and typically talk in terms of units of microstrain (1/1000th units). Think about what can withstand greater tensile strain: the shaft of a tennis racket, the shaft of a golf club, or the shaft of a fiberglass pole?

4.2.3 Stiffness and Mechanical Strength

Engineers study the mechanical behavior of a material by loading a small sample in a material testing system (MTS), which simultaneously measures the force (**F**) and displacement (**d**) of the material as it is deformed at various rates. The resulting graph is called a load-deformation curve (▣ Fig. 4.3) and like we saw in ▶ Fig. 3.5. **F-d** graph scan be converted with other measurements (material area and initial length) to obtain a stress–strain graph. Load-deformation graphs have several variables and regions of interest. The **elastic region** is the initial linear region of the graph where the slope corresponds to the **stiffness** or **Young's modulus** of

elasticity of the material. **Stiffness** or Young's modulus is technically defined as the ratio of stress to strain in the elastic region of the curve but is often approximated by the ratio of load to deformation (ignoring the change in dimension of the material). If the test were stopped within the elastic region, the material would return to its initial shape. If the material was perfectly elastic, the force at a given deformation during restitution (unloading) would be the same as in loading. We will see later that connective tissues adapt to either be effective springs, losing small amounts of stored in them during deformation or lose higher amounts of stored energy so they act as shock absorbers/dampers.

Beyond the linear elastic region is the plastic region, where increases in deformation occur with minimal, nonlinear, and complex changes in load due to failure of material integrity (▣ Figs. 4.3 and 4.4). The yield point or elastic limit is the point on the graph separating the elastic and plastic regions. When the material is deformed beyond the yield point, the material will not return to its initial dimensions. In biological materials, normal physiological loading occurs within the elastic

▣ **Fig. 4.4** The typical load-deformation (elongation) curve for human tendon is more complex than for many materials. Initial elongation is resisted by small force increases in the toe region, followed by the elastic region. Much of the physiological loading of tendons in normal movement are likely within the toe region (<5% strain: Maganaris and Paul 2000), but elongation beyond the elastic limit damages the tendon

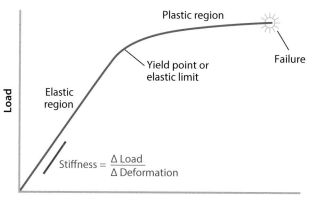

▣ **Fig. 4.3** The regions and key variables in a load-deformation graph of a material

4

region, and deformations near and beyond the elastic limit are associated with microstructural damage to the tissue. Another important variable calculated from these measurements is the mechanical strength of the material.

Activity: Mechanical (Failure) Strength

Two strong materials are nylon and steel. Nylon strings in a tennis racket can be elongated a great deal (high strain and a lower stiffness) compared to steel strings. Steel is a stiff and strong material. Take a paper clip and apply a bending load. Did the paper clip break? Bend it back the opposite way, and repeat counting the number or bends before the paper clip breaks. Most people cannot apply enough force in one shearing effort to break a paper clip, but over several loadings, the steel weakens, and you can get a sense of the total mechanical work/energy you had to exert to break the paper clip.

The **mechanical strength** of a material is the measurement of the maximum force or total **mechanical energy** the material can absorb before failure. The energy absorbed and mechanical work done on the material can be measured by the area under the load-deformation graph. Within the plastic region, the pattern of failure of the material can vary, and the definition of failure can vary based on the interest of the research. Conditioning and rehabilitation professionals might be interested in the yield strength (force at the end of the elastic region) of healthy and healing ligaments. Sports medicine professionals may be more interested in the ultimate strength that is largest force or stress the material can withstand. Sometimes it is of interest to know the total amount of strain energy (see ▶ Chap. 6) the material will absorb before it breaks because of the residual forces that remain after ultimate strength. This is failure strength and represents how much total loading the material can absorb before it is broken. This text will be specific in regard to the term *strength*, so that when used alone the term will refer to muscular strength and the mechanical strengths of materials will be identified by their relevant adjective (yield, ultimate, or failure).

4.2.4 Viscoelasticity

Biological tissues are structurally complex and also have complex mechanical behavior in response to loading. First, biological tissues are **anisotropic**, which means that their mechanical strength properties are different for each major direction of loading. Second, the nature of the protein fibers and amount of calcification all determine the mechanical response. Third, most soft

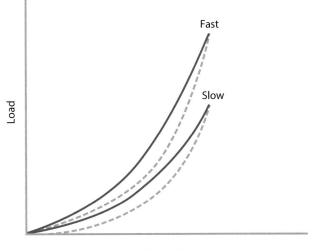

□ Fig. 4.5 Load-deformation curves for the tendon stretched at a fast and a slow rate to the same length. Force at any length in loading (solid line) is higher than in unloading (dashed line). A viscoelastic material has different stiffness at different rates of deformation. A faster stretch results in greater force in the tendon for a given length compared to a slow stretch. Hysteresis is the work and energy lost in the restitution of the tendon and can be visualized as the area between loading and unloading of the tendon

connective tissue components of the muscle, tendons, and ligaments have another region in their load-deformation graph. For example, when a sample of the tendon is stretched at a constant rate, the response illustrated in □ Fig. 4.4 is typical. Note that the response of the material is more complex (nonlinear) than the Hookean elasticity illustrated in ▶ Fig. 3.5. The initial increase in deformation with little increase in force before the elastic region is called the *toe region*. The toe region corresponds to the straightening of the wavy collagen fiber in connective tissue (Carlstedt and Nordin 1989). After the toe region, the slope of the elastic region will vary depending on the rate of stretch. This means that tendons (and other biological tissues) are not perfectly elastic but are **viscoelastic**.

Viscoelastic means that the stress and strain in a material are dependent on the rate of loading because of damping effects of the material, so the timing of the force application affects the strain response of the material. □ Figure 4.5 illustrates the response of a ligament that is stretched to a set length at two speeds, slow and fast. Note that a high rate of stretch results in a higher stiffness than a slow stretch. Muscles and tendons also have increasing stiffness with increasing rates of stretch. The viscoelasticity of muscles and tendons has great functional significance. A slow stretch will result in a small increase in passive resistance (high compliance) from the muscle, while the muscle will provide a fast increase in passive resistance (high stiffness) to a rapid stretch. This is one of the reasons that stretching exercises should be performed slowly, to minimize the increase in force in the muscle–tendon unit (MTU) for a given amount of stretch. The solid lines of the graph represent the loading response of the ligament, while the dashed lines represent the mechanical response of the tissue as the load is released (unloading).

Activity: Viscoelasticity

An extreme example of viscoelastic behavior that serves as a good demonstration for teaching stretching exercises is the behavior of Silly Putty™. Roll the putty into a cylinder, which serves as a model of the muscle. The putty has low stiffness at slow stretching rates, so it gradually increases in length under low force conditions and has a plastic response. Now stretch the putty model quickly, and note the much higher stiffness. This high stiffness makes the force in the putty get quite high at short lengths and often results in the putty breaking. You may be familiar with this complex (different) behavior of the material because the shape can be molded into a stable shape like a ball, but when the ball is loaded quickly (thrown at a wall or the floor), it will bounce rather than flatten out.

There are other important properties of viscoelastic materials: **creep**, **stress relaxation**, and **hysteresis**. **Creep** is the gradual elongation (increasing strain) of a material over time when placed under a constant tensile stress. **Stress relaxation** is the decrease in stress over time when a material is elongated to a set length. For example, holding a static stretch at a specific joint position results in a gradual decrease in tension in the muscle from stress relaxation. If you leave a free weight hanging from a nylon cord, you might return several days later to find the elongation (creep) in the cord has stretched it beyond initial length. Creep and stress relaxation are nonlinear responses and have important implications for stretching (see application box on flexibility and stretching) and risk of injury in repetitive tasks. For example, long-held or repetitive work postures that stretch ligaments, reducing their mechanical and proprioceptive effectiveness, likely increase joint laxity and logically the risk of injury (Solomonow 2004).

Application: Stress Relaxation

Guitar players will know that steel strings do not lose tension (consequently their tuning) as quickly as nylon strings; this phenomenon is not related to strength but to viscoelasticity. Steel guitar strings are much more elastic (stiffer) and have negligible viscoelastic properties compared to nylon strings. In a similar fashion, nylon tennis strings lose tension over time. Skilled players who prefer a higher tension to grab the ball for making greater spin will often need to cut out and replace nylon strings before they break. Gut strings are more elastic than nylon and tend to break before there is substantial stress relaxation. Similarly, when a static stretch holds a muscle group in an extended position for a long period of time, the tension in the stretched muscle group decreases over time. This stress relaxation occurs quickly (most within the first 15 seconds), with diminishing amounts of relaxation with longer amounts of time (see Knudson 1998). How might a coach set up a stretching routine that maximizes stress relaxation of the athlete's muscles? If many people dislike holding stretched positions for a long period to time, how might kinesiology professionals program stretching to get optimal compliance and muscle stress relaxation?

Hysteresis is the property of viscoelastic materials of having a different unloading response than its loading response (�’ Fig. 4.5). Hysteresis also provides a measure of the amount of energy lost because the material is not perfectly elastic. The area between the loading and unloading lines is the energy lost in the recovery from that stretch. We will learn in ▶ Chap. 6 that energy and work are related and that mechanical work is defined as force times displacement ($\mathbf{F} \cdot \mathbf{d}$), so work can be visualized as an area under a force-displacement graph. If you want to visualize the failure strength (work) of the material in �’ Fig. 4.3, imagine or shade in the total area above zero and below the load-deformation graph.

All these mechanical response variables of biological materials depend on precise measurements and characteristics of the samples. The example mechanical strengths and strains mentioned in the next section represent typical values from the literature. Do not assume these are exact values because factors like training, age, and disease all affect the variability of the mechanical response of tissues. Methodological factors like how the human tissues were stored (fresh, frozen), attached to the machine, or preconditioned (like a warm-up before testing) all affect the results. Remember that the rate of loading has a strong effect on the stiffness, strain, and strength of biological materials. This has led some scholars to develop impact devices (rather than typical engineering materials testing machines) to document mechanical properties of human tissues (Bates et al. 2017). The following sections will emphasize more the mechanical strengths of tissues in different directions and how these are likely related to training strategies and risk of common injuries.

4.3 Biomechanics of the Passive Muscle–Tendon Unit (MTU)

The mechanical response of the MTU to passive stretching exercise is viscoelastic, so the response of the tissue depends on the time or rate of stretch. At a high rate of passive stretch (fast), the MTU is stiffer than when it is slowly stretched. This is the primary reason why slow, static stretching exercises are preferred over ballistic stretching techniques. A slow stretch results in less passive tension in the muscle for a given amount of elongation compared to a faster stretch. The load in an MTU during other movement conditions is even more complicated because the load can vary widely with activation, previous muscle action, and kind of muscle action. All these variables affect how load is distributed in the active and passive components of the MTU. Keep in mind that the Hill model of the muscle has a contractile component that modulates tension with activation, as well as two passive tension elements: the parallel elastic and the series elastic components. The mechanical behavior of activated muscle is presented in the upcoming section on the Mechanical Characteristics of Muscle.

The tendon is the connective tissue that links the muscle to the bone and strongly affects how muscles are used or injured in movement. The tendon is a well-vascularized tissue whose mechanical response is primarily related to the protein fiber *collagen*. The parallel arrangement of collagen fibers in tendon and cross-links between fibers makes the tendon about three times stronger in tension than the muscle. The ultimate strength of the tendon is usually about 100 MPa (megapascals) or 14,500 lbs/in^2 (Kirkendall and Garrett 1997). Even though the diameter of tendons is often smaller than the associated muscle belly, their great tensile strength makes tendon rupture injuries rare. Acute overloading of the MTU usually results in **strains** (sports medicine term for over-stretched muscle, not mechanical strain). Muscle strains are disruptions of the muscle fibers that can occur throughout the muscle but most often occur at the muscle–tendon junction (Garrett 1996).

In creating movement, a long tendon can act as an efficient spring in fast bouncing movements in many animals (Alexander 1992) because the stiffness of the muscle belly can exceed tendon stiffness in high states of activation. A muscle with a short tendon transfers force to the bone more quickly because there is less slack to be taken out of the tendon. The intrinsic muscles of the hand are well suited to the fast finger movements of a violinist because of their short tendons. The Achilles tendon provides some compliance to smooth out the forces of the large calf muscle group (soleus and gastrocnemius) but also stores and returns considerable energy. Conversely, the connective tissues of cartilage and your heel pad have greater hysteresis, so the extra energy they absorb allows them to act as shock absorbers rather than elastic springs.

Application: Flexibility and Stretching

A common health-related fitness component is *flexibility*. Flexibility is defined as "the intrinsic property of body tissues, which determines the range of motion achievable without injury at a joint or group of joints" (Holt et al. 1996:172). Static flexibility or extensibility refers to the usual linear or angular measurements of the actual limits of range of motion (ROM) in a joint or joint complex. Static flexibility measurements have elements of subjectivity because of variations in testers and patient tolerance of stretch. Clinical professions (AT/PT/OT) use goniometers and inclinometers to measure joint ROM and refer to these rotations as osteokinematics. They also qualitatively assess the tiny linear, gliding motions along joint surfaces (arthrokinematics).

The kinematic (linear or angular) measure of the extensibility or ROM of a muscle group at a joint or joints is different from the tissue resistance. The kinetic variables often measured in biomechanical studies of flexibility are passive tension and its rate of increase (muscle–tendon stiffness). The largest force or moment measured to stretch a muscle group to a person's point of discomfort has been considered a measure of passive tension at the point of that person's subjective stretch tolerance. The slope of this increase in force/moment is the stiffness of the MTU. ROM/extensibility is correlated with peak resistance or stretch tolerance (Kay et al. 2015), but ROM/extensibility is poorly correlated with stiffness in stretching (Whitting et al. 2013). Inactivity and immobilization have been shown to decrease static range of motion and increase muscle group stiffness (Akeson et al. 1987; Heerkens et al. 1986). Flexibility remains a complex concept with different biomechanical variables that describe aspects of this ability to move without injury.

Static ROM/extensibility is related to variations in body position because the passive tension increases in each MTU, especially in multiarticular MTUs (passive insufficiency). This is why there are strict rules for body positioning in flexibility testing. What muscles of the leg and thigh are unloaded when students try to cheat by bending their knees in a sit-and-reach test? What calf muscle is unloaded in a seated toe-touch stretch when the ankle is plantar flexed?

Many students might be surprised to find that recent research has challenged the association of flexibility with health (Nuzzo 2019) and acute benefits of pre-activity stretching on performance (Knudson 1999;

Magnusson and Renstrom 2006; Shrier 2004) or injury (Knudson et al. 2000; Shrier 1999; Magnusson and Renstrom 2006). Muscle activation and muscular strength are significantly decreased for 15 and 60 minutes, respectively, following stretching (Fowles et al. 2000). The dose of stretching that significantly decreases strength may be between 20 and 40 seconds (Knudson and Noffal 2005; Simic et al. 2013). Injury risk in subsequent maximal muscular actions may be reduced, however, by generalized warm-up (Fradkin et al. 2006; Knudson 2007). Several studies have reported that large percentages of collegiate coaches and sports medicine staff still use stretching in warm-up protocols (Judge et al. 2020).

Stretching is a common practice in physical conditioning and sports. Stretching exercises must be carefully prescribed to focus tension on the MTUs and not the ligaments that maintain joint integrity (Knudson 1998). Long-term stretching programs likely increase static range of motion by stimulating the production of new sarcomeres in muscle fibers (De Deyne 2001) and neuromuscular factors (Guissard and Duchateau 2006). Stretching program factors influencing long-term range of motion have been reviewed by Thomas et al. 2018).

4.4 Biomechanics of Bone

Unlike muscle, the primary loads experienced by most bones are compressive. The mechanical response of the bone to compression, tension, and other complex loads depends on the complex structure of bones. Remember that bones are living tissues with blood supplies, made of a high percentage of water (25% of bone mass), and having considerable deposits of calcium salts and other minerals. The strength of the bone depends strongly on its density of mineral deposits and collagen fibers and is also strongly related to dietary habits and physical activity. The loading of bones in physical activity results in greater osteoblast activity, laying down the bone. Immobilization or inactivity will result in dramatic decreases in bone density, stiffness, and mechanical strength. A German scientist is credited with the discovery that bones remodel (lay down greater mineral deposits) according to the mechanical stress in that area of the bone. This laying down of more bone mass where it is stressed and reabsorption of the bone in the absence of stress is called **Wolff's Law**. Bone remodeling is well illustrated by the formation of the bone around the threads of screws in the hip prosthetic in the X-ray in ◘ Fig. 4.6a.

◘ **Fig. 4.6**   **a** X-ray of a fractured femur with a metal plate repair. Note the remodeling of the bone around the screws that transfer load to the plate. Reprinted with permission from Nordin and Frankel (2001). **b** Radiograph of a patient with a hybrid right total hip arthroplasty without periprosthetic lucency. (Reprinted with permission from Springer Nature (2020))

4

Application: Osteoporosis

Considerable research is currently being directed at developing exercise machines as countermeasures for the significant bone density loss in extended space flight. A microgravity environment substantially decreases the loading of the large muscles and bones of the lower extremity, resulting in loss of bone and muscle mass. There is also interest in exercise as a preventative and remedial strategy for increasing the bone mass of postmenopausal women. The strong link between the positive stresses of exercise on bone density, however, is often complicated by such other things like diet and hormonal factors. In the late 1980s, researchers were surprised to find that elite women athletes were at greater risk for **stress fractures** because they had the bone density of women two to three times their age. **Stress fractures** are very small breaks in the cortical (see below) bone that result from physical activity without adequate rest. What was discovered was that overtraining and the very low body fat that resulted in amenorrhea also affected estrogen levels that tended to decrease bone mass. This effect was stronger than the bone growth stimulus of the physical activity. High-level women athletes in many sports must be careful in monitoring training, diet, and body fat to maintain bone mass. Kinesiology professionals must be watchful for signs of these problems. While this was initially labeled the female athlete triad, attributed to disordered eating, amenorrhea, and osteoporosis, it is now known as relative energy deficiency in sport (Monuntjoy et al. 2014) because males are also affected.

The macroscopic structure of the bone shows a dense, external layer called cortical (compact) bone and the less-dense internal cancellous (spongy) bone. The mechanical response of the bone is dependent on this "sandwich" construction of cortical and cancellous bone. This design of a strong and stiff material with a weaker and more flexible interior (like fiberglass) results in a composite material that is strong for a given weight (Nordin and Frankel 2001). This is much like a surf board constructed of fiberglass bonded over a foam core. Cortical bone is stiffer (maximum strain about 2%), while cancellous bone is less stiff and can withstand greater strain (7%) before failure. In general, this design results in ultimate strengths of the bone of about 200 MPa (29,000 lbs/in^2) in compression, 125 MPa (18,000 lbs/in^2) in tension, and 65 MPa (9500 lbs/in^2) in shear (Hayes 1986). This means that an excessive bending load on the femur like in ◘ Fig. 4.2b would most likely cause a fracture to begin on the lateral aspect that is under tensile loading. Using sports rules to protect athletes from lateral blows (like blocking rules in American football) is wise because the bone is weakest under shearing loads.

It is also important to understand that the ultimate mechanical strength of the bone depends on nutritional, hormonal, and physical activity factors. Research done with an elite powerlifter found that the ultimate compressive strength of a lumbar vertebral body (more than 36,000 N or 4 tons) estimated from bone mineral measurements was twice that of the previous maximal value. More recent studies of drop jump training in prepubescent children has demonstrated that bone density can be increased, but it is unclear if peak forces, rates of loading, or repetitions are the training stimulus for the increases in bone mass (Bauer et al. 2001). More research on the osteogenic effects of various kinds of loading and exercise programs could help physical educators design programs that help school children build bone mass. The following section will outline the mechanical response of ligaments to loading.

4.5 Biomechanics of Ligaments

Ligaments are tough connective tissues that connect bones to guide and limit joint motion as well as provide important proprioceptive and kinesthetic afferent signals (Solomonow 2004). Most joints are not perfect hinges with a constant axis of rotation, so they tend to have small accessory motions and moving axes of rotation that stress ligaments in several directions. The collagen fibers within ligaments are not arranged in parallel like tendons but in a variety of directions. Normal physiological loading of most ligaments is 2–5% of tensile strain, which corresponds to a load of ≈500 N (112 lbs) in the human anterior cruciate ligament (Carlstedt and Nordin 1989). Except for "spring" ligaments that have a large percentage of elastin fibers (ligamentum flavum in the spine), which can stretch more than 50% of their resting length, the maximum strain of most ligaments and tendons is about 8–10% (Rigby et al. 1959).

Like bone, ligaments and tendons remodel according to the stresses they are subjected to. A long-term increase in the mechanical strength of articular cartilage with the loads of regular physical activity has also been observed (Arokoski et al. 2000). Inactivity, however, results in major decreases in the mechanical strength of ligaments and tendon, with reconditioning to regain this strength taking longer than deconditioning (Carlstedt and Nordin 1989). The ability of the musculoskeletal system to adapt tissue mechanical properties to the loads of physical activity does not guarantee a low risk of injury. There is likely a higher risk of tissue overload when deconditioned individuals participate in vigorous activity or when trained individuals push the envelope, training beyond the tissue's ability to adapt during the rest periods between training bouts. We will see in the next section that muscle mechanical properties also change in response to activity and inactivity.

4.6 Major Mechanical Characteristics of Muscle

Previously we discussed the passive tension in an MTU as it is passively stretched. Now it is time to examine the tensile forces the MTU experiences in the wide variety of muscle actions, lengths, and other active conditions encountered in movement. The tensile force potential of a MTU varies and can be described by three mechanical characteristics. These characteristics deal with the variations in muscle force because of differences in velocity, length, and the time relative to activation.

4.6.1 Force–Velocity Relationship

The **force–velocity relationship** explains how the force of activated muscle varies with velocity. This may be the most important mechanical characteristic since muscle strength expression in all three muscle actions (eccentric, isometric, concentric) are reflected in the graph.

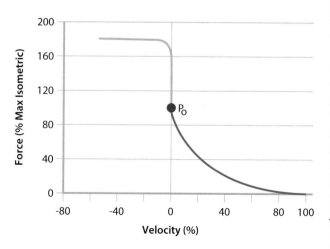

◻ Fig. 4.7 The *in vitro* force–velocity relationship of the muscle. Muscle force potential rapidly decreases with increasing velocity of shortening (concentric action), while the force within the muscle increases with increasing velocity of lengthening (eccentric action). Muscle velocity expressed as a percentage of maximal shorting and force as a percentage of isometric (P_O) maximum

We will see that the force or tension a muscle can create varies widely across actions and across the many speeds of movement. The discovery and formula describing this fundamental relationship in concentric conditions are also attributed to A. V. Hill. Hill made careful measurements of the velocity of shortening when a preparation of maximally stimulated frog muscle was released from isometric conditions. These studies of isolated preparations of the muscle are performed in what we term *in vitro* (Latin for "in glass") conditions. ◻ Figure 4.7 illustrates the shape of the complete force–velocity relationship of skeletal muscle. The force–velocity curve essentially states that the force the muscle can create decreases with increasing velocity of shortening (concentric actions), while the force the muscle can resist increases with increasing velocity of lengthening (eccentric actions). The force in isometric conditions is labeled P_0 in Hill's equation. The right side of the graph corresponds to how the tension potential of the muscle rapidly decreases with increases in speed of concentric shortening. Also note, however, that increasing negative velocities (to the left of isometric) show how muscle tension rises in faster eccentric muscle actions. In isolated muscle preparations, the forces that the muscle can resist in fast eccentric actions can be almost twice the maximum isometric force (Alexander 2002). It turns out the extent of damage done to a muscle eccentrically overstretched is strongly related to the peak force during the stretch (Stauber 2004). In athletics they say the sprinter "pulled" their hamstring, but this muscle strain injury is often result of large forces (high mechanical stress) from a too intense eccentric muscle action. We will see later that neuromuscular factors focus these high eccentric forces in smaller numbers of muscle fibers, also contributing to injury risk.

If the force capability of an *in vitro* muscle preparation varies with velocity, can this behavior be generalized to a whole MTU, muscle groups, or the body in normal movement? Researchers have been quite interested in this question, and the answer is clearly yes, although the shape of the relationship is more linear with the slope mediated by other factors. The torque a muscle group can create depends on the previous action, activation, rate of force development, and the combination of the characteristics of the muscles acting at that and nearby joints. Despite these complications, the *in vivo* (in the living animal) torque–angular velocity relationship of muscle groups usually matches the shape of the *in vitro* curve. These *in vivo* torque–angular velocity relationships are established by testing at many angular velocities on **isokinetic** and specialized dynamometers. These studies tend to show that in repeated isokinetic testing the peak eccentric torques are higher than peak isometric torques,

4

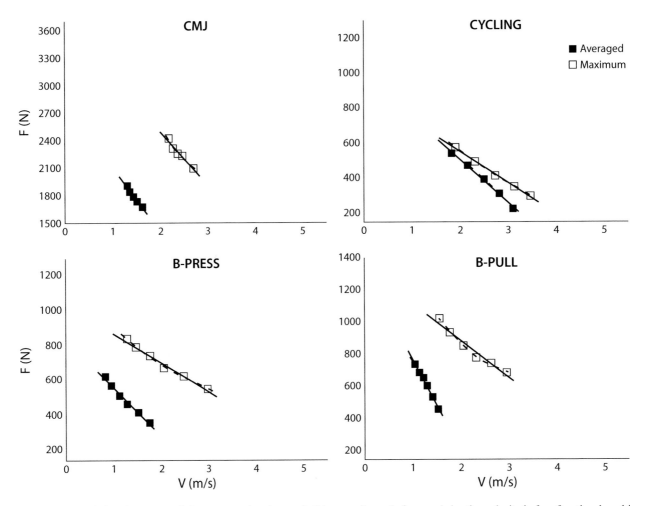

□ **Fig. 4.8**   Associations between peak (open squares) and mean (solid squares) muscle force and shorting velocity in four functional, multi-segment human movements. (Adapted permission of Zivkovic et al. 2017)

but not to the extent of isolated muscle preparations (Holder-Powell and Rutherford 1999), while concentric torques decline with varying slopes with increasing speed of shortening (De Koning et al. 1985; Gulch 1994; Pinniger et al. 2000). Expressions of maximal muscular strength for the whole body, as well as the upper and lower extremities usually result in linear (□ Fig. 4.8) declines in muscular force with increasing concentric velocity (Zivkovic et al. 2017).

This general shape of a muscle's potential maximum tension has many implications for human movement. First, it is not possible for muscles to create large forces at high speeds of shortening. Muscles can create high tensions to initiate motion, but as the speed of shortening increases, their ability to create force (maintain acceleration) decreases. Second, the force potential of muscles at small speeds of motion (in the middle of the graph) depends strongly on isometric muscular strength (Zatsiorsky and Kraemer 2006). This means that muscular strength will be a factor in most movements, but this

influence will vary depending on the speed and direction (moving or braking) the muscles are used. Third, the inverse relationship between muscle force and velocity of shortening means you cannot exert high forces at high speeds of shortening, and this has a direct bearing on muscular power. In ► Chap. 6 we will study mechanical power flows and look more closely at the right mix of force and velocity that creates peak muscular power output. This also means that isometric strength and muscle speed are really two different muscular abilities.

So there are major implications for human movement because of the functional relationship between muscle force and velocity. What about training? Does training alter the relationship between muscle force and velocity, or does the force–velocity relationship remain fairly stable and determine how you train muscle? It turns out that we cannot change the nature (shape) of the force–velocity relationship with training, but we can shift the graph upward to improve performance (De Koning et al. 1985; Fitts and Widrick 1996). Weight

Fig. 4.9 Training shifts the force–velocity curve upward and is specific to the kind of training. Heavy weight training primarily shifts the curve upward for isometric and slow concentric actions, while speed training improves muscle forces at higher concentric speeds

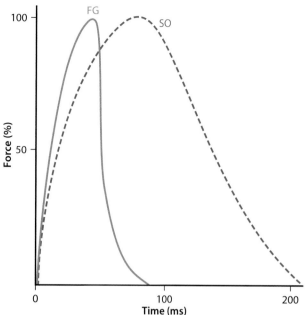

Fig. 4.10 The twitch response of fast-twitch (FG) and slow-twitch (SO) muscle fibers. Force output is essentially identical for equal cross-sectional areas, but there are dramatic differences in the rise and decay of tension be- tween fiber types that affect the potential speed of movement

training with high loads and few repetitions primarily shifts the force–velocity curve up near isometric conditions (Fig. 4.9), while fast lifting of light loads shifts the curve up near V_{max}, which is the maximum velocity of shortening for a muscle. Athletes training to maximize throwing speed will train differently based on the load and speed of the implements in their sport. Athletes putting the shot will do higher weight and low repetition lifting, compared to athletes that throw lighter objects like a javelin, softball, or baseball, who would train with lower weights and higher speeds of movement. One of the best books that integrates the biomechanics of movement and muscle mechanics in strength and conditioning is by Zatsiorsky and Kraemer (2006).

Another area where the force–velocity relationship shows dramatic differences in muscle performance is related to muscle fiber types. Skeletal muscle fibers fall on a continuum between slow twitch (Type I) and fast twitch (Type II). Type I are also called slow oxidative (SO) because of their high oxidative glycolysis capacity (considerable mitochrondion, myoglobin, triglycerides, and capillary density). Type II fibers are also called fast glycolytic (FG) because of their greater anaerobic energy capacity (considerable intramuscular ATP and glycolytic enzymes). Muscle fibers with intermediate levels are usually called FOG (fast-oxidative–glycolytic) fibers. Muscle fiber types have been classified in many ways (Scott et al. 2001), but biomechanics often focuses on the twitch response and velocity of shortening characteristics of fiber types. This is because the force

potential of fast and slow-twitch fibers per given physiological cross-sectional area is about the same. The timing that the muscle fibers create force and speed of shortening, however, is dramatically different. This fact has major implications for high-speed and high-power movements.

The easiest way to illustrate these differences is to look at the **twitch** response of different fiber types. If an *in vitro* muscle fiber is stimulated one time, the fiber will respond with a twitch. The rate of tension development and decay of the twitch depend on the fiber type of the fiber. Figure 4.10 illustrates a schematic of the twitch responses of several fiber types. A fiber at the slow end of the fiber type continuum gradually rises to peak tension in between 60 and 120 milliseconds (about a tenth of a second). A fiber at the high end of the continuum (FG) would quickly create a peak tension in 20 to 50 ms. This means that the muscle with a greater percentage of FG fibers can create a greater velocity of shortening than a similar (same number of sarcomeres) one with predominantly SO muscle fibers. Muscles with higher percentages of SO fibers will have a clear advantage in long-duration, endurance-related events.

Human muscles are a mix of fiber types. There are no significant differences in fiber types of muscles across gender, but within the body, the antigravity muscles (postural muscles that primarily resist the torque created by gravity) like the soleus, erector spinae, and

4

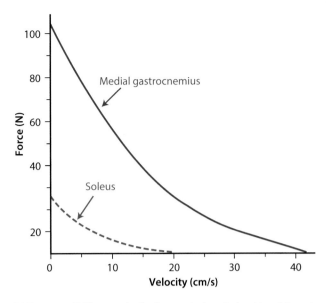

◨ Fig. 4.11 Differences in the force–velocity relationship of the primarily fast-twitch medial gastrocnemius and primarily slow-twitch soleus of the cat. (Reprinted, by permission, from Edgerton, Roy, Gregor, and Rugg, (1986))

abdominals tend to have a higher percentage of slow fibers than fast fibers. The fiber type distribution of elite athletes in many sports has been well documented. There is also interest in the trainability and plasticity of fiber types (Fitts and Widrick 1996; Kraemer et al. 1996). ◨ Figure 4.11 illustrates the force–velocity relationship in the predominantly slow-twitch soleus and predominantly fast-twitch medial gastrocnemius muscles in a cat. This fiber distribution and mechanical behavior are likely similar to humans. If the gastrocnemius was a more significant contributor to high-speed movements in sport, how might exercise position and technique be used to emphasize the gastrocnemius over the soleus?

Interdisciplinary Issue: Speed
Running speed in an important ability in many sports. The force–velocity relationship suggests that as muscles shorten concentrically faster, they can create less tension to continue to increase velocity. What are the main factors that determine sprinting speed? Do muscle mechanical properties dominate sprinting performance, or can technique make major improvements in to running speed? Several lines of research suggest that elite sprinting ability may be more related to muscular and structural factors than technique. Near top running speed, stride rate appears to be the limiting factor rather than stride length (Luthanen and Komi 1978; Chapman and Caldwell 1983; Mero et al. 1992). In the 100 meter dash, running speed is clearly corre-

lated with percentage of fast-twitch fibers (Mero et al. 1981) and the length of muscle fascicles (Abe et al. 2000; Kumagai et al. 2000) in high-level sprinters. Athletes with longer fascicles are faster. Future research into genetic predisposition to fiber dominance and trainability might be combined with biomechanical research to help improve the selection and training of sprinters.

Application: Domains of Muscular Strength
Therapists, athletes, and coaches often refer to a functional characteristic called **muscular strength**. While muscular strength is commonly measured in weight training with one-repetition maxima (1RM is the maximum weight a person can lift only one time), most researchers define muscular strength in isometric conditions at a specific joint angle to eliminate the many mechanical factors affecting muscle force (Clarke 1966; Knuttgen and Kraemer 1987; Atha 1981). Many fitness test batteries include tests for components called *muscular strength* and *muscular endurance*. Early physical education research demonstrated that muscular strength has several domains of functional expression. Statistical analysis of fitness testing demonstrated that muscular strength is expressed as static (isometric), dynamic (slow to moderate movements), and explosive for fast movement (Jackson and Frankiewicz 1975; Myers et al. 1993). This corresponds closely to the major changes in force capability in the force–velocity relationship. Other experts often include another domain of muscular strength related to eccentric actions: stopping strength (Zatsiorsky and Kraemer 2006). Functional muscular strength is also complicated by the fact that the force a muscle group can express also depends on the inertia of the resistance. The peak force that can be created in a basketball chest pass is nowhere near peak bench press isometric strength because of the small inertia of the ball. The ball is easily accelerated (because of its low mass), and the force the muscles can create at high shortening velocities rapidly declines, so the peak force that can be applied to the ball is much less than with a more massive object. So the force–velocity property of skeletal muscle and other biomechanical factors is manifested in several functional "strengths." Kinesiology professionals need to be aware of how these various muscular performance parameters correspond to the movements of their clients. Professionals should use muscular strength terminology correctly to prevent the spread of inaccurate information and interpret the literature carefully because of the many meanings of the word *strength*.

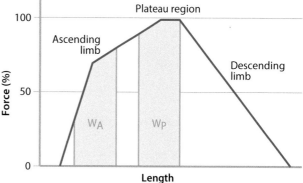

Fig. 4.12 The force–length relationship of human skeletal muscle. The active component follows an inverted "U" pattern according to the number of potential cross-bridges as muscle length changes. Passive tension increases as the muscle is stretched beyond its resting length (L_0). The total tension potential of the muscle is the sum of the active and passive components of tension

Fig. 4.13 The three regions of the active component of the length–tension relationship. Differences in the work ($W = \mathbf{F} \cdot \mathbf{d}$) the muscle can do in the ascending limb (W_A) versus the plateau region (W_P) are illustrated. Shortened muscle would have to work over a greater displacement to create the same work that a shorter action in the midrange of motion. Work can be visualized as the area under a force–displacement graph

4.6.2 Force–Length Relationship

The length of a muscle also strongly affects the ability of the muscle to create tension. The **force–length relationship** documents how muscle tension varies at different muscle lengths. The variation in potential muscle tension at different muscle lengths, like the force–velocity relationship, also has a dramatic effect on how movement is created. We will see that the force–length relationship is as influential on the torque a muscle group can make as the geometry (moment arm) of the muscles and joint (Rassier et al. 1999).

Remember that the tension a muscle can create has both active and passive sources, so the length–tension graph of muscle will have both of these components. Figure 4.12 illustrates the force–length relationship for a skeletal muscle fiber. The active component of the force–length relationship (dashed line) has a logical association with the potential numbers of cross-bridges between the actin and myosin filaments in the sliding filament theory. Peak muscle force can be generated when there are the most cross-bridges. This is called resting length (L_0) and usually corresponds to a point near the middle of the range of motion. Potential active muscle tension decreases for shorter or longer muscle lengths because fewer cross-bridges are available for binding. The passive tension component (solid line) shows that passive tension increases very slowly near L_0 but dramatically increases as the muscle is elongated. Passive muscle tension usually does not dramatically contribute to movements in the middle portion of the range of motion but does contribute to motion in various neuromuscular disorders (Salsich et al. 2000) and when muscles are stretched near the end of the range of motion. The exact shape of the force–length relationship slightly

varies between muscles because of differences in active (fiber area, angle of pennation) and passive (tendon) tension components (Gareis et al. 1992).

The active tension component of the force–length relationship is usually classified into three regions (Fig. 4.13). The ascending limb represents the decreasing force output of the muscle as it is shortened beyond resting length. Movements that require a muscle group to shorten considerably will not be able to create maximal muscle forces. The plateau region represents the high muscle force region, typically in the midrange of the anatomical range of motion. Movements initiated near the plateau region will have the potential to create maximal muscle forces. The descending limb represents the decreasing active tension a muscle can make as it is elongated beyond resting length. At extremes of the descending limb, the dramatic increases in passive tension provide the muscle force to bring a stretched muscle back to shorter lengths, even though there are virtually no potential cross-bridge attachment sites. Biomechanical research has begun to demonstrate that muscles adapt to chronic locomotor movement demands and the coordination of muscles may be organized around muscles suited to work on the ascending, plateau, or descending limb of the force–length curve (Rassier et al. 1999; Maganaris 2001). It is clear that the length of muscles influences how the central nervous system coordinates their actions (Nichols 1994).

Activity: Force–Length Relationship

Active insufficiency is the decreased tension of a multiarticular muscle when it is shortened across one or more of its joints. Vigorously shake the hand of a partner. Fully flex your wrist, and try to create the same large grip force in the handshake. What happened to

4

your strength? Which limb of the force–length relationship creates this phenomenon? Is your grip strength as weak when you wrist is hyperextended as you shake another person's hand?

The implications for a muscle working on the ascending limb versus the plateau region of the force–length curve are dramatic. The mechanical work that a muscle fiber can create for a given range of motion can be visualized as the area under the graph because work is force times displacement. Note the difference in mechanical work (area under **F-d** graph) that can be created if the muscle fiber works in the ascending limb instead of near the plateau region (⬛ Fig. 4.13). These effects also interact with the force–velocity relationship to determine how muscle forces create movement throughout the range of motion. These mechanical characteristics also interact with the time delay in the rise and fall of muscle tension, the force–time relationship.

4.6.3 Force–Time Relationship

Another important mechanical characteristic of muscle is related to the temporal delay in the rise and fall of muscle tension. The **force–time relationship** primarily refers to the delay in the development of muscle tension of the whole MTU and can be expressed as the time from the **motor action potential** (electrical signal of depolarization of the fiber that makes of the electromyographic or EMG signal) to the initial rise in muscle tension or some relevant tension value (size of load or maximum). We will see later the decline in muscle tension after muscle stimulation ceases is slower than the rise time.

The time delay that represents the force–time relationship in the rise of muscle tension can be split into two parts. The first part of the delay is related to the rise in muscle stimulation sometimes called active state or excitation dynamics. In fast and high-force movements, the neuromuscular system can be trained to rapidly increase (down to about 20 milliseconds [ms]) muscle stimulation. The second part of the delay involves the actual build-up of tension that is sometimes called contraction dynamics. Recall that the contraction dynamics of different fiber types was about 20 ms for FG and 120 ms for SO fibers. When many muscle fibers are repeatedly stimulated, the fusion of many twitches means the rise in tension takes even longer. The length of time depends strongly on the cognitive effort of the subject, training, kind of muscle action, and the activation history of the muscle group. ⬛ Figure 4.14 shows a

⬛ Fig. 4.14 The rectified electromyographic (REMG) signal from the quadriceps and the force of knee extension in an isometric action. The delay between the activation (REMG) and the build-up of force in the whole muscle is the *electromechanical delay* and represents the force–time relationship of the muscle. It takes 250 to 400 ms for peak force to be achieved after initial activation of the muscle group

schematic of rectified electromyography (measure the electrical activation of muscle) and the force of an isometric grip force. Note that peak isometric force took about 500 ms. Revisit ⬛ Fig. 2.14 for another example of the electromechanical delay (delay from rectified EMG to whole muscle force). Typical delays in peak tension of whole muscle groups (the force–time relationship) can be quite variable. Peak force can be developed in as little as 100 ms and up to over a second for maximal muscular strength efforts. The force–time relationship is often referred to as the **electromechanical delay** in electromyographic (EMG) studies. This delay is an important thing to keep in mind when looking at EMG plots and trying to relate the timing of muscle force rise and fall to the movement. Recall that the rise in muscle tension is also affected by the stiffness of the connective tissue components of muscle (passive tension from SEC and PEC), so the size of the electromechanical delay is affected by the slack or tension in the connective tissue (Muraoka et al. 2004). In ▶ Chap. 5 we will see that kinematics provides a precise description of how motion builds to a peak velocity and where this occurs relative to the accelerations and force that create motion.

Application: Strength Curves
The torque-generating capacity of a muscle primarily depends on its physiological cross-sectional area, moment arm, and muscle length (Murray et al. 2000). The maximum torque that can be created by a muscle group through the range of motion does not always have a shape that matches the *in vitro* force–length relationship of muscle fibers. This is because muscles with different areas, moment arms, and length properties

are summed and often overcome some antagonistic muscle activity in maximal exertions (Kellis and Baltzopoulos 1997).There is also some evidence that the number of sarcomeres in muscle fibers may adapt to strength training and interact with muscle moment arms to affect were the peak torque occurs in the range of motion (Koh 1995). "Strength curves" of muscle are often documented by multiple measurements of the isometric or isokinetic torque capability of a muscle group throughout the range of motion (Kulig et al. 1984). The torque-angle graphs created in isokinetic testing also can be interpreted as strength curves for muscle groups. The peak torque created by a muscle group tends to shift later in the range of motion as the speed of rotation increases, and this shift may be related to the interaction of active and passive sources of tension (Kawakami et al. 2002). How the shape of these strength curves indicates various musculoskeletal pathologies is controversial (Perrin 1993). Knowledge of the angles where muscle groups create peak torques or where torque output is very low is useful in studying movement. Postures and stances that put muscle groups near their peak torque point in the range of motion maximize their potential contribution to motion or stability (Zatsiorsky and Kraemer 2006). In combative sports, an opponent put in an extreme joint position may be easily immobilized (active insufficiency, poor leverage, or pain from the stretched position). Accommodating resistance exercise machines (Nautilus™ was one of the first) are usually designed to match the average strength curve of the muscle group or movement. These machines are designed to stay near-maximal resistance (match the strength curve) throughout the range of motion (Smith 1982), but this is a difficult objective because of individual differences (Stone et al. 2002). There will be more discussion of strength curves and their application in ▶ Chap. 7.

This delay in the development of muscle tension has implications for the coordination and regulation of movement. It turns out that deactivation of muscle (timing of the decay of muscle force) also affects the coordination of movements (Neptune and Kautz 2001), although this section will limit the discussion of the force–time relationship to a rise in muscle tension.

Kinesiology professionals need to know about these temporal limitations so they understand the creation of fast movements and can provide instruction or cues consistent with what the mover's body does. For example, it is important for coaches to remember that when they see high-speed movement in the body, the forces and torques that created that movement preceded the peak speeds of motion they observed. Clinicians need to recognize that even slow movements result from muscle action forces that precede the movement being observed. A coach that provides urging to increase effort late in the movement is missing the greater potential for acceleration earlier in the movement and is asking the performer to increase effort when it will not be able to have an effect. Muscles are often preactivated before to prepare for a forceful event, like the activation of plantar flexors and knee extensors before a person lands from a jump. A delay in the rise of muscle forces is even more critical in movements that cannot be pre-programmed due to uncertain environmental conditions. To make the largest muscle forces at the initiation of an intended movement, the neuromuscular system often uses a carefully timed movement and muscle activation strategy. This strategy is called the stretch-shortening cycle and will be discussed in the following section.

4.7 Stretch-Shortening Cycle (SSC)

The mechanical characteristics of skeletal muscle have such a major effect on the force and speed of muscle actions that the central nervous system has a preferred muscle action strategy to maximize performance in most fast movements. This strategy is most beneficial in high-effort events but is also usually selected in submaximal movements. Most normal movements unconsciously begin a **stretch-shortening cycle** (SSC): a countermovement away from the intended direction of motion that is slowed down (braked) with eccentric muscle action that is immediately followed by concentric action in the direction of interest. Sometimes this is also referred to as a stretch-shorten cycle. This active bounce out of an eccentric braking of a countermovement results in potentiation (increase) of force in the following concentric action if there is minimal temporal delay between the two muscle actions (Wilson et al. 1991; Elliott et al. 1999). In normal movements, muscles are also used in shortening-stretch cycles where the muscle undergoes concentric shortening followed by eccentric elongation as the muscle torque decreases below the resistance torque (Rassier et al. 1999); however it is the stretch-shortening cycle that is the most important history-dependent mechanical property of the muscle. Previous muscle action creates history-dependent responses like passive stretch-induced strength deficits, SSC, and post-activation potentiation (PAP).

4

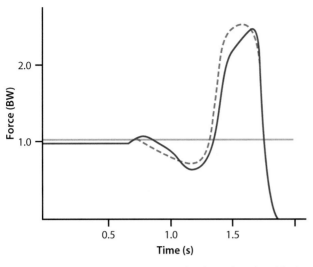

◨ **Fig. 4.15** Vertical jump ground reaction forces in units of bodyweight (BW) for two athletes. Athlete A (dashed line) has a greater rate of force development compared to athlete B (solid line). The rate of force development is the slope of the graphs when vertical forces are building above 1 BW. Do not interpret the up and down motion of the graph as motion of the jumper's body; it represents the sum of the vertical forces the athlete makes against the ground

Application: Rate of Force Development

Biomechanists often measure force or torque output of muscle groups or movements with dynamometers. One variable derived from these force–time graphs is the rate of force development (F/t), which measures how quickly the force rises. A high rate of force development is necessary for fast movements. The vertical ground reaction forces of the vertical jump of two athletes are illustrated in ◨ Fig. 4.15. Note that both athletes create the same peak vertical **ground reaction force**, but athlete A (dashed line) has a higher rate of force development (steeper slope) than athlete B (solid line). This allows athlete A to create a larger vertical impulse and jump higher. The ability to rapidly increase the active state and consequently muscle force has been demonstrated to contribute strongly to vertical jump performance (Bobbert and van Zandwijk 1999). Rate of force development is even more important in running jumps, where muscle actions and ground contact times are much shorter (50 to 200 ms) than in a vertical jump or MVC (◨ Fig. 4.14). Training the neuromuscular system to rapidly recruit motor units is very important in these kinds of movements (Aagaard 2003). What kinds of muscle fibers help athletes create a quick rise in muscle force? What kinds of muscle actions allow muscles to create the largest tensions? The next two sections will show how the neuromuscular system activates muscles and coordinates movements to make high rates of muscle force development possible.

Early research on frog muscle by Cavagna et al. (1965) demonstrated that concentric muscle work was potentiated (increased) when preceded by active stretch (eccentric action). This phenomenon is known as the **stretch-shortening cycle** or stretch-shorten cycle and has been extensively studied by Paavo Komi (1984, 1986). The use of fiber-optic tendon force sensors and measures of MTU length has allowed Komi and his colleagues to create approximate *in vivo* torque–angular velocity diagrams. Many biomechanical variables have been examined to study the mechanisms of the SSC benefit. The benefits of the SSC in movements are dependent on what variables and when these variables are calculated (Bird and Hudson 1998) and the resistance moved (Cronin et al. 2001b). The important point is that well-timed SSC can make dramatic improvements in maximal muscular performance. In moderate/slow SSC movements like a vertical jump performance can be enhanced 10–20% (see the SSC Activity and Lab Activity 4) and faster stretches up to 40–50% or more (Komi 1984; Bober et al. 1987; Cronin et al. 2001a, b).

The mechanisms of the beneficial effects of SSC coordination are of considerable interest to biomechanics scholars. There are four likely sources of the greater muscle force in the concentric phase of an SSC: contractile potentiation, reflex potentiation, storage and reutilization of elastic energy, and the time available for force development (Komi 1986; van Ingen Schenau et al. 1997). Contractile potentiation of muscle force is one of the several variations in muscle force potential based on previous muscles actions. These phenomena are called history-dependent behaviors (see Herzog et al. 2000; Sale 2002). Shortening actions tend to depress force output of subsequent muscle actions, while eccentric actions tend to increase concentric actions that immediately follow. Force potentiation of muscle is also dependent on muscle length (Edman et al. 1997).

Activity: Stretch-Shortening Cycle (SSC)

The benefits of stretch-shortening cycle muscle actions are most apparent in vigorous, full-effort movements. To see the size of the benefit for performance, execute several overarm throws for distance on a flat, smooth field. Measure the distance of your maximal effort throw with your feet still and body facing the direction of your throw. Have someone help you determine about how far your trunk and arm backswing was in the normal throw. Measure the distance of a primarily concentric action beginning from a static position that matches your reversal trunk and arm position in the normal throw. Calculate the benefit of the SSC (prestretch augmentation) in the throw as (normal–concentric)/concentric. Compare your results with those of others, the lab activity, and research on vertical jumping (Walshe et al. 1996; Kubo et al. 1999). The benefit of the SSC to other faster movements is likely to be even higher than in vertical jumping (Komi and Gollhofer 1997). What factors might affect amount of prestretch augmentation? What might be the prestretch augmentation in other movements with different loads?

Another mechanism for the beneficial effect of the SSC is a greater contribution from the myotatic or stretch reflex (see the section on "Proprioception of Muscle Action and Movement"). Muscle spindles are proprioceptors of muscle length and are particularly sensitive to fast stretch. When muscles are rapidly stretched, like in an SSC movement, muscle spindles activate a short reflex loop that strongly activates the muscle being stretched. Studies of athletes have shown greater activation of muscles in the concentric phase of an SSC movement compared to untrained subjects (Komi and Golhoffer 1997). The lack of a precise value for the electromechanical delay (the force–time relationship) makes it unclear if stretch reflexes contribute to greater muscle forces in the late eccentric phase or the following concentric phase. The contribution of reflexes to the SSC remains controversial and is an important area of study.

One of the most controversial issues is the role of elastic energy stored in the eccentric phase, which can be subsequently recovered in the concentric phase of an SSC. There has been considerable interest in the potential metabolic energy savings in the reutilization of stored elastic energy in SSC movements. It may be more accurate to say elastic mechanisms in the SSC are preventing energy loss or maintaining muscle efficiency (Ettema 2001), rather than an energy-saving mechanism. Animal studies (e.g., wallabies and kangaroo rats) have been used to look at the extremes of evolutionary adaptation in muscle–tendon units related to SSC movement economy (Griffiths 1989, 1991; Biewener 1998; Biewener and Roberts 2000). A special issue of the *Journal of Applied Biomechanics* was devoted to the role of stored elastic energy in the human vertical jump (Gregor 1997). Recent in vivo studies of the human gastrocnemius muscle in SSC movements have shown that the compliant tendon allows the muscle fibers to act in near isometric conditions at joint reversal and while the whole muscle shortens to allow elastic recoil of the tendinous structures to do more positive work (Kubo, et al. 2000; Kurokawa et al. 2001). The interaction of tendon and muscle must be documented to fully understand the benefits of the SSC action of muscles (Finni et al. 2000). Training studies indicate that another mechanism of increased jumping

could be related to more effective eccentric phases of SSC movements (Cormie et al. 2010).

Another mechanism for the beneficial effect of SSC coordination is related to the timing of force development. Recall that the rate of force development and the force–time relationship have dramatic effect on high-speed and high-power movements. The idea is that if the concentric movement can begin with near-maximal force and the slack taken out of the elastic elements of the MTU, the initial acceleration and eventual velocity of the movement will be maximized. While this is logical, the interaction of other biomechanical factors (force–length relationship, architecture, and leverage) makes it difficult to examine this hypothesis. Interested students should see papers on this issue in the vertical jump (Bobbert et al. 1996; Bobbert and van Zandwijk 1999), and sprint starts (Kraan et al. 2001).

The most influential mechanism for the beneficial effect of an SSC will likely depend on the movement. Some events like the foot strike in sprinting or running jump (100 to 200 ms) require high rates of force development that are not possible from rest due to the force–time relationship. These high-speed events require a well-trained SSC technique and likely have a different mix of the four factors than a standing vertical jump. Remember, however, that even slower movements will normally be coordinated with SSC muscle actions because of their greater mechanical efficiency (more muscle force for less metabolic cost when muscles act in quasi-isometric conditions and recover more energy stored in the muscle-tendon unit.

Plyometric (plyo = more metric = length) is another form of training will likely increase the athlete's ability to tolerate higher eccentric muscle forces and increase the potentiation of initial concentric forces (Komi 1986). Plyometrics are most beneficial for athletes in high-speed and dynamic sports/activities. There has been considerable research on the biomechanics of lower-body drop jumping plyometrics (Bobbert 1990). Early studies showed that jumpers tend to spontaneously adopt one of two techniques (Bobbert et al. 1986) in drop jumping exercises. Research has also focused on technique adaptations due to the compliance of the landing surface (Sanders and Allen 1993), the effect of landing position (Kovacs et al. 1999), and what might be the optimal drop height (Lees and Fahmi 1994). Less research has been conducted on the biomechanics of upper body plyometrics (Newton et al. 1996; Knudson 2001). Loads for plyometric exercises are controversial, with loads ranging between 30 and 70% of isometric muscular strength because maximum muscular performance varies with technique and the movement (Kaneko et al. 1983; Wilson et al. 1993; Izquierdo et al. 1999; Cronin et al. 2001a; Newton et al. 1997).

4

Plyometrics are not usually recommended for novice and individuals with limited training experience. Even though eccentric muscle actions are common in natural movements, intense unaccustomed eccentric activity is clearly associated with overload and potential damage. Eccentric-induced muscle fiber damage has been extensively studied and appears to be related to excessive strain in sarcomeres (see the review by Lieber and Friden 1999) rather than the high forces of eccentric actions. Kinesiology professionals should carefully monitor plyometric technique and exercise intensity to minimize the risk of injury with vigorous training. Now we will turn to the neuromuscular factors that determine how we create internal forces from muscle activation.

4.8 Neuromuscular Control

The mechanical response of muscles also strongly depends on how the muscles are activated. The neuromuscular control of movement by the central nervous system is an active area of study where biomechanical research methods have been particularly useful. This section will summarize the important structures and their functions in the activation of muscles to regulate muscle forces and movement.

4.8.1 The Functional Unit of Control: Motor Units

The coordination and regulation of muscle force to create movement are of considerable interest to many scholars. At the structural end of the neuromuscular control process are the functional units of the control of muscles: motor units. A **motor unit** is one alpha motor neuron and all the muscle fibers it innervates. A muscle may have from a few to several hundred motor units. The activation of a motor axon results in stimulation of all the fibers of that motor unit and the resulting twitch. All fibers of a motor unit have this synchronized, "all-or-nothing" response. Pioneering work in the neurophysiology of muscle activation was done in the early twentieth century by Sherrington, Arian, and Denny-Brown (Burke 1986).

4.8.2 Regulation of Muscle Force

If the muscle fibers of a motor unit twitch in unison, how then does a whole muscle generate a smooth increase in tension? The precise regulation of muscle tension results from two processes: the **recruitment** of different motor units and the regulation of motor unit **firing rate**.

Recruitment is the activation of different motor units within a muscle. Physiological research has determined three important properties of recruitment of motor units. First, motor units tend to be organized in pools or task groups (Burke 1986). Second, motor units tend to be recruited in an asynchronous fashion. Different motor units are stimulated at slightly different times, staggering the twitches to help smooth out the rise in tension. There is evidence that some motor unit synchronization develops to increase rate of force development (Semmler 2002), but too much synchronous recruitment results in pulses of tension/tremor that is associated with disease (Parkinson's) or extreme fatigue (final repetition of an exhaustive set of weight lifting). The recruitment of motor units is likely more complex than these general trends since serial and transverse connections between parallel architecture muscles allow active fibers to modify the tension and length of nearby fibers (Sheard 2000).

The third organizational principle of recruitment has been called orderly recruitment or the **size principle** (Denny-Brown and Pennybacker 1938; Henneman et al. 1965). It turns out that motor units tend to have specialized innervation and homogeneous fiber types, so motor units take on the characteristics of a particular fiber type. A small motor unit consists of a motor axon with limited myelination and primarily SO muscle fibers, while a large motor unit has a large motor axon (considerable myelination) and primarily FG fibers. The recruitment of a large motor unit by the brain results in the quickest message and build-up in tension, while recruitment of a small motor unit has a slower nerve conduction velocity and a gradual tension build-up (Fig. 4.16).

In essence, the size principle says that motor units are recruited progressively from small (slow twitch) to large (fast twitch). A gradual increase in muscle force would result from recruitment of SO dominant motor units followed by FOG and FG dominant units, and motor units would be de-recruited in reverse order if the force is to gradually decline. This holds true for most movements, but there is the ability to increase firing rate of large motor units within the size principle to move quickly or rapidly build-up forces (Bawa 2002; Burke 1986). Have you ever picked up a light object (empty suitcase) when you expected a heavy one? If so, you likely activated many pools of large and small motor units immediately and nearly threw the object. Athletes in events requiring high rates of force development (jumping, throwing) will need to train their ability to speed up and override the size principle and activate many motor units rapidly. There are also many other factors that complicate the interpretation that the size principle is an invariant in

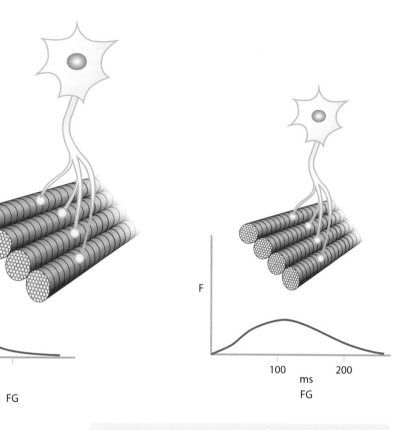

Fig. 4.16 Schematic of the differences in the size of motor units. Typical twitch response, size of the motor nerve, and typical recruitment are illustrated

motor control (Enoka 2015). One example is that recruitment tends to be patterned for specific movements (Sale 1987; Desmedt and Godaux 1977), so the hamstring muscles may not be recruited the same in a jump, squat, or knee flexion exercise. Activation of muscles also varies across the kinds of muscle actions (Enoka 1996; Gandevia 1999; Gielen 1999) and can be mediated by fatigue and sensory feedback (Enoka 2015).

Firing rate or *rate coding* is the repeated stimulation of a particular motor unit over time. To create the muscle forces for normal movements, the frequency (times per second [Hz]) that motor units are usually rate coded is between 10 and 30 Hz, while FG motor units have a faster relaxation time and can be rate coded between 30 and 60 Hz (Sale 1992).

Interdisciplinary Issue: The Control of Movement
With hundreds of muscles, each with hundreds of motor units that must be repeatedly stimulated, to coordinate in a whole-body movement, can the brain centrally control all those messages? If the brain could send all those messages in a preprogrammed fashion, could it also monitor and evaluate efferent sensory and proprioceptive information and adjust the movement? Early motor learning research and theory focused on the brain's central control of movement or a motor program. More recent research is based on a Bernstein

or dynamical systems perspective (Feldman et al. 1998; Schmidt and Wrisberg 2007), where more general control strategies interact with sensory feedback. Since kinetic variables (torques, forces, EMG) can be measured or calculated using biomechanics, many motor control scholars are interested in looking at these variables to uncover clues as to how movement is coordinated and regulated. Some biomechanists are interested in the control of movement, so here is an ideal area for interdisciplinary research with scholars from motor control, development, and other subdisciplines of kinesiology.

The repeated stimulation of a motor unit increases the twitch force above the level of a single twitch (up to 10 times) because the tension in the fibers begins at a higher level, before the decay or relaxation in tension. Since recruitment tends to be asynchronous and firing rates vary with motor unit size, the twitches of the motor units in a whole muscle combine and fill in variations over time, resulting in smooth changes in tension. When muscle is artificially stimulated for research or training purposes to elicit maximal force, the frequency used is usually higher than 60 Hz to make sure that motor unit twitches fuse into a **tetanus**. A **tetanus** is the summation of individual twitches into a smooth increase in muscle tension.

4

■ **Fig. 4.17** Raw EMG of eight leg muscles in five cycles of bicycle pedaling. The initiation of the downstroke of the first pedal cycle is illustrated by the dashed line region highlighted by the red arrow. (Reprinted from Laplaud et al., Journal of Electromyography and Kinesiology © (2006), with permission from Elsevier)

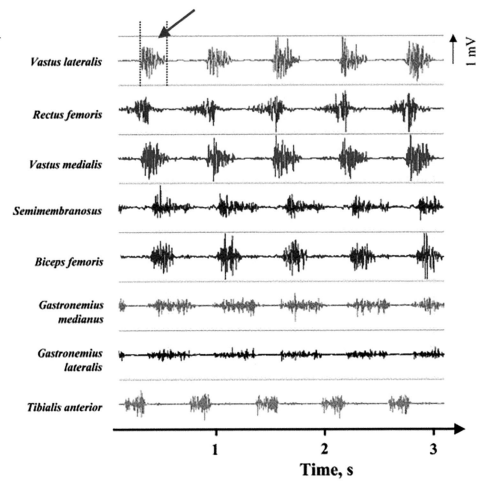

Both recruitment and firing rate have a dramatic influence on the range of muscle forces that can be created. How recruitment and firing rate interact to increase muscle forces is quite complex, but it appears that recruitment dominates for forces up to 50% of maximum with increasing importance of firing rate (Enoka 2015). The combined effect of recruitment and firing rate of motor units is reflected in the size, density, and complexity of the eletromyographic (EMG) signal. Special indwelling EMG electrode techniques are used to study the recruitment of individual motor units (Basmajian and DeLuca 1985).

Recall that in ▶ Chap. 2 we learned how EMG research has shown that at the whole muscle level muscles are activated to in complex synergies to achieve movement or stabilization tasks. Muscles are activated in short bursts that coordinate with other forces (external and segmental interactions) to create human movement. ■ Figure 4.17 shows the lower extremity muscle activation in several pedal strokes in cycling. Compare the short burst pattern of activation in 4.17 to the sustained activity of an MVC in ■ Fig. 4.14. Physical medicine professionals often take advantage of this flexibility

of the neuromuscular system by training muscle actions that compensate for physical limitations from disease or injury. Motor learning scholars are interested in EMG and the activation of muscles as clues to neuromuscular strategies in learning movements. While there has not been extensive research in this area, it appears that changes in EMG with practice/training depend on the nature of the task (Gabriel and Boucher 2000). As people learn submaximal movements, the duration of EMG bursts decreases; there are decreases in extraneous and coactivation of muscles and a reduction in EMG magnitude as the body learns to use other forces (inertial and gravitational) to efficiently create the movement (Englehorn 1983; Moore and Marteniuk 1986; Newell et al. 1989). Maximal effort movements are believed to be more reliant on changes in the magnitude and rise time of activation than the duration of muscle activation (Gottlieb et al. 1989). In maximal high-speed movements, the magnitude and rate of increase in activation tend to increase (Darling and Cooke 1987; Corcos et al. 1993; Gabriel and Boucher 2000) with practice. The activation and cooperative actions of muscles to create skilled human movement are very complex phenomena.

Unfortunately, athletes are often stereotyped as dumb jocks with gifted physical abilities. How much of movement ability do physical characteristics like muscular strength, speed, and coordination contribute to performance compared to neuromuscular abilities (a good motor brain)? Think about the ability your favorite athlete would have if he/she had a stroke that affected part of their motor cortex. In training and conditioning, there are several areas of research where there is evidence that the effects of training on muscle activation by the central nervous system are underrated. First, it is well known that the majority of the initial gains in strength training (first month) are related to the neural drive rather than hypertrophy (see Sale 1992). Second, it is known that both normal and injured subjects are not usually able to achieve true maximum muscle force in a maximal voluntary contraction. This is called **muscle inhibition** and is studied using an electrical stimulation method called twitch interpolation technique (Brondino et al. 2002). Another area of neuromuscular research relates to the inability to express bilateral muscular strength (both arms or legs) equal to the sum of the unilateral strength of each extremity. This phenomenon has been called the bilateral deficit but the decrements (3–20%) are not always observed (Jakobi and Cafarelli 1998). Interest in the biomechanics of the vertical jump has made this movement a good model for examining a potential bilateral deficit (Challis 1998). How might a professional try to differentiate true differences in muscular strength between sides of the body and a bilateral deficit? If there truly is a bilateral deficit that limits the neuromuscular activation of two extremities, how should you train for bilateral movements (bilaterally, stronger or weaker limb)?

4.8.3 Proprioception of Muscle Action and Movement

Considerable sensory information about the body and its environment are used in the regulation of movement. While persons use all their senses to gather information about the status or effectiveness of their movements, there are musculoskeletal receptors that provide information to the brain to help produce movement. These receptors of information about the motion and force in muscles and joints are called **proprioceptors**. While we usually do not consciously attend to this information, this information and the various reflexes they initiate are important in the organization of movement. A reflex is an involuntary response initiated by some sensory stimulus. Reflexes are only initiated if the sensory stimulus is above some threshold.

There are many proprioceptive receptors that monitor aspects of movement. Information about joint position is provided by four kinds of receptors. The vestibular system of the inner ear provides information about the head's orientation with respect to gravity. This section will summarize the important MTU proprioceptors that provide information on muscle length (muscle spindles) and force (Golgi tendon organs). Human movement performance relies on an integration of all sensory organs, and training can be quite effective in utilizing or overriding various sensory or reflex responses. A dancer spinning in the transverse plane can reduce dizziness (from motion in inner ear fluid) by spotting—rotating the neck opposite to the spin to keep the eyes fixed on a point followed by a quick counterrotation to the spin so as to find that point again. Athletes in "muscular strength" sports not only train their muscle tissue to shift the force–velocity relationship upward; they also train their central nervous system to more effectively activate more motor units and override the inhibitory (relaxation) effect of Golgi tendon organs.

When muscle is activated, the tension that is developed is sensed by Golgi tendon organs. Golgi tendon organs are located at the musculotendinous junction and have an inhibitory effect on the creation of tension in the muscle. Golgi tendon organs connect to the motor neurons of that muscle and can relax a muscle to protect it from excessive loading. The intensity of this autogenic inhibition varies, and its functional significance in movement is controversial (Chalmers 2002). If an active muscle were forcibly stretched by an external force, the Golgi tendon organs would likely relax that muscle to decrease the tension and protect the muscle. Much of high-speed and high muscular strength performance is training the central nervous system to override this safety feature of the neuromuscular system. The rare occurrence of a parent lifting part of an automobile off a trapped child is an extreme example of overriding Golgi tendon organ inhibition from the emotion and adrenaline. Famous strong men and women, past and present, train to overcome the effects of this inhibition and pain of extremely high muscular forces. The action of Golgi tendon organs is also obvious when muscles suddenly stop creating tension. Good examples are the collapse of a person's arm in a close wrist wrestling match (fatigue causes the person to lose the ability to override inhibition) or the buckling of a leg during the great loading of the takeoff leg in running jumps.

4

Muscle spindles are sensory receptors located between muscle fibers that sense the length and speed of lengthening or shortening. Muscle spindles are sensitive to stretch and send excitatory messages to activate the muscle and protect it from stretch-related injury. Muscle spindles are sensitive to slow stretching of muscle but provide the largest response to rapid stretches. The rapid activation of a quickly stretched muscle (100–200 ms) from muscle spindles is due to a short reflex arc. Muscle spindle activity is responsible for this **myotatic reflex** or stretch reflex. The use of a small rubber reflex hammer by physicians allows them to check the stretch reflex responses of patients. The large numbers of spindles and their innervation allow them to be sensitive and reset throughout the range of motion. Recall that a stretch reflex is one possible mechanism for the benefit of an SSC. Stretch reflexes may contribute to the eccentric braking action of muscles in follow-throughs. In performing stretching exercises, the rate of stretch should be minimized to prevent activation of stretch reflex responses from muscle spindles.

Interdisciplinary Issue: Muscle Inhibition and Disinhibition

The neuromuscular aspects of training and detraining are fertile areas for the cooperation of scholars interested in human movement. Rehabilitation (physical therapists, athletic trainers, etc.) and strength and conditioning professionals, as well as neurophysiologists, and biomechanists all might be involved in understanding the inhibition of muscle activation following an injury. They might also collaborate on research questions if the neuromuscular changes in strength development are similar in strength redevelopment following injury or disuse.

The other important neuromuscular effect of muscle spindles is inhibition of the antagonist (opposing muscle action) muscle when the muscle of interest is shortening. This phenomenon is call **reciprocal inhibition**. Relaxation of the opposing muscle of a shortening muscle contributes to efficient movement. In lifting a drink to your mouth, the initial shortening of the biceps brachii inhibits triceps brachii activity that would make the biceps work harder than necessary. Reciprocal inhibition is often overridden by the central nervous system when coactivation of muscles on both sides of a joint is needed (muscle synergy) to push or move in a specific direction. Reciprocal inhibition also plays a role in several stretching techniques designed to utilize neuromuscular responses to facilitate stretching. The contract–relax–agonist–contract technique of proprioceptive neuromuscular facilitation (PNF) is designed to use reciprocal inhibition to relax the muscle being stretched by activating the opposite muscle group (Hutton 1993; Knudson 1998). For example, in stretching the hamstrings, the stretcher activates the hip flexors to help relax the hip extensors being stretched. The intricacies of proprioceptors in neuromuscular control (Enoka 2015; Taylor and Prochazka 1981) are relevant to all movement professionals but are of special interest to those who deal with disorders of the neuromuscular system (e.g., neurologists, physical therapists). More research is needed to understand how the central nervous system weighs these many neuromuscular factors is activating muscles to create and control skilled movement.

4.9 Force–Motion Principle

Creation of human movement is largely driven by muscular forces interacting with other forces external to the body. The **force–motion principle** states that it takes unbalanced forces (and the subsequent torques they induce) to create or modify motion. To know what size and direction of force, like from muscular actions, to change, recall that a **free-body diagram** of the biomechanical system is usually employed. We are not in a position to perform quantitative calculations to determine the exact motion created at this point in the text, but this section will provide examples of the qualitative application of the force–motion principle in improving human movement. Later on, in ▶ Chaps. 6 and 7, we will explore Newton's laws of motion and the quantitative methods used in biomechanics to explore the forces and moments that create and modify human movement.

Kinesiology professionals often work in the area of physical conditioning or rehabilitation to improve function. Function can be high-level sport performance or remediation of the effects of an injury, disuse, or aging. If muscle forces are the primary motors (hip extensors in running faster) and brakes (plantar flexors in landing from a jump), the force–motion principle suggests that muscle groups that primarily contribute to the motion of interest should be trained. Remember that this is more complex task than consulting an anatomy book, requiring the integration of several kinds of biomechanics research. So how can we know what exercises, technique (speed, body position), or load to prescribe in conditioning or rehabilitation?

Imagine a physical education teacher working with students on their upper body muscular strength. A particular student is working toward improving his score on a pull-up test in the fitness unit. The forces in a pull-up exercise can be simplified into two vertical forces: the downward gravitation force of bodyweight and an upward force created by concentric muscle actions at the elbows, shoulders, and back. The considerable isometric actions of the grip, shoulder girdle, and trunk do not appear to obviously limit this youngster's performance.

Fig. 4.18 The force–motion principle can be applied in a situation where a gymnast is having difficulty in performing inverted splits. The two forces that may limit the split are the passive tension resistance of the hip muscles or inadequate strength of the hip abductors. The coach must decide which forces limit this athlete's performance

You note that this student's bodyweight is not excessive, so losing weight is not an appropriate choice. The teacher decides to work on exercises that train the elbow flexors, as well as the shoulder adductors and extensors. The teacher will likely prescribe exercises like lat pulls, arm curls, and rowing to increase the student's ability to pull downward with a force larger than his bodyweight. Which of these exercises to emphasize more would require some library/online research on the biomechanics of these exercises and pull-ups.

Suppose a coach is interested in helping a young gymnast improve her "splits" position in a cartwheel or other arm support stunts (Fig. 4.18). The gymnast can easily overcome the passive muscular tension in the hip adductors to create a split in a seated position, but the downward force creating this static position is large (weight of the upper body) compared to the weight of the leg that assists the split in the inverted body position (◘ Fig. 4.18). The force–motion principle suggests that the balance of forces at the hips must be downward to create the split in the dynamic action of the stunt. In other words, the forces of gravity and hip abductors must create a torque equal to the upward torque created by the passive tension in the hip adductors. If the gymnast is having trouble with this stunt, the two biomechanical solutions that could be considered are stretching the hip adductors (to decrease passive muscle tension resistance) and increase the muscular strength or activation of the hip abductors in this particular phase of the stunts.

These examples of the force–motion principle have been kept simple for several reasons. First, we are only beginning our journey to an understanding of biomechanics. Second, the force–motion principle deals with a complex and deeper level of mechanics (kinetics) that explains the causes of motion. Third, the complexity of the biomechanical and neuromuscular system makes inference of muscle actions complicated. The variability in the forces and kinematics of human movement, therefore, has been of interest to a variety of scholars (see the ► Chap. 3 Interdisciplinary Issue on Variability). The application of biomechanical principles is also not easy, because they interact with each other and also with factors related to the task, individual differences, or the movement environment. The rest of the book will provide challenges to the perception that the causes of and solutions to human movement problems are simple and introduce you to the variables and principles of biomechanics that are used to answer questions about the causes of movement.

4.10 Summary

Forces applied to the musculoskeletal system create loads in tissues. Loads are named based on their direction and line of action relative to the structure. Several mechanical variables are used to document the mechanical effect of these loads on the body. How hard forces act on tissue is mechanical stress, while tissue deformation is measured by strain. Simultaneous measurement of force applied to a tissue and its deformation allow biomechanists to determine the stiffness and mechanical strength of biological specimens. Musculoskeletal tissues are viscoelastic, meaning their response is complex and depends on length and timing of loading. Viscoelastic deformation depends on the rate of loading, and the material will lose some energy (hysteresis) stored in it when returning to normal shape. Bones are strongest in compression, while ligaments and tendon are strongest in tension. Three major mechanical characteristics of muscle that affect the tension skeletal muscles can create are the force–velocity, force–length, and force–time relationships. An important neuromuscular strategy used to maximize the initial muscle forces in most movements is the rapid reversal of a countermovement with an eccentric muscle action into a concentric action. This strategy is called the stretch-shortening cycle. The creation of muscular force is controlled by recruitment of motor units and modulating their firing rate. Several musculotendon proprioceptors provide length and tension information to the central nervous system to help regulate muscle actions. The force–motion principle was applied to hypothesize muscles to focus training on in exercise prescription for fitness and how muscular passive tension may affect a gymnastic stunt.

❓ Review Questions

1. What are the major kinds of mechanical loads experienced by the muscle, tendon, and bone?

2. What are the mechanical variables that can be determined from a load-deformation curve, and what do they tell us about the response of a material to loading?

3. Compare and contrast the mechanical strength of the muscle, tendon, ligaments, and bone. How does this correspond to the incidence of various musculoskeletal injuries?

4. How does the passive behavior of the muscle–tendon unit affect the prescription of stretching exercises?

5. What are the functional implications of the force–velocity relationship of skeletal muscle for strength training?

6. What is the force–motion principle, and give an example of its application in a specific human movement?

7. Use the force–length relationship to describe how the active and passive components of muscle tension vary in the range of motion.

8. Compare and contrast the force–time relationship of the muscle with the force–time principle of biomechanics.

9. When might increasing the time of force application not benefit the development of movement speed and why?

10. How does the brain control muscle force, and how is muscle fiber type related?

11. What is the stretch-shortening cycle, and in what kinds of movements is it most important?

12. What are the two major proprioceptors in the muscle that monitor length and force?

13. Explain how range of motion affects the speed, accuracy, and force potential of movement. Give an example of when the range of motion principle is mediated by other mechanical factors.

14. What biomechanical properties help contribute to the beneficial effect of continuous passive movement therapy (i.e., the slow, assisted motion of an injured limb) following surgery?

15. Write a complete description of a stretching or conditioning exercise. Identify the likely muscle actions and forces contributing to the movement.

❗ Key Terms

Bending · Creep · Compression · Degrees of freedom · Electromechanical delay · Energy (mechanical) · Firing rate (rate coding) · Force–length relationship · Force-motion principle · Force-time relationship (electromechanical delay) · Force–velocity relationship · Hysteresis · Isokinetic · Load · Motor unit · Muscle inhibition · Myotatic reflex · Proprioceptors · Reciprocal inhibition · Recruitment · Shear · Strain · Strength (mechanical) · Stress · Stress fracture · Stress relaxation · Stretch-shortening cycle · Tetanus

⭐ Suggested Reading

Alexander, R. M. (1992). *The human machine*. New York: Columbia University Press.

Biewener, A. A., & Roberts, T. J. (2000). Muscle and tendon contributions to force, work, and elastic energy savings: A comparative perspective. *Exercise and Sport Sciences Reviews, 28*, 99–107.

De Luca, C. J. (1997). The use of surface electromyography in biomechanics. *Journal of Applied Biomechanics, 13*, 135–163.

Enoka, R. (2015). *The neuromechanical basis of human movement* (5th ed.). Champaign: Human Kinetics.

Fung, Y. C. (1993). *Biomechanics: Mechanical properties of living tissues* (2nd ed.). New York: Springer-Verlag.

Guissard, N., & Duchateau, J. (2006). Neural aspects of muscle stretching. *Exercise and Sport Sciences Reviews, 34*, 154–158.

Kawakami, Y., & Fukunaga, T. (2006). New insights into in vivo human skeletal muscle function. *Exercise and Sport Sciences Reviews, 34*, 16–21.

Komi, P. V. (Ed.). (1992). *Strength and power in sport*. London: Blackwell Scientific Publications.

Meyers, D. C., et al. (1993). The dimensions of human physical performance: Factor analysis of strength, stamina, flexibility, and body composition measures. *Human Performance, 6*, 309–344.

Patel, T. J., & Lieber, R. L. (1997). Force trans-mission in skeletal muscle: From actomyosin to external tendons. *Exercise and Sport Sciences Reviews, 25*, 321–364.

Rassier, D. E., et al. (1999). Length dependence of active force production in skeletal muscle. *Journal of Applied Physiology, 86*, 1445–1457.

Vogel, S. (2001). *Prime mover: A natural history of muscle*. New York: W.W. Norton & Co.

Whiting, W. C., & Zernicke, R. F. (2008). *Biomechanics of musculoskeletal injury* (2nd ed.). Champaign: Human Kinetics.

Zatsiorksy, V. M., & Kraemer, W. J. (2006). *Science and practice of strength training* (2nd ed.). Champaign: Human Kinetics.

Zatsiorsky, V. M., & Prilutsky, B. I. (2012). *Biomechanics of skeletal muscles*. Champaign: Human Kinetics.

📶 Internet Resources

▶ Memoriam of Y.C. "Bert" Fung. *wisyr.org/2019/12/26/in-memoriam-yuan-cheng-bert-fung-father-of-biomechanics-1919-2019/*

▶ Muscle Physiology Online–UC San Diego website with a comprehensive review of physiological and biomechanical aspects of muscle. muscle. *ucsd.edu/musintro/jump.shtml*

▶ Stretching-Physics Girl. *youtube.com/watch?v=1JgBp7dX4AU*

▶ Visible Human Project–National Library of Medicine. *nlm.nih.gov/research/visible/visible_human.html*

Linear and Angular Kinematics

Key Questions

➤ **What do you need to know to begin?**
Knowledge of musculoskeletal anatomy and algebra.

➤ **Why do you need to know this material?**
Understanding of actual human movement kinematics informs general recommendations for technique, coordination, and the motion of projectiles.

➤ **What is the key idea?**
Biokinematic variables and principles are essential tools in documenting movement technique and how it can be improved.

Kinematics is the accurate description (measurement) of motion and is essential to understanding the biomechanics of human motion. Qualitative anatomical descriptions of joint rotations could be also considered kinematics, but biomechanics uses precise mathematical measurements of motions of living things. Recall from ► Chap. 3 that kinematics is subdivided according to the kinds of measurements used, either linear or angular. Whatever the form of measurement, biomechanical studies of the kinematics of all kinds of people provide valuable knowledge on desirable movement technique. Biomechanics has a long history of kinematic measurements of human motion (Cappozzo et al. 1990). Precise kinematic measurements are sometimes used for the calculation of more complex, kinetic variables. This chapter will introduce key kinematic variables in documenting both linear and angular human motions. The principles of biomechanics explored in this chapter are **Optimal Projection** and the **Coordination Continuum**.

5.1 Linear Motion

Motion is change in position with respect to some **frame of reference**. In mathematical terms, linear motion is simple to define algebraically: Subtract a measurement of initial position from the final position of interest. The simplest linear motion variable is a scalar called **distance** (*l*). The use of the symbol *l* may be easy to remember if you associate it with the length an object travels irrespective of the direction(s) of motion. Typical units of distance are meters and feet. Imagine an outdoor adventurer leaves base camp and climbs for 4 hours through rough terrain along the path illustrated in ◘ Fig. 5.1. If her final position traced a 1.3 km climb measured relative to the base camp (0 km) with a pedometer, the distance she climbed was 1.3 km (final position–initial position). Note that 1.3 km (kilometers) is equal to 1300 meters. The odometer in your car works in a similar fashion, counting the revolutions (angular motion) of the tires to generate a measurement of the distance (linear variable) the car travels irrespective if you are moving forward or backing up. Because distance is a scalar, your odometer does not tell you in what direction you are driving on the one-way street! Scalar quantities are always positive numbers and students should remember that the units of measurement (often meters or feet in linear motion) need to be specified.

The corresponding vector quantity to distance is **displacement (d)**. As a vector, displacements usually have an adjective that describes their direction (horizontal displacement and vertical displacement), and mathematically, these directions are usually defined using right-angle directions, which are convenient for the purpose of description. Making vector calculations on these axes also allows signs (+ and −) to represent displacement in opposite directions in a particular direction. For most two-dimensional (2D planar) analyses of human movements in a vertical plane, like in ◘ Fig. 5.1, the directions used are horizontal and vertical, so displacements are calculated as final position minus initial position in that particular dimension. The usual convention is that motions to the right on the x-axis and upward along the y-axis are positive, with motion in the opposite directions negative. Since displacement is a vector quantity, if motion upward and to the right is defined as positive, motion downward and motion to the left are negative displacements.

Assuming ◘ Fig. 5.1 is drawn to the scale shown, it looks like the climber had a positive

© Springer Nature Switzerland AG 2021
D. Knudson, *Fundamentals of Biomechanics*, https://doi.org/10.1007/978-3-030-51838-7_5

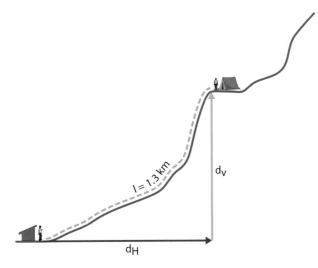

5

□ Fig. 5.1 An outdoor adventurer climbs from base camp to a camp following the illustrated path. The distance the climber covers is 1.3 km. Her displacement is 0.8 km horizontally and 0.7 km vertically

0.8 km of horizontal displacement and 0.7 km of vertical displacement. This eyeballing of the horizontal and vertical components of the hike will be fairly accurate because displacement is a vector. Vectors can be conveniently represented by combinations of right-angle components, like the horizontal and vertical displacements in this example. If our adventurer were stranded in a blizzard and a helicopter had to lower a rescuer from a height of 0.71 km above base camp, what would be the rescuer's vertical displacement to the climber? The vertical displacement of the rescuer would be −0.01 km or −10 meters (final vertical position minus initial vertical position: 0.7–0.71 = −0.01 km).

Biomechanists most often use measures of displacement rather than distance because they carry directional knowledge that is crucial to calculation of other kinematic and kinetic variables. There are a couple of subtleties to these examples. First, the analysis is a simple 2D model of truly 3D reality. Second, the human body is modeled as a **point mass**. In other words, we know nothing about the orientation of the body or body segment motions; we just confine the analysis to the whole body mass acting at one point in space. Finally, an absolute frame of reference was used, when we are interested in the displacement relative to a moving object—like the helicopter, a relative frame of reference can be used.

In other biomechanical studies of human motion, the models and frames of reference can get more complicated as the **model** of the biomechanical object/system gets more complex and more dimensions of motion are considered. The analysis might not be focused on whole-body movement but how much a muscle is shortening between two attachments. Biomechanics primar-

ily use **rigid body** models of anatomical segments that are combined to create a **linked segment** model for part or all of the body. Scientists use model complexity and planes of motion that match the needs of kinematic description of the movement and study goals.

A three-dimensional (3D) analysis of the small accessory gliding motions of the knee joint would likely measure along anatomically relevant axes like proximal–distal, mediolateral, and anteroposterior. Three-dimensional kinematic measurements in biomechanics require considerable numbers of markers, spatial calibration, and mathematical complexity for completion. **Degrees of freedom** represent the kinematic complexity of a biomechanical model. The degrees of freedom (dof) correspond to the number of kinematic measurements needed to completely describe the position and motion of an object. A point mass model moving in a 2D plane has only two dof, so the motion of the object can be described with an x (horizontal) and a y (vertical) coordinate. The dimensionless nature of this point in space means there is no side of the object or rotation to consider like we do later in this chapter. A rigid body model of a single body segment in 2D has three dof: two linear motions and one angle to describe its orientation in the plane.

The 3D motion of a body segment has six dof, because there are three linear coordinates (x, y, z) and three angles (to define the orientation of the segment) that must be specified. For example, physical therapy likes to describe the six dof for the lower leg at the knee joint using the terms osteokinematics (three anatomical joint rotations) and arthrokinematics (three small gliding or linear motions along joint surfaces). The mathematical complexity of 3D kinematics is much greater than the 2D kinematics illustrated in this text. Good sources for a more detailed description of kinematics in biomechanics are available (Allard et al. 1995; Zatsiorsky 1998). The field of biomechanics is striving to develop standards for reporting joint kinematics so that data can be exchanged and easily applied in various professional settings (Wu and Cavanagh 1995).

The concept of **frame of reference** is, in essence, where you are measuring or observing the motion from. Reference frames in biomechanics are either absolute or relative. An absolute or global frame of reference is assumed essentially motionless, like the apparent horizontal and vertical motion we experience relative to the earth and its gravitational field (as in □ Fig. 5.1). A relative frame of reference is measuring from a point that is clearly moving, like the motion of the foot relative to the hip or the plant foot relative to the soccer ball. There is no one frame of reference that is best, because the biomechanical description that is most relevant depends on the purpose of the analysis. Sometimes it is important to know how far one part of the body is from another moving part (position of body part like the knee or center of mass relative to the limits of the base of support).

This point of motion being relative to your frame of reference is important for several reasons. First, the appearance and amount of motion depends on where the motion is observed or measured from. You could always answer a question about a distance as some arbitrary number from an "unknown point of reference," but the accuracy of that answer may be good for only partial credit. Second, the many ways to describe the motion are much like the different anatomical terms that are sometimes used for the identical motion. Finally, this is a metaphor for an intellectually mature kinesiology professional who knows there is not one single way of seeing or measuring human motion because your frame of reference affects what you see and the numbers reported. The next section will examine higher-order kinematic variables that are associated with the rates of change of an object's linear motion. It will be important to understand that these new variables are also dependent on the model and frame of reference used for their calculation.

5.1.1 Speed and Velocity

Speed is how fast an object is moving without regard to direction. Speed is a scalar quantity like distance, and most people have an accurate intuitive understanding of speed. **Speed** (s) is defined as the rate of change of distance ($s = l/t$), so typical units are m/s, ft/s, km/hr, or miles/hr. It is very important to note that our algebraic formula for speed (l/t), and other kinematic time derivatives to come, means "the change in the numerator divided by the change in the denominator." This means that the calculated speed is an *average* value for the time interval used for the calculation. If you went jogging across town (5 miles) and arrived at the turnaround point in 30 minutes, your average speed would be (5 miles/0.5 hours) or an average speed of 10 miles per hour. You likely had intervals where you ran faster or slower than 10 mph, so we will see how representative or accurate the kinematic calculation is depends on the size of time interval and the accuracy of your linear and time measurements. A radar gun measures a closer estimate of the top speed of a thrown ball than a stopwatch timing of the throw trajectory. The reflection of the radar beam picked up the instrument are collected quickly when the ball passes along the axis of the radar beam, but the mean speed calculated with a stopwatch time is likely lower because of the large time interval and slowing of the ball from gravity (upward motion) and air resistance.

Since biomechanical studies have used both the English and metric systems of measurements, students need to be able to convert speeds from one system to the other. Speeds reported in m/s can be converted to speeds that make sense to American drivers (mph) by essentially doubling them (mph = m/s • 2.23). Speeds in ft/s can be converted to m/s by multiplying by 0.30, and km/hour can be converted to m/s by multiplying by 0.278. Other conversion factors can be found in Appendix B. ◻ Table 5.1 lists some top speeds in sports and other human movements that have been reported in the biomechanics literature or world records. Sports medicine professionals document also walking speed in monitoring rehabilitation where precise timing accuracy is not so strict (see Application: Speed box).

Be sure to remember that speed is also relative to frame of reference. The motion of the runner in ◻ Fig. 5.2 can correctly be described as 8 meters per second (m/s), 1 m/s, or −1.5 m/s. Runner A is moving 8 m/s relative to the starting line, 1 m/s faster than runner B, and 1.5 m/s slower than runner C. All are correct kinematic descriptions of the speed of runner A.

Application: Speed

One of the most important athletic abilities in many sports is running speed. Coaches often quip that "luck follows speed" because a fast athlete can arrive in a crucial situation before their opponent. Coaches that often have a good understanding of speed are in cross-country and track. The careful timing of various intervals of a race, commonly referred to as pace, can be easily converted average speeds over that interval. Pace (time to run a specific distance) can be a good kinematic variable to know, but its value depends on the duration of the interval, how the athlete's speed changes over that interval, and the accuracy of the timing. How accurate do you think stopwatch measures of time and, consequently, speed are in track? What would a two-tenths-of-a-second error mean in walking (200 s), jogging (100 s), or sprinting (30 s) a lap on a 220 m track? Average speeds for these events are 1.1, 2.2, and 7.3 m/s, with potential errors of 0.1, 0.2, and 0.7%. Less than 1%! That sounds good, but what about shorter events like a 100 m dash or the "hang time" (time the ball is in the air) of a punt in football? American football has long used the 40 yard dash as a measure of speed, ability, and potential for athletes despite little proof of its value (see Ostarello 1986; Maisel 1998). A 40 yard/meter race is primarily a measure of running horizontal acceleration, where athletes may not even get to top running speed. If you measured time by freezing and counting frames of video, how much more accurate would a 40 meter dash timing be? One must be careful because video cameras have many different frame rates (15, 30, 60 pictures per second). Kinematic timing can quite accurate with high-speed video cameras (250, 500, 1000, and more pictures per

5

■ **Table 5.1**   Typical peak speeds in human movement

Speed	m/s	mph
Bar in a bench press	0.25	0.6
Muscle shortening	0.5	1.1
Walking	1.1–1.8	2.5–4.0
Vertical jump	2.3	5.1
Free throw	7.0	15.7
Sprinting	12.0	26.8
Tennis forehand	20.0	44.6
Batting	31.3	70.0
Soccer kick	44.8	100
Baseball pitch	47.0	105
Tennis serve	70.4	157
Golf drive	94.6	211

second).If the current world record for 100 m is 9.79 seconds and elite runners cover 40 m (43.7 yards) out of the starting blocks in about 4.7 seconds, should you believe media guides that say that a certain freshman recruit at Biomechanical State University ran the forty (40 yard dash) in 4.3 seconds?

Velocity is the vector corresponding to speed. The vector nature of velocity makes it more complicated than speed, so many people incorrectly use the words interchangeably and have incorrect notions about velocity. Velocity is essentially the speed of an object, *in a particular direction*. Velocity is the rate of change of displacement ($V = \mathbf{d}/t$), so its units are the same as speed and are usually qualified by a directional adjective (i.e., horizontal, vertical, resultant). Note that when the adjective "angular" is not used, the term *velocity* refers to linear velocity. If you hear a coach say a pitcher has "good velocity," the coach is not using biomechanical terminology correctly. A good question to ask in this

■ **Fig. 5.2**   The speed of runner A depends on the frame of reference of the measurement. Runner A can be described as moving at 8 m/s relative to the track or −1.5 m/s relative to runner C

Fig. 5.3 The horizontal plane running path (dashed line) of a physical education student changing practice stations. The linear displacements can be measured along a water fountain axis and equipment axis (child's frame of reference is likely oriented to these two landmarks: access to the basketballs and water)

situation is: "That's interesting. When and in what direction was the pitch velocity so good?" The "when" in this sentence indicates the understanding that even if a direction is specified, the velocity will have many values for different points in time (like the sprint velocity in ■ Fig. 5.4).

The phrase "rate of change" is very important because velocity defines how quickly position is *changing* in the specified direction (displacement). Most students might recognize this phrase as the same one used to describe the derivative or the slope of a graph (like the hand dynamometer example in ▶ Fig. 3.4).

Remember to think about velocity as a speed, but in a particular/specified direction. A simple example of the velocity of human movement is illustrated by the path (dotted line in ■ Fig. 5.3) of a physical education student in a horizontal plane as he changes exercise stations in a circuit-training program. The directions used in this analysis are a fixed reference frame (horizontal plane) that is relevant to young students: the equipment axis and water axis.

The student's movement from his initial position (I) to the final position (F) can be vectorially represented by displacements along the equipment axis ($\mathbf{d}_E$) and along the water axis ($\mathbf{d}_W$). Note that the definition of these axes is arbitrary since the student must combine displacements in both directions to arrive at the basketballs or a drink. The net displacements for this student's movement are positive, because the final position measurements are larger than the initial positions. Let's assume

that $\mathbf{d}_E = 8$ m and $\mathbf{d}_W = 2$ m and that the time it took this student to change stations was 10 seconds. The average velocity along the water axis would be $\mathbf{V}_W = \mathbf{d}_W/t = 2/10 = 0.2$ m/s. Note that the motion of most interest to the student is the negative displacement ($-\mathbf{d}_W$), which permits a quick trip to the water fountain. The average velocity along the equipment axis would be 0.8 m/s ($\mathbf{V}_E = \mathbf{d}_L/t = 8/10$). Right-angle trigonometry can then be used to calculate the magnitude and direction of the resultant displacement ($\mathbf{d}_R$) and then the average velocity of the student. We will use right-angle trigonometry in ▶ Chap. 6 to analyze the effect of force vectors. By the way, if your right-angle trigonometry is a little rusty, check out Appendix D for a refresher.

Remember that calculations of speed and velocity using algebra are *average* velocities over the time interval used. It is important to realize that the smaller the time interval, the greater the potential accuracy of kinematic calculations. In the previous example, for instance, smaller time intervals of measurement would have detected the negative velocity (to get a drink) and positive velocity of the student in the water direction. Biomechanics research often uses high-speed film or video imaging (Gruen 1997) to make kinematic measurements over very small time intervals (200 or thousands of pictures per second). The use of calculus allows for kinematic calculations ($\mathbf{v} = \mathbf{d}_d/\mathbf{d}_t$) to be made to *instantaneous* values for any point in time of interest. If kinematic calculations are based over a too large time interval, you may not be getting information much

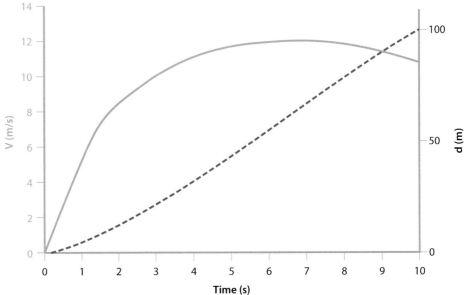

◘ Fig. 5.4 The horizontal displacement-time (dashed curve) and horizontal velocity-time (solid curve) graphs for the 100 m dash of an elite male sprinter

better than the time or pace of a whole race, or you may even get the unusual result of zero velocity because the race finished where it started.

Graphs of kinematic variables versus time are extremely useful in showing the motion and patterns within that movement. Because human movement occurs across time, biokinematic variables like displacement, velocity, and acceleration are usually plotted versus time, although there are other graphs that are of value. ◘ Figure 5.4 illustrates the horizontal displacement and velocity graphs for an elite male sprinter in a 100 m dash. Graphs of the speed over a longer race precisely document how the athlete runs the race. Notice that the athlete first approaches top speed at about the 40 to 50 meter mark. You can compare your velocity profile to ◘ Fig. 5.4 and to those of other sprinters in Lab Activity 5.

5.1.2 Acceleration

The second derivative of displacement with respect to time, or the rate of change of velocity, is **acceleration**. Acceleration is how quickly velocity is changing (remember that velocity changes when speed or direction changes). This vector nature of velocity and acceleration means that it is important to *think of acceleration as an unbalanced force in a particular direction*. The acceleration of an object can speed it up, slow it down, change its direction, or both. It is incorrect to assume that "acceleration" means an object is speeding up. The use of the term "deceleration" should normally be avoided because it implies that the object is slowing down and does not take into account changes in direction.

Let's look at an example that illustrates why it is not good to assume the direction of motion when studying acceleration. Imagine a person is swimming laps, as illustrated in ◘ Fig. 5.5. Motion to the right is designated positive, and the swimmer has a relatively constant velocity (zero horizontal acceleration) in the middle of the pool and as she approaches the wall. As her hand touches the wall, there is a negative acceleration that first slows her down and then speeds her up in the negative direction to begin swimming again. Thinking of the acceleration at the wall as a push in the negative direction is correct throughout the turn. As the swimmer touches the other wall, there is a positive acceleration that decreases her negative velocity, and if she keeps pushing (hasn't had enough exercise), it will increase her velocity in the positive direction back into the pool. The algebraic definition of acceleration (**a**) is **V**/t, so typical units of acceleration are m/s² and ft/s². Another convenient way to express acceleration is in units of gravitational acceleration (*g*'s). When you jump off a box, you experience (in flight) one *g* of acceleration, which is about −9.81 m/s/s or −32.2 ft/s/s (negative because gravity force always acts downward). This means that, in the absence of significant air resistance or other vertical force, your vertical velocity will change 9.81 m/s every second in the negative direction. Note that this means you slow down 9.81 m/s every second on the way up and speed up 9.81 m/s every second on the way down. Gravitational acceleration units (g) are used for large acceleration events like a big change of direction on a roller coaster (4 g), the shockwaves in the lower leg following heel strike in running (5 g), a tennis shot (50 g), or head acceleration in a football tackle (40–200 g). When a person is put under sustained (several seconds

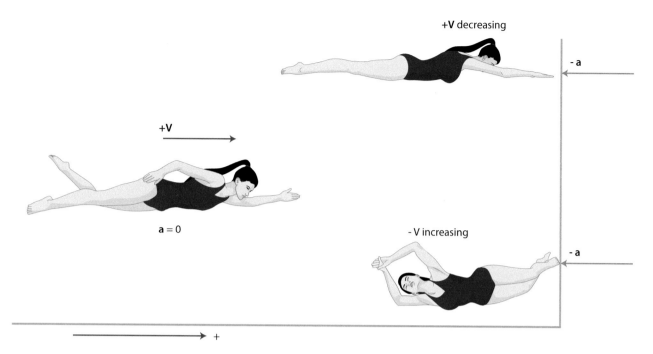

Fig. 5.5 The motion and accelerations of swimmers as they change direction in lap swimming. If motion to the right is designated positive, the swimmer experiences a negative acceleration as they make the turn at the pool wall. The negative acceleration first slows positive velocity and then begins to build negative velocity to start swimming in the negative direction. It is important to associate signs and accelerations with directions

instead of an instant, like the previous examples) high-level acceleration like in jet fighters (5–9 g), pilots must wear a pressure suit and perform whole-body isometric muscle actions to prevent blacking out from the blood shifting in their body.

Acceleration due to gravity always acts in the same direction (toward the center of the earth) and may cause speeding up or slowing down depending on the direction of motion. Remember to think of acceleration as a push in a direction or a tendency to change velocity, not as speed or velocity. The vertical acceleration of a ball at peak flight in the toss of a tennis serve is −1 g, not zero. The vertical velocity may be instantaneously zero, but the constant pull of gravity is what prevents it from staying up there.

Let's see how big the horizontal acceleration of a sprinter is in getting out of the blocks and into good running position. This is an easy example because the rules require that the sprinter have an initial horizontal velocity of zero. If video measurements of the sprinter showed that they passed the 10 m point at 1.9 seconds with a horizontal velocity of 7 m/s, what would be the runner's horizontal acceleration? The sprinter's change in velocity was 7 m/s (7−0), so the sprinter's horizontal acceleration was $\mathbf{a} = \mathbf{V}/t = 7/1.9 = 3.7$ m/s/s. If the sprinter could maintain this acceleration for 3 seconds, how fast would he be running?

Close examination of the displacement, velocity, and acceleration graphs of an object's motion is an excellent exercise in qualitative understanding of linear kinematics. Examine the pattern of horizontal acceleration in the 100 m sprint mentioned earlier (see **Fig. 5.4). Note that there are essentially three phases of acceleration in this race that roughly correspond to the slope of the velocity graph. There is a positive acceleration phase (increasing positive velocity), a phase of near zero acceleration (flat velocity graph), and a negative acceleration phase (decreasing positive velocity). Most sprinters struggle to prevent running speed from declining at the end of races over 60 m. Elite female sprinters have similar velocity graphs in 100 m races. What physiological factors might account for the inability of people to maintain peak speed in sprinting?

Note that the largest accelerations (largest rates of change of velocity) do not occur at the largest or peak velocities. Peak velocity must occur when acceleration is zero. Coaches often refer to quickness as the ability to react and move fast over short distances, while speed is the ability to cover moderate distances in a very short time. Based on the velocity graph in **Fig. 5.4, how might you design running tests to differentiate speed and quickness?

Acceleration is the kinematic (motion description) variable that is closest to a kinetic variable (explanation of motion). Kinesiology professionals need to remember that the pushes (forces) that create accelerations precede the peak speeds (where accelerations are zero) they eventually create. Coaches observing movement cannot see acceleration, but they can perceive changes in speed or direction that can be interpreted as acceleration that happened just before these events. Just remember that by the time the coach perceives the acceleration, the muscular and body actions which created those forces occurred just before the motion changes you are able to see.

5.1.3 Uniformly Accelerated Motion

In rare instances the forces acting on an object are constant and therefore create a constant acceleration in the direction of the resultant force. The best example of this special condition is the force of earth's gravity acting on projectiles. A projectile is an object launched into the air that has no self-propelled force capability (◘ Fig. 5.6). A person's body in jumping acts as a projectile, just like the ball a tennis player tosses is a projectile. Many human projectile movements have vertical velocities that are sufficiently small so that the effects of air resistance in the vertical direction can be ignored

(see ▶ Chap. 8). Without fluid forces in the vertical direction, projectile motion is uniformly accelerated by one force, the force of gravity. There are exceptions, of course (e.g., skydiver, badminton shuttle), but for the majority of human projectiles, we can take advantage of the special conditions of vertical motion to simplify kinematic description of the motion. The Italian Galileo Galilei is often credited with discovering the constant nature of gravitational acceleration using some of the first accurate of measurements of objects falling and rolling down inclines. This section will briefly summarize these mathematical descriptions, but will emphasize several important facts about this kind of motion and how this can help determine optimal angles of projection in sports.

When an object is thrown or kicked without significant air resistance in the vertical direction, the path or **trajectory** will be some form of a parabola. The uniform nature of the vertical force of gravity creates a linear change in vertical velocity and a second-order change in vertical displacement. The constant force of gravity also assures that the time it takes to reach peak vertical displacement (where vertical velocity is equal to zero) will be equal to the time it takes for the object to fall to the same height that it was released from. The magnitude of the vertical velocity when the object falls back to the

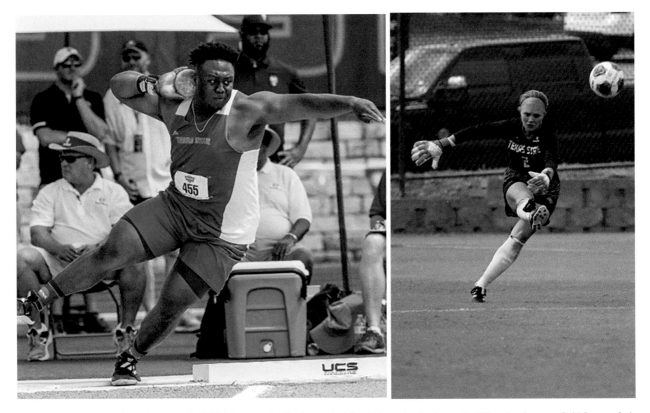

◘ **Fig. 5.6**   The shot (left) and a soccer ball (right) are projectiles because they are not self-propelled when thrown or kicked. At slower vertical velocities of release, the balls experience uniformly accelerated

(-9.8 m/s/s) motion in the vertical direction because fluid forces of air resistance are small (see ▶ Chap. 8). Photo used by permission of Texas State University Department of Athletics same position of release will be the same as the velocity of release. A golf ball tossed vertically at shoulder height at 10 m/s (to kill time while waiting to play through) will be caught at the same shoulder level at a vertical velocity of −10 m/s. The velocity is negative because the motion is opposite of the toss, but is the same magnitude as the velocity of release. Think about the 1 *g* of acceleration acting on this golf ball and these facts about uniformly accelerated motion to estimate how many seconds the ball will be in flight.

This uniformly changing vertical motion of a projectile can be determined at any given instant in time using three formulas and the kinematic variables of displacement, velocity, acceleration, and time. My physics classmates and I memorized these by calling them VAT, SAT, and VAS (provides the order of the kinematic variables within the formulas). The various kinematic variables are obvious, except for "S," which is another common symbol for displacement. The final vertical velocity of a projectile can be uniquely determined if you know the initial velocity (V_i) and the time of flight of interest (VAT: $V_f^2 = V_i + at$). Vertical displacement is also uniquely determined by initial velocity and time of flight (SAT: $d = V_i t + 0.5at^2$). Finally, final velocity can be determined from initial velocity and a known displacement (VAS: $V_f^2 = V_i^2 + 2ad$).

Let's consider a quick example of using these facts before we examine the implications for the best angles of projecting objects. Historically great jumpers in the National Basketball Association like Michael Jordan or David Thompson are credited with standing vertical jumps about twice as high (1.02 m or 40 inches) as skilled college males. This outstanding jumping ability is not an exaggeration (Krug and LeVeau 1999). Given that the vertical velocity is zero at the peak of the jump and the jump height, we can calculate the takeoff velocity of our elite jumper by applying VAS. Solving for V_i in the equation:

We select the velocity to be positive when taking the square root because the initial velocity is opposite to gravity, which acts in the negative direction. If we wanted to calculate his hang time, we could calculate the time of the fall with SAT and double it because the time up and time down are equal:

$$d = V_i t + 0.5at^2$$
$$-1.02 = 0 + 0.5(-9.81)t^2 \quad t = 0.456\,s$$

So the total fight time is 0.912 seconds. If you know what your vertical jump is, you can repeat this process and compare your takeoff velocity and hang time to that of elite jumpers. The power of these empirical relationships is that you can use the mathematics as models for simulations of projectiles. If you substitute in rea-

sonable values for two variables, you get good predictions of kinematics for any instant in time. If you wanted to know when a particular height was reached, what two equations could you use? Could you calculate how much higher you could jump if you increased your takeoff velocity by 10%?

So we can see that uniformly accelerated motion equations can be quite useful in modeling the vertical kinematics of projectiles. The final important point about uniformly accelerated motion, which reinforces the directional nature of vectors, is that, once the object is released, the vertical component of a projectile's velocity is independent of its horizontal velocity. The extreme example given in many physics books is that a bullet dropped the same instant another is fired horizontally would strike level ground (pretty far away!) at the same time. Given constant gravitational conditions, the height of release and initial vertical velocity uniquely determine the time of flight of the projectile. The range or horizontal distance the object will travel depends on this time of flight and the horizontal velocity. Athletes may increase the distance they can throw by increasing the height of release (buying time against gravity), vertical velocity, and horizontal velocity. The optimal combination of these depends on the biomechanics of the movement, not just the kinematics or trajectory of uniformly accelerated motion. The next section will summarize a few general rules that come from the integration of biomechanical models and kinematic studies of projectile activities. These rules are the basis for the Optimal Projection Principle of biomechanics.

5.2 Optimal Projection Principle

For most sports and human movements involving projectiles, there is a range of angles that results in best performance. The **Optimal Projection Principle** refers to the range of angle(s) that an object is projected to achieve a particular goal. This section will outline some general rules for optimal projections that can be easily applied by coaches and teachers. These optimal angles are "rules of thumb" that are consistent with the biomechanical research on projectiles, so don't take the term "optimal" too literally. Finding true optimal angles of projection requires integration of descriptive studies of athletes at all ability levels (e.g., Bartlett et al. 1996), the laws of physics (like uniformly accelerated motion and the effects of air resistance; see ▸ Chap. 8), and modeling studies that incorporate the biomechanical effects of various release parameters. Determining an exact optimal angle of projection for the unique characteristics of a particular athlete and in a particular environment may be possible using a combination of experimental data

Fig. 5.7   The three variables that determine the release parameters of a projectile in two dimensions: height of release and the horizontal and vertical velocities of release

and modeling (Hubbard et al. 2001; Li 2012); however, this is exceedingly difficult and would cease to be best in different conditions (increase in wind, fatigue from numerous throw, etc.). There will be, however, some general trends or rules for teaching and coaching projectile events where biomechanical research has shown that certain factors dominate the response of the situation and favor certain release angles.

In most instances, a two-dimensional point mass model of a projectile is used to describe the compromise between the height of release and the vertical and horizontal components of release velocity (Fig. 5.7). If a ball was kicked and then landed at the same height, and the air resistance was negligible, the optimal angle of projection for producing maximum horizontal displacement would be 45°. Forty-five degrees above the horizontal is the perfect mix of horizontal and vertical velocity to maximize horizontal displacement. Angles above 45° create shorter ranges because the extra flight time from larger vertical velocity cannot overcome the

loss in horizontal velocity. Angles smaller than 45° cause loss of flight time (lower vertical velocity) that cannot be overcome by the larger horizontal velocity. Try the activity below to explore optimal angles of projection.

Activity: Angles of Projection

Use a garden hose to water the grass and try various angles of projection of the water. The air resistance on the water should be small if you do not try to project the water too far. Experiment and find the angle that maximizes the distance the water is thrown. First see if the optimal angle is about 45°, when the water falls back to the height that it comes out of the hose. What happens to the optimal angle during long-distance sprinkling as the height of release increases?

Note how the optimal angle of projection changes from 45° when the height of release is above and below the target.

Before we can look at generalizations about how these factors interact to apply the Optimal Projection Principle, the various goals of projections must be analyzed. The mechanical objectives of projectiles are displacement, speed, and a combination of displacement and speed. The goal of an archer is accuracy in displacing an arrow to the target. The basketball shooter in Figure 5.7 strives for the right mix of ball speed and displacement to score. A soccer goalie punting the ball out of trouble in his end of the field focuses on ball speed rather than kicking the ball to a particular location.

When projectile displacement or accuracy is the most important factor, the range of optimal angles of projection is small. In tennis, for example, Brody (1987) has shown that the vertical angle of projection (angular "window" for a serve going in) depends on many factors but is usually less than 4°. The goal of a tennis serve is the right combination of displacement and ball speed, but traditionally the sport and its statistics have emphasized the importance of consistency (accuracy) so as to keep the opponent guessing. In a tennis serve, the height of projection above the target, the net barrier, the spin on the ball, the objective of serving deep into the service box, and other factors favor angles of projection at or above the horizontal (Elliott 1983). Elite servers can hit high-speed serves 3° below the horizontal, but the optimal serving angle for the majority of players is between 0 and 15° above the horizontal (Owens and Lee 1969; Elliott 1983).

This leads us to our first generalization of the Optimal Projection Principle. *In most throwing or striking events, when a mix of maximum horizontal speed and*

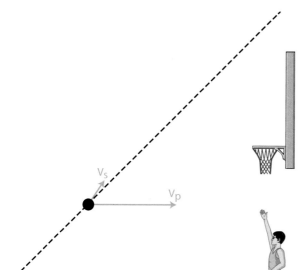

Fig. 5.8 Coaches can visually estimate the initial path of thrown balls to check for optimal projection. The initial path of the ball can be estimated relative to an imaginary 45° angle. When throwing for distance, many small children select very high angles of release that do not maximize the distance of the throw

displacement are of interest, the optimal angle of projection tends to be below 45°. The higher point of release and dramatic effect of air resistance on most sport balls makes lower angles of release more effective. Coaches observing softball or baseball players throwing should look for initial angles of release between 28 and 40° above the horizontal (Dowell 1978). Coaches should be able to detect the initial angle of a throw by comparing the initial flight of the ball with a visual estimate of 45° angle (Fig. 5.8). Note that there is a larger range of optimal or desirable angles that must accommodate differences in the performer and the situation. Increasing the height of release (a tall player) will tend to shift the optimal angle downward in the range of angles, while higher speeds of release (gifted players) will allow higher angles in the range to be effectively used. What do you think would happen to the optimal angles of release of a javelin given the height of release and speed of approach differences of an L5-disabled athlete in a wheelchair compared to an able-bodied athlete?

There are a few exceptions to this generalization, which usually occur due to the special environmental or biomechanical conditions of an event. In long jumping, for example, the short duration of takeoff on the board limits the development of vertical velocity, so that takeoff angles are usually between 18 and 23° (Hay et al. 1986; Linthorne et al. 2005). In the standing long jump, jumpers prefer slightly higher takeoff angles with relatively small decreases in performance (Wakai and

Linthorne 2005). We will see in ▶ Chap. 8 that the effect of air resistance can quickly become dominant on the optimal release parameters for many activities. In football place-kicking, the lower-than-45° generalization applies (optimal angles are usually between 25 and 35°), but the efficient way the ball can be punted and the tactical importance of time during a punt make the optimal angle of release about 50°. With the wind at the punter's back, he might kick above 50° while using a flatter kick against a wind. The backspin put on various golf shots is another example of variations in the angle of release because of the desirable effects of spin on fluid forces and the bounce of the ball. Most long-distance clubs have low pitches, which agrees with our principle of a low angle of release, but a golfer might choose a club with more loft in situations where he wants higher trajectory and spin rate to keep a ball on the green.

The next generalization relates to projectiles with the goal of upward displacement from the height of release. *The optimal angle of projection for tasks emphasizing vertical displacement or a mix of vertical displacement and time of flight tends to be above 45°.* Examples of these movements are the high jump and basketball shooting. Most basketball players (not the giants of the NBA) release a jump shot below the position of the basket. Considerable research has shown that the optimal angle of projection for basketball jump and set shots is between 49 and 55° (see Knudson 1993). This angle generally corresponds to the arc where the minimum speed may be put on the ball to reach the goal, which is consistent with a high-accuracy task (Nakano et al. 2020). Ironically, a common error of beginning shooters is to use a very flat trajectory that requires greater ball speed and may not even permit an angle of entry so that the ball can pass cleanly through the hoop! Coaches that can identify appropriate shot trajectories can help players improve more quickly (Fig. 5.9). The optimal angles of release in basketball are clearly not "high-arc" shots, but are slightly greater than 45° and match the typical shooting conditions in recreational basketball. Of course, real-world optimal shooting trajectories in basketball are more complex and include shots off the backboard (Huston and Grau 2003), defenders, fatigue, and many other factors. Recent work has also indicated that individual optimal angles of projection for individuals from free throw to 3-point distances are between 42 and 50 ° (Inaba et al. 2019). Much taller, stronger, and faster athletes (elite-level performers) are more likely to have high release points in jump shots and have release angles closer to this lower range near 45 degrees. There may even be body-type variations (recall from ▶ Chap. 2 that scholars in anthropometry examine the influence of body physical properties on performance). Taller athletes with longer limbs and greater heights of release

5

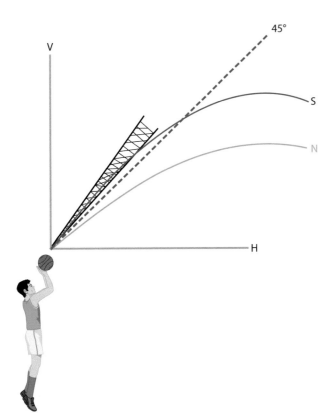

45°

V

S

N

H

Fig. 5.9 The optimal projection angles for most basketball jump shots for young basketball players are between 49 and 55° above the horizontal (hatched). These initial trajectories represent the right mix of low ball speed and a good angle of entry into the hoop. Novice shooters (N) often choose a low angle of release. Skilled shooters (S) really do not shoot with high arcs, but with initial trajectories that are in the optimal range and tailored to the conditions of the particular shot

have an advantage in kicking, throwing, and striking sports. Shorter athletes have an advantage in strength and tumbling sports.

The optimal projection principles involves several generalizations about desirable initial angles of projection. These general rules are likely to be effective for most performers. Care must be taken in applying these principles in special populations. The biomechanical characteristics of elite (international caliber) athletes or wheelchair athletes are likely to affect the optimal angle of projection. Wheelchair basketball athletes are limited by a lower point of release and a lack of lower extremities to contribute to ball speed so higher angles of research are best for them. Extremely tall and excellent jumpers in elite basketball may have lower angles of release that are near optimal for their jump shots. Kinesiology professionals should be aware that biomechanical and environmental factors interact to affect the optimal angle of projection. For example, a stronger athlete might use an angle of release slightly lower than expected but which is close to optimal for her. Her extra strength allows her to release the implement at a higher

point without losing projectile speed so that she is able to use a slightly lower angle of release. Professionals coaching projectile sports must keep up on the biomechanical research related to optimal conditions for their athletes.

5.3 Angular Motion

Angular kinematics is the description (measurement) of angular motion or rotation. Angular kinematics is particularly appropriate for the study of human movement because the major anatomical motions of most human joints can be described using one, two, or three orthogonal (right-angle) rotations. Angular kinematics should also be somewhat easy for biomechanics students because for every linear kinematic variable, there is a corresponding angular kinematic variable. It will even be easy to distinguish angular from linear kinematics because the adjective "angular" or a Greek letter symbol is used instead of the Arabic letters used for linear kinematics.

Angular displacement (θ: theta) is the vector quantity representing the change in angular position of an object. Angular displacements are measured in degrees, radians (dimensionless ratio unit equal to 57.3°), and revolutions (360°). The usual convention to keep directions straight and be consistent with our 2D linear kinematic directional conventions and calculations is to consider counterclockwise rotations as positive. Angular displacement measured with a **goniometer** is one way to measure **static flexibility** of a joint or joint complex in sports medicine. As in linear kinematics, the frames of reference for these angular measurements can vary. Some tests define complete knee joint extension as 0°, while other tests refer to that position as 180°. For a review of several physical therapy static flexibility tests, see Norkin and White (2017).

In analyzing the curl-up exercise shown in ☐ Fig. 5.10, the angle between the thoracic spine and the floor is often used. This exercise is usually limited to the first 30 to 40° above the horizontal to limit the involvement of the hip flexors (Knudson 1999). The angular displacement of the thoracic spine in the eccentric phase would be −38° (final angle minus initial angle: 0–38 = −38°). This trunk angle is often called an **absolute angle** because it is measured relative to an "unmoving" earth frame of reference. **Relative angles** are defined between two segments that can both move. Examples of relative angles in biomechanics are joint angles. The knee angle (θ_K) in ☐ Fig. 5.10 is a relative angle that would tell if the person is changing the positioning of both legs in the exercise. Note this angular convention would consider complete knee extension as approximately 180 degrees (3.14 radians).

Another example of three absolute angles is illustrated for the frontal plane in the stance phase of running (■ Fig. 5.11). The trunk angle is measured to the vertical (yellow), while a pelvis and upper leg angles (red) are approximated using anterior superior iliac spine (ASIS) markers and the midline of the patella.

Notice the subtraction of these two absolute angles to the vertical would give you a relative angle measurement of hip adduction/abduction. This increased pelvic drop and hip adduction could create unequal stresses in the knee joint and increase the risk of anterior cruciate ligament (ACL) injury.

■ **Fig. 5.10** The angular kinematics of a curl-up exercise can be measured as the angle between the horizontal and a thoracic spine segment. This is an example of an absolute angle (global frame of reference) because it defines the angle of an object relative to external space. The knee joint angle (θ_K) is a relative angle (local frame of reference) because both the leg and thigh can move

5.3.1 Angular Velocity

Angular velocity (ω: omega) is the rate of change of angular displacement and is usually expressed in degrees per second or radians per second. The formula for angular velocity is $\omega = \theta/t$, and calculations would be the same as for a linear velocity, except the displacements are angular measurements. Angular velocities are vectors are drawn by the right-hand rule, where the flexed fingers of your right hand represent the rotation of interest and the extended thumb would be along the axis of rotation and would indicate the direction of the angular velocity vector. This book does not go over detailed examples of strict vector algebra mathematical technique, but will employ a curved arrow just to illustrate angular velocities and moments/torques (■ Fig. 5.12).

Biomechanics emphasizes angular velocities over angular speed because vectors provide important direc-

■ **Fig. 5.11** Frontal plane angles measured in midstance for two runners. Note the runner in panel (**b**) has more hip adduction (relative angle) and left hip drip than the runner in panel (**a**). (Reprinted from Dingenen et al., (2019), page 187, copyright (2019), with permission of Elsevier)

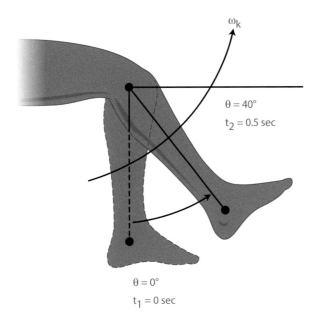

ω_k

$\theta = 40°$
$t_2 = 0.5$ sec

$\theta = 0°$
$t_1 = 0$ sec

▫ Fig. 5.12 The average angular velocity (ω) of the first half of a knee extension exercise can be calculated from the change in angular displacement divided by the change in time (θ/t)

▫ Table 5.2 Typical peak angular speeds in human movement		
	Speed deg/s	**rad/s**
Knee extension:		
sit-to-stand	150	2.6
Trunk extension:		
vertical jump	170	3.0
Elbow flexion:		
arm curl	200	3.5
Knee extension:		
vertical jump	800	14.0
Ankle extension:		
vertical jump	860	15.0
Wrist flexion:		
baseball pitching	1000	17.5
Radio/ulnar pronation:		
tennis serve	1400	24.4
Knee extension:		
soccer kick	2000	34.9
Shoulder flexion:		
softball pitch	5000	87.3
Shoulder internal rotation:		
pitching	7400	129.1

tional information that scalars do not. If relative angles are calculated between anatomical segments at joint, the angular velocities calculated can represent the speed of flexion or extension and other anatomical rotations. Biomechanical research often indirectly calculates joint angles from the linear coordinates (measurements) derived from video images or directly from electrogoniometers attached to participants performing the experimental movements. It is also useful for kinesiology professionals to be knowledgeable about the typical angular velocities of joint movements. This allows professionals to understand the similarity between skills and determine appropriate training and rehabilitation exercises. ▫ Table 5.2 lists typical peak joint angular speeds for a variety of human movements.

Let's calculate the average angular velocity of a typical knee extension exercise and compare it to the peak angular velocity in the table. ▫ Figure 5.12 illustrates the exercise and the data for the example. The subject extends a knee, taking their leg from a vertical orientation to the middle of the range of motion. If we measure the angle of the lower leg from the vertical, the exerciser has moved their leg 40° in a 0.5 second period of time. The average knee extension angular velocity can be calculated as follows: $\omega_K = \theta/t = 40/0.5 = 80$ deg/s. The angular velocity is positive because the rotation is counterclockwise. So the exercise averages 80° per second of knee extension velocity over the half-second time interval, but the instantaneous peak angular velocity at the position shown in the figure is likely larger than that (perhaps 150 or 160 deg/s). The peak knee extension angular velocity in this exercise likely occurs in the mid-

range of the movement, and the knee extension velocity must then slow to zero at the end of the range of motion. This illustrates some limitations of free-weight and rehabilitation exercises. There is a range of angular velocities (which have an effect on the linear force–velocity relationship of the muscles), and there must be a decrease in the angular velocity of the movement at the end of the range of motion. This negative acceleration (if the direction of motion is positive) protects the joints and ligaments, but is not specific to many events where peak speed is achieved near the release of an object where other movements can gradually slow the body in the follow-through.

5.3.2 Angular Acceleration

The rate of change of angular velocity is angular acceleration ($\alpha = \omega/t$). Angular acceleration is symbolized by the Greek letter alpha (α). The typical units of angular acceleration are deg/s/s and radians/s/s. Like linear acceleration, it is best to think angular acceleration as an unbalanced moment or rotary effect. An angular acceleration of −200 rad/s/s means that there is an unbalanced clockwise moment of force tending to rotate the object being studied. The angular acceleration of an isokinetic dynamometer in the middle of the range of motion should be zero because the machine is designed to match or balance the torque created by the person, constraining the rotation so the arm of the machine should be moving at a near constant angular velocity.

Angular kinematics graphs are particularly useful for providing precise descriptions of how joint rotation patterns occur in human movements. ◘ Figure 5.13 illustrates the angular displacement and angular velocity of a simple elbow extension and flexion movement in the sagittal plane. Imagine that the data represented a student tired of studying exercise physiology, who reached forward to grab a refreshing, 48 ounce *Fundamentals of Biomechanics* text! Note as we look at the kinematic information in these graphs that the complexity of a very simple movement grows as we look at the higher-order time derivatives (velocity and acceleration).

Interdisciplinary Issue: Specificity

One of the most significant principles discovered by early kinesiology research is the principle of specificity. Specificity applies to the various components of fitness, training response, and motor skills. Motor learning research clearly indicates that there is specificity of motor skills, but there is potential transfer of ability between similar motor skills. In strength and conditioning, the principle of specificity states that the exercises prescribed should be specific, as close as possible to the movement that is to be improved. Biomechanics research that measures the angular kinematics of various sports and activities can be used to assess the motion similarity and potential specificity of training exercises. Given the peak angular velocities in ◘ Table 5.2, how specific are most weight training or isokinetic exercise movements that are limited to 500° per second or slower to the dynamic movements of sports or the motion of the elderly? The peak speed of joint rotations is just one kinematic aspect of movement specificity. What other control, learning, psychological, or other factors affect the specificity of an exercise for a particular movement?

Interdisciplinary Issue: Isokinetic Dynamometers

Isokinetic dynamometers were developed by J. Perrine in the 1960s. His Cybex machine could be set at different angular velocities and would accommodate the resistance to the torque applied by a subject to prevent angular acceleration beyond the set speed. Since that time, isokinetic testing of virtually every muscle group has become a widely accepted and gold-standard measure of muscular strength in clinical and research settings. Isokinetic dynamometers have been influential in documenting the balance of strength between opposing muscle groups (Grace 1985).There are a journal (*Isokinetics and Exercise Science*) and several books (e.g., Brown 2000; Perrin 1993) that focus on the many uses of isokinetic testing. Isokinetic machines, however, are not perfectly isokinetic throughout the range of motion, because there has to be an acceleration to the set speed at the beginning of a movement that often results in a torque overshoot as the machine negatively accelerates the limb to the set angular velocity (Winter et al. 1981) as well as another negative acceleration at the end of the range of motion. The effects of inertia (Iossifidou and Baltzopoulos 2000), shifting of the limb in the seat/restraints (Arampatzis et al. 2004), and muscular co-contraction (Kellis and Baltzopoulos 1998) are other issues being investigated that affect the validity of isokinetic testing. It is important to note that the muscle group is not truly shortening or lengthening in an isokinetic fashion. Muscle fascicle-shortening velocity is not constant (Ichinose, Kawakami et al. 2000) in isokinetic dynamometry even when the arm of the machine is rotating at a constant angular velocity. This is because linear motion of points on rotating segments does not directly correspond to angular motion in isokinetic (Hinson et al. 1979) or other joint motions.

The elbow angular displacement data show an elbow extended (positive angular displacement) from about a 37° to about an 146° elbow angle to grasp the book. The extension movement took about 0.6 seconds, but flexion with the book occurred more slowly. Since the elbow angle is defined on the anterior aspect of a subject's arm, larger numbers mean elbow extension. The corresponding angular velocity–time graph represents the speed of extension (positive ω) or the speed of flexion (negative ω). The elbow extension angular velocity peaks at about 300 deg/s (0.27 sec) and gradually slows. The velocity of elbow flexion increases and decreases more gradually than the elbow extension.

The elbow angular acceleration would be the slope of the angular velocity graph. Think of the elbow angular acceleration as an unbalanced push toward extension or flexion. Examine the angular velocity graph and note the general phases of acceleration. When are there general upward or downward trends or changes in the angular velocity graph? Movements like this often have three major phases. The extension movement was initiated by a phase of positive acceleration, indicated by an

5

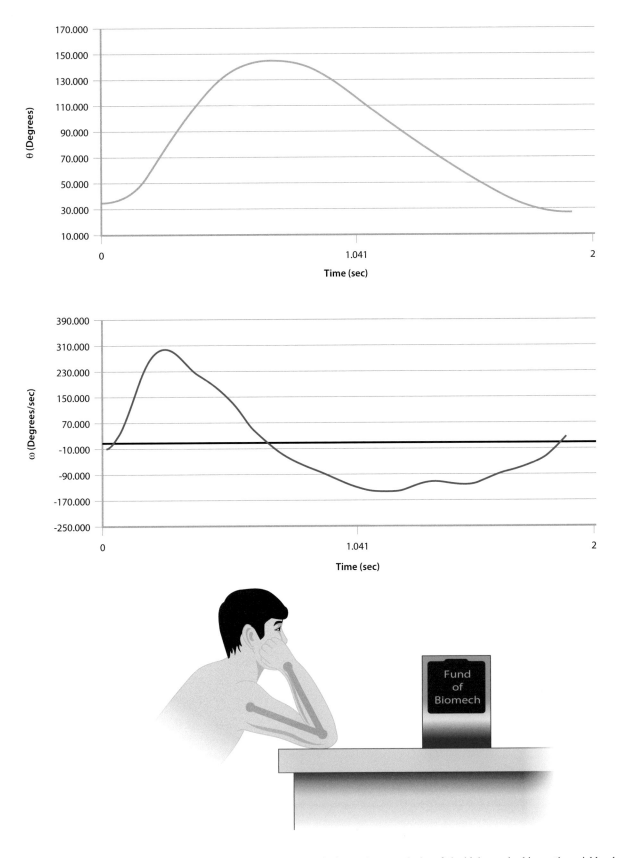

◻ Fig. 5.13 The angular displacement (anterior angle that begins at about 35 degrees) and angular velocity of a simple elbow extension/flexion movement to grab a book. See the text for an explanation of the increasing complexity of the higher-order kinematic variables describing the motion increasing angular velocity. The second phase is a negative acceleration (downward movement of the angular velocity graph) that first slows elbow extension and then initiates elbow flexion. The third phase is a small positive angular acceleration that slows elbow flexion as the book nears the person's head. These three phases of angular acceleration correspond to typical muscle activation in this movement. This movement would usually be created by a tri-phasic pattern of bursts from the elbow extensors, flexors, and extensors. Accelerations (linear and angular) are the kinematic variables closest to the causes (kinetics) of the motion and are more complex than lower-order kinematic variables like angular displacements.

🔲 Figure 5.14 plots the ankle angle, angular velocity, plantar flexor torque, and REMG for the gastrocnemius muscle in a concentric-only and an SSC hop. Notice how only the SSC has a negative angular velocity (describing essentially the speed of the eccentric stretch of the calf muscles) and the dramatic difference in the pattern and size of the plantar flexor torque created following the SSC coordination.

Angular and linear kinematics give scientists important tools to describe and understand exactly how movements occur. Remember to treat the linear and angular measurements separately: like the old saying goes, "don't mix apples and oranges." A good example is a rotating object (like a golf club). As the club is rotated in the swing plane around an axis between the player's two hands, a point near the club head travels a larger distance compared to a point near the center (club shaft or center of mass). How can two points make the same revolutions per minute and travel at different speeds? Easy, if you notice the last sentence mixes or compares angular and linear kinematic variables. In linear kinetics we will look at the trigonometric functions that allow linear measurements to be mapped to angular.

Biomechanists usually calculate angular kinematic variables from linear coordinates of body segments with trigonometry. There is another simple formula that converts linear to angular kinematics in special conditions. It is useful to illustrate why the body tends to extend segments just prior to high-speed release or impact events in sports. The tangential linear velocity of a point on a rotating object, *relative to its axis of rotation*, can be calculated as the product of its angular velocity and the distance from the axis to the point (called the radius): $V = \omega \cdot r$. The special condition for using this formula is to use angular velocity in radians/second. Using a dimensionless unit like rad/s, you can multiply a radius measured in meters and get a linear velocity in meters/second.

The most important point is to notice that the angular velocity and the radius are equally important in creating linear velocity. To hit a golf ball harder, you can either use a longer club or rotate the club faster. We will see in ▶ Chap. 7 that angular kinetic analysis can help us decide which of these two options is best for a particular situation. In most throwing events, the arm is extended late in the throw to increase the linear velocity of a projectile. Angular kinetics is necessary to understand why this extension or increase in the radius of segments is delayed to just before release.

The golf swing is also interesting because the advantage of swinging a long club with a large range of motion to create great club head speed (🔲 Table 5.1) is mediated by the need for accuracy. Fast club head speeds are not effective if the orientation (angle) and club velocity (direction) do not align with the target at impact with the ball. The teaching cues used for golfers to strive for accuracy is to strive to hit on a "swing plane" and generally not take the club shaft past horizontal in the backswing. Three-dimensional kinematic studies of skilled golfers have indicated that they typically have three patterns of downswing that approximate club rotation in a swing plane (🔲 Fig. 5.15). Three-dimensional kinematic studies of skilled performers provide good preliminary information about effective technique and subtle technique variations.

5.4 Coordination Continuum Principle

Many kinesiology professionals are interested in the coordination of movement. Coordination is commonly understood as the sequence and timing of body actions used in a movement. Unfortunately, there is no universally agreed-upon definition or way to study coordination in the kinesiology literature. A wide variety of approaches has been proposed to describe the coordination of movement. Some approaches focus on the kinematics of the joint or segmental motions (Hudson 1986; Kreighbaum and Bartels 1996), while others are based on the joint forces and torques (kinetics) that create the subsequent movement (Chapman and Sanderson 1990; Putnam 1991, 1993; Roberts 1991; Zajac 1991; Prilutsky 2000). This section presents the **Coordination Continuum Principle**, which is adapted from two kinematic approaches to defining coordination (Hudson 1986; Kreighbaum and Bartels 1996), because teachers and coaches most often modify the spatial and temporal aspects of movement. While teaching cues that focus on muscular effort may be used occasionally, much of the teaching, coaching, and instruction in movement and

5

▫ Fig. 5.14 Ankle angle, angular velocity, torque, and rectified EMG in a concentric-only (PFJ: plantar flexion jump) and SSC hop exercise (RJ: rebound jump). (Figure reprinted permission of Sugisaki et al. (2005))

Fig. 5.15 Three styles of club motion that approximate the "swing plane" of a golf swing: Planar style (**a**), spiral style (**b**), and the reverse style (**c**). (Data reprinted by permission of Kwon and Han (2016))

rehabilitation exercise remains in the positioning and motions of the body.

Kinematic coordination of movements can be pictured as a continuum ranging from simultaneous body actions to sequential actions. The **Coordination Continuum Principle** states that movements requiring the generation of high forces tend to utilize simultaneous segmental movements, while lower-force and high-speed movements are more effective with more sequential movement coordination. A person lifting a heavy box simultaneously extends the hips, knees, and ankles (**Fig. 5.16**). In overarm throwing, people usually use a more sequential action of the whole kinematic chain, beginning with the legs, followed by trunk and arm motions.

Because the kinematics of coordination falls on a continuum and the speed and forces of movement vary widely, it is not always easy to determine what coordination pattern is best. In vertical jumping, resistance is moderate and the objective is to maximize height of takeoff and vertical velocity. While a vertical jump looks like a simultaneous movement, biomechanical studies show that the kinematics and kinetics of different jumpers have simultaneous and sequential characteristics (Hudson 1986; Bobbert and van Ingen Schenau 1988; Aragon-Vargas and Gross 1997). Kinesiology professionals need to remember that coordination is not an either/or situation in many activities. Until there is more research determining the most effective range of technique or coordination, there will be a little art to the teaching and coaching of movements not at the extremes of the continuum.

The motor development of high-speed throwing and striking skills tends to begin with restricted degrees of freedom and simultaneous actions. Children throwing, striking, or kicking tend to make initial attempts with simultaneous actions of only a few joints. Skill develops

Fig. 5.16 Coordination to move a heavy load usually involves simultaneous joint motions like in this squat lift

with the use of more segments and greater sequential action. In high-speed throwing, for example, the sequential or "differential" rotation of the pelvis and upper trunk in

5

◘ Fig. 5.17 Poor sequential coordination in throwing and striking results in slow segment speeds at impact that can be visually identified by slow ball speeds, lack of eccentric loading of distal segments, or limited movement in the follow-through (like this volleyball spike)

the transverse plane is a late-developing milestone of high-skill throwing (Roberton and Halverson 1984). It is critical that physical educators know the proper sequential actions in these low-force and high-speed movements. Kinematic studies help identify these patterns of motion in movement skills. Unfortunately, the relative youth of biomechanics means that kinematic documentation of coordination in the wide variety of human movement in all populations of people (age, sex, disability) and environmental conditions (wind, elevation, equipment, competitors) is not complete. Early biomechanics research studies (1970s and 1980s) emphasized advanced and elite, young male performers, leaving little knowledge on how sex, culture, disability, lower skill levels, age, or other environmental factors affect movement kinematics. This limited picture of a diverse movement world is similar to what psychologists have described as the "WEIRD" (Western, Educated, Industrialized, Rich, Democratic) research in their field (Henrich et al. 2010).

Suppose a junior high volleyball coach is working with a tall athlete on spiking. The potential attacker lacks a strong overarm pattern and cannot get much speed on the ball (◘ Fig. 5.17). The kinematics of the preparatory action lacks intensity, stretch, and timing. At impact the player's elbow and upper arm are well forward of his shoulder. The coach suspects that his overarm throwing pattern is still immature and must be developed before skilled spiking is possible. This coach has integrated biomechanical and motor development information to determine the best course of action to help this player improve. The lack of ball speed (kinematics) and muscle stretch-shortening cycles within a sequential coordination are biomechanical factors missing in this athlete. The forward elbow position at impact is a motor development indicator of an immature trunk and arm

action within an overarm pattern. How coaches work on this problem may vary, but one good strategy would be to simplify the movement and work on throwing the volleyball. Sequential rotation of the trunk, arm, forearm, and wrist is the focus of training.

Strength and conditioning and physical rehabilitation (AT, OT, PT) professionals closely monitor training technique, because body position and motion in exercises dramatically affect muscular actions and risk of injury. In strength training, resistances are near maximal, so coordination in most exercises tends to be simultaneous. Imagine someone performing a squat exercise with a heavy weight. Is the safest technique to simultaneously flex the hips and knees in the eccentric phase and then simultaneously extend in the concentric phase? If the resistance is lighter (bodyweight), like in standing up out of a chair, after a person leans forward to put their upper bodyweight over their feet, do the major joints of the body simultaneously act to stand? Would the coordination change if the person is weakened from injury or the elderly from lack of activity? In the next chapter, we will examine variations in conditioning for intense, high-speed movements that are different than high-force (strength) movements. Do you think dynamic, high-speed movements will also have simultaneous coordination or will the coordination shift a little toward sequential? Why?

5.5 Summary

A key branch of biomechanics is kinematics, the precise description or measurement of human motion. Human motion is measured relative to some frame of reference and is usually expressed in linear (meters, feet) or angular (radians, degrees) units. Angular kinematics are particularly appropriate in biomechanics because these can be easily adapted to document the rotations of the major joints of the body. There are many kinematic variables that can be used to document the human motion. Simple kinematic variables are scalars, while others are vector quantities that take into account the direction of motion. The first and second time derivatives (rates of change) of displacement measurements are velocity and acceleration, respectively. The Optimal Projection Principle states that sporting events involving projectiles have a range of desirable initial angles of projection appropriate for most performers, although exact "optimal" conditions are difficult to define. The kinematic timing of segment motions falls on a Coordination Continuum from simultaneous to sequential movement. High-force movements use more simultaneous joint rotations, while high-speed movements use more sequential joint rotations.

? Review Questions

1. What is a frame of reference and why is it important in kinematic measurements?
2. Compare and contrast the scalar and vector linear kinematic variables.
3. Explain the difference between calculation of average and instantaneous velocities, and how does the length of the time interval used affect the accuracy of a velocity calculation?
4. Use the velocity graph in ◘ Fig. 5.4 to calculate the average acceleration of the sprinter in the first and the last 10 m intervals.
5. A patient lifts a dumbbell 1.2 m in 1.5 s and lowers it back to the original position in 2.0 s. Calculate the average vertical velocity of the concentric and eccentric phases of the lift.
6. Explain why linear and angular accelerations should be thought of as pushes in a particular direction rather than speeding up or slowing down.
7. Why are angular kinematics particularly well suited for the analysis of human movement?
8. From the anatomical position a person abducts their shoulder to 30° above the horizontal. What is the angular displacement of this movement with the usual directional (sign) convention?
9. A soccer player attempting to steal the ball from an opponent was extending her knee at 50 deg/s when her foot struck the opponent's shin pads. If the player's knee was stopped (0 deg/s) within 0.2 seconds, what angular acceleration did the knee experience?
10. A golfer drops a ball to replace a lost ball. If the ball had an initial vertical velocity of 0 m/s and had a vertical velocity before impact of −15.7 m/s exactly 1.6 seconds later, what was the vertical acceleration of the ball?
11. A softball coach is concerned that her team is not throwing at less than 70% speed in warm-up drills. How could she estimate or measure the speeds of the warm-up throws to make sure her players are not throwing too hard?
12. A biomechanist uses video images to measure the position of a box in the sagittal plane relative to a worker's toes during lifting. Which coordinate (x or y) usually corresponds to the height of the box and the horizontal position of the box relative to the foot?
13. Use the formula for calculating linear velocity from angular velocity ($V = \omega \cdot r$) to calculate the velocity of a golf club relative to the player's hands (axis of rotation). Assume the radius is 1.5 m and the angular velocity of the club was 2000 deg/s. Hint: remember to use the correct units for angular velocity of radians. Think about how this example would relate to how push

distance would turn into speed based on different diameters of the wheels on a wheelchair.

14. Give an example of a fixed and a relative frame of reference for defining joint angular kinematics. Which frame of reference is better for defining anatomical rotations versus rotations in space?
15. What is the vertical acceleration of a volleyball at the peak of its flight after the ball is tossed upward in a jump serve?
16. What would you estimate to be the "window" of successful launch angles (Optimal Projection Principle) to the horizontal for a soccer penalty kick and why?
17. The Coordination Continuum Principle would predict what coordination for an adult lifting 10 pounds from the floor? A preschooler? Elderly in assisted living?
18. Kinematics of motion can be modified by looking at what force acting on the object/system?
19. Calculate the average horizontal velocity of a sprinter over the first 20 meters ($\Delta t = 3.0$ s) and over the first 60 m ($\Delta t = 7.0$ s). Does the change from V_{20} to V_{60} agree with the chapter discussion of acceleration in sprinting?

! Key Terms

Absolute frame of reference · Angle · Acceleration · Coordination continuum principle · Degrees of freedom · Displacement · Distance · Goniometer · Isokinetic · Linked segment model · Optimal projection principle · Point mass (particle) · Relative frame of reference · Rigid body · Speed · Static Flexibility · Trajectory · Velocity

★ Suggested Reading

Cappozzo, A., et al. (1990). *Biolocomotion: A century of research using moving pictures*. Rome: Promograph.

Hudson, J. L. (1986). Coordination of segments in the vertical jump. *Medicine and Science in Sports and Exercise, 18*, 242–251.

Li, L., & Holmes, M. (2014). *Gait biometrics: Basic patterns, role of neurological disorders and effects of physical activity*. Hauppauge: Nova Science Publishers.

Norkin, C. C., & White, D. J. (2017). *Measurement of joint motion–A guide to goniometry* (7th ed.). Philadelphia: F. A. Davis.

Zatsiorsky, V. M. (1998). *Kinematics of human motion*. Champaign: Human Kinetics.

🔊 Internet Resources

▶ "2D Video Analysis" Videos at the BiomechancsMMU Youtube Channel. *youtube.com/channel/UCfOgUG_55Wf0OyteQvh2a3Q*

▶ BBC TWO Four-minute story on uniformly accelerated motion in NASAs Worlds Largest vacuum chamber. *youtube.com/watch?v=E43-CfukEgs&feature=youtu.be*

▶ Khan Academy–2D Projectiles. *khanacademy.org/science/physics/two-dimensional-motion*

► Kinovea 2D video app. *kinovea.org*
► NBC News Learn–Video lessons on the Science of Golf, Science of NFL Football, and Science of the Winter Olympics. *nbclearn. com/portal/site/learn/resources*
► Vectors and 2D Motion I Crash Course Physics–PBS video lesson on kinematics. *klru.pbslearningmedia.org/resource/vectors-2dmotion-crashcourse-1004/vectors-2dmotion-crash-course-physics/*
► Vicon Biomechanics and Sports Science Showreel 2018. *youtube. com/watch?v=uPn26JbRN4g*

Linear Kinetics

Key Questions

➤ **What do you need to know to begin?**
Knowledge of musculoskeletal anatomy, algebra, and kinematics.

➤ **Why do you need to know this material?**
Understanding the causes of linear motion informs general recommendations for technique and training.

➤ **What is the key idea?**
Biokinetic variables and principles for analysis of linear motion are essential tools in establishing the causes of human movement.

In the previous chapter, we learned that kinematic principles and measurements of motion could be used to provide knowledge for improving human movement. This chapter will summarize the important laws of kinetics that show how forces overcome inertia and create human motion. Studying the causes of linear motion is the branch of mechanics known as linear kinetics. Identifying the causes of motion may be the most useful kind of mechanical knowledge for determining what potential changes could be used to improve human movement. The biomechanical principles that will be discussed in this chapter are **Inertia, Force–Time,** and **Segmental Interaction**.

6.1 Laws of Kinetics

Linear kinetics of classical mechanics (Newton's Laws of Motion) provide precise ways to document the causes of the linear motion of all objects. The specific mechanical variables and mathematical expression a biomechanist will choose to use in analyzing the causes of linear motion often depends on the nature of the movement and analysis. When instantaneous effects of forces are of interest, we will see one form of Newton's 2nd Law of Motion ($\mathbf{F} = m\mathbf{a}$) will be used. When studying movements and the effects of forces over intervals of time, Newton's original language (words and mathematics) of his 2nd Law (Impulse–Momentum Relationship) is usually used. Later in the chapter, a third approach to studying the causes of motion using the Work–Energy Relationship will be presented. This chapter summarizes these concepts in the context of human movement. Most importantly, we will see how these laws can be applied to human motion in the biomechanical principles of force–time, inertia, and segmental interaction principles.

6.2 Newton's Laws of Motion

Arguably, some of the most important discoveries of mechanics are the three laws of motion developed by the Englishman, Sir Isaac Newton. Newton is famous for many influential scientific discoveries, including developments in calculus, the Law of Universal Gravitation, and the Laws of Motion of classical mechanics. The importance of his laws cannot be overemphasized in our context, for they are the keys to understanding how human movement occurs. The publication of these laws in his 1686 book (*Principles of Natural Philosophy* when translated to English) marked one of the rare occasions of scientific breakthrough. Thousands of years of dominance of the incorrect mechanical views of the Greek philosopher Aristotle were overturned forever. Newton's laws of motion determine the motion of biomechanical systems and objects (balls, planes, rockets, planets,. ..) and only are replaced by more complex relativistic mechanics at extreme speeds near the speed of light (like electromagnetic signals sent by radios/cell phones).

6.2.1 Newton's First Law and First Impressions

Newton's first law is called the Law of Inertia because it outlines a key property of the matter related to motion. Newton stated that all objects have the inherent property to resist a change in their state of motion.

D. Knudson, *Fundamentals of Biomechanics*, https://doi.org/10.1007/978-3-030-51838-7_6

His first law is usually stated something like this: objects tend to stay at rest or in uniform motion unless acted upon by an unbalanced force. A player sitting and "warming the bench" has just as much inertia as a teammate of equal mass running at a constant velocity on the court. It is vitally important that kinesiology professionals recognize the effect inertia and Newton's first law have on movement technique. The linear measure of inertia (◘ Fig. 6.1) is mass and has units of kg in the SI system and slugs in the English system. This section is an initial introduction to the fascinating world of kinetics and will demonstrate how our first impressions of how things work from casual observation are often incorrect.

Understanding kinetics, like Newton's first law, is both simple and difficult. They are simple because there are only a few physical laws that govern all human movement, can be easily understood and demonstrated using simple algebra, and contain only a few variables. The study of biomechanics can be difficult, however, because the laws of mechanics indicate causes of motion that are often counterintuitive for many people. This is because the observations of everyday life often lead to incorrect assumptions about the nature of the world and motion. Many children and adults have incorrect notions about inertia, and this view of the true nature of motion has its own "cognitive inertia," which is hard to displace. The natural state of objects in motion is to slow down, right? Wrong! The natural state of motion is to continue whatever it is doing! Newton's first law shows that objects tend to resist changes in motion and that things only seem to naturally slow down because forces like friction and air or water resistance that tend to slow an object's motion (Remember the force–motion principle). Most objects around us appear at rest, so isn't there something natural about being apparently motionless? The answer is yes, if the object is initially at rest! The same object in linear motion has the same natural or inertial tendency to keep moving. In short, the mass (and consequently its linear inertia) of an object is the same whether it is motionless or moving.

We also live in a world where most people take atmospheric pressure for granted. They are aware that high winds can create very large forces, but would not believe that in still air there can be hundreds of pounds of resultant force on both sides of a house window (or a person) due to the pressure of the atmosphere all around us. The true nature of mechanics in our world often becomes more apparent under extreme conditions. The pressure of the sea of air we live in becomes real when a home explodes or implodes from a passing tornado, or a fast-moving weather system brings a change in pressure that makes a person's injured knee ache. People interested in scuba diving need to be knowledgeable about water pressure differences and the timing of these changes when they dive.

So casual observation can often lead to incorrect assumptions about the laws of mechanics. We equate forces with objects in contact or a collision between two objects. Yet we live our lives exercising our muscles against the consistent force of gravity, a force that acts at quite a distance whether we are touching the ground or not. We also tend to equate the velocity (speed and direction) of an object with the force that made it. In this chapter, we will see that a force that acts on an object does not have to be acting in the direction of the resultant motion of the object (◘ Fig. 6.2). It is the skilled person that creates muscle forces with size and direction to precisely combine with other (external) forces to balance on a bike or throw the ball in the correct direction.

Casual visual observation also has many examples of perceptual illusions about the physical realities of our world. Our brains work with our eyes to give us a mental image of physical objects in the world and their motion, so that most people routinely mistake this constructed mental image for the actual object. The color of objects is also an illusion based on the wavelengths of light that are reflected from an object's surface. So what about touch? The solidity of objects is also a perceptual illusion because the vast majority of the volume in atoms is "empty" space. The forces and hardness we feel when we touch things are the magnetic forces of electrons on the two surfaces repelling each other, while the material

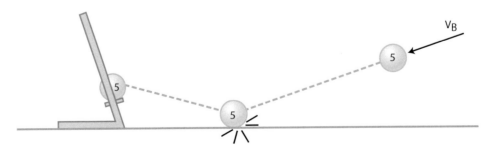

◘ **Fig. 6.1**   All objects have the inherent property of inertia, the resistance to a change in the state of motion. The measure of linear motion inertia is mass. A medicine ball has the same resistance to acceleration (5 kg of mass) in all conditions of motion (moving: dashed line and resting in the rack), assuming it does not travel near the speed of light

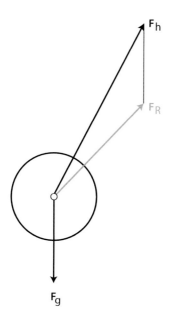

Fig. 6.2 Force and motion do not always act in the same direction. This free-body diagram of the forces and resultant force ($\mathbf{F_R}$) on a basketball before release illustrates how a skilled player applies a force to the ball ($\mathbf{F_h}$) that combines with the force of gravity ($\mathbf{F_g}$) to create the desired effect. The ball will be accelerated in the direction of the resultant force $\mathbf{F_R}$

strength of an object we bend is related to its physical structure and chemical bonding. We also have a distorted perception of time and the present. We rely on light waves bouncing off objects and toward our eyes. This time delay is not a problem at all, unless we want to observe very high-speed or distant objects like in astronomy. There are many other examples of our cognitive molding or construction of the nature of reality, but the important point is that there is a long history of careful scientific measurements which demonstrate that certain laws of mechanics represent the true nature of objects and their motion. These laws provide a simple structure that should be used for understanding and modifying motion, rather than erroneous perceptions about the nature of things. Newton's first law is the basis for the Inertia Principle in applying biomechanics.

Interdisciplinary Issue: Body Composition
A considerable body of kinesiology research has focused on the percentage of fat and lean mass in the human body. There are metabolic, mechanical, psychological, and health effects of the amount and location of fat mass. In sports performance, fat mass can be both an advantage (increased inertia for a football lineman or sumo wrestler) and a disadvantage. Increasing lean body mass usually benefits sports performance, although greater mass means increasing inertia, which could decrease agility and quickness. When coaches are asked by athletes "How much should I weigh?"

they should answer carefully, focusing the athlete's attention first on healthy body composition. Then the coach can discuss with the athlete the potential risks and benefits of changes in body composition that might hold performance benefits in their sport. How changes in an athlete's inertia affect their sport performance should not be evaluated without regard to broader health issues. Athletes trying to "cut" weight to make a certain weight class or try to gain too much body fat, both are taking risks with their long-term health for sport-term sport performance.

6.2.2 Newton's Second Law

Newton's second law is arguably the most important law of motion because it shows how the forces that create motion (kinetics) are linked to the motion (kinematics). The second law is called the **Law of Momentum** or **Law of Acceleration**, depending on how the mathematics is written. The most common approach is the instantaneous version $\mathbf{F} = \mathbf{ma}$. This is the law of acceleration, which describes motion (acceleration) for any instant in time. The formula correctly written is $\Sigma\mathbf{F} = m \cdot \mathbf{a}$, and states that the acceleration an object experiences is proportional to the resultant force, is in the same direction, and is inversely proportional to the mass. The larger the unbalanced force in a particular direction, the greater the acceleration of the object in that direction. With increasing mass, the inertia of the object will decrease the acceleration if the force doesn't change. This expression gives rise to the units of force and are therefore, called Newtons in his honor.

Let's look at an example using skaters in the push-off and glide phases during ice skating (**Fig. 6.3**). If the skaters have a mass of 59 kg and the horizontal forces are known (measured), we can calculate the hori-

Fig. 6.3 Friction forces acting on ice skaters during push-off and gliding. Newton' Second Law of Motion applied in the horizontal direction (see text) will determine the horizontal acceleration of the skater

6

zontal acceleration of the skater. During push-off, the net horizontal force is +200 N because air resistance is negligible, so the skater's horizontal acceleration is: $\Sigma F = m \cdot a$, 200 = 59a, so a = 3.4 m/s/s. The skater has a positive acceleration and would tend to speed up 3.4 m/s every second if she could maintain her push-off force this much over any horizontal air resistance. In the glide phase, the friction force is now a resistance rather than a propulsive force. During glide the skater's acceleration is −0.08 m/s/s because: $\Sigma F = m \cdot a$, −5 = 59a, so a = −0.08 m/s/s.

The kinesiology professional can qualitatively break down movements with Newton's second law. Large changes in the speed or direction (acceleration) of a person means that large forces must have been applied just before the observed changes in state of motion. If an athletic contest hinges on the agility of an athlete in a crucial play, the coach should select the lightest and quickest player. An athlete with a small mass is easier to accelerate than an athlete with a larger mass, provided they can create similar forces relative to their body mass. If a smaller player is being overpowered by a larger opponent, the coach can substitute a larger more massive player to defend against this opponent (again if this more massive athlete has similar relative strength). Note that increasing force or decreasing mass are both equally important in creating acceleration and subsequent movement.

Newton's second law plays a critical role in quantitative biomechanics. Biomechanists wanting to study the net forces that create human motion take acceleration and body segment mass measurements and apply $\Sigma F = ma$ in all relevant directions This working backward from kinematics to the resultant kinetics is called **inverse dynamics**. Other scientists build complex computer models of biomechanical systems and use **direct dynamics**, essentially calculating the motion from the "what-if" kinetics and body configurations they input.

6.2.3 Newton's Third Law

Newton's third law of motion is called the **Law of Reaction**, because it is most often translated as: for every action, there is an equal and opposite reaction. For every force exerted, there is an equal and opposite force being exerted. If a patient exerts a sideways force of +150 N on an elastic cord, there has to be −150-N reaction force of the cord acting on the patient's hand (◘ Fig. 6.4). The key insight by Newton that people often miss is that a force is really a mutual interaction between two bodies. It may seem strange that if you push horizontally against a wall, the wall is simultaneously pushing back toward you, but it is. This is *not* to say that a force on a free-body diagram should be represented by two vectors, but a person must understand that the effect of a force is not just on one object. A free body diagram is *one* object or mechanical system and the forces acting on it, so the double vectors in ◘ Figs. 6.4 and 6.5 can sometimes be confusing because they are illustrating both objects (patient and elastic band) and are not true free body diagrams. If someone ever did not seem to kiss you back after a date, you can always take some comfort in the fact that at least in mechanical terms they did.

An important implication of the law of reaction is how reaction forces can change the direction of motion opposite to our applied force when we exert our force on objects with larger forces or inertia (◘ Fig. 6.5a). During push-off in running an athlete exerts downward

◘ **Fig. 6.4** Newton's third law states that all forces have an equal and opposite reaction forces on the other object, like in this elastic exercise. The −150-N (**F$_A$**) force created by the person on the elastic cord coincides with a 150-N reaction force (**F$_B$**) exerted on the person by the cord

$F_A = -150 \, N$

$F_B = 150 \, N$

and backward push with the foot, which creates a **ground reaction force** to propel the body upward and forward. The extreme mass of the earth easily overcomes our inertia, and the ground reaction force accelerates our body in the opposite direction of force applied to the ground. Another example would be eccentric muscle actions where we use our muscles as brakes, pushing in the opposite direction to another force. The force exerted by the tackler in ◘ Fig. 6.5b likely ends up being an eccentric muscle action as the inertia and ground reaction forces created by the runner are too great for the falling defender. Remember that when we push or pull, this force is exerted on some other object and the object pushes or pulls back on us too!

There are several kinds of force-measuring devices used in biomechanics to study how forces modify movement. Two important devices are the **force platform** (or force plates) and pressure sensor arrays. A force plate is a rigid platform that measures the forces and torques in all three dimensions applied to the surface of the platform (Schieb 1987). Force plates are often mounted in a floor to measure the ground reaction forces that are equal and opposite to the forces people make against the ground (see ◘ Fig. 6.5). Since the 1980s, miniaturization of a variety of force sensors has allowed for rapid development of arrays of small-force sensors that allow measurement of the distribution of forces (and pressure because the area of the many small sensors are known) on a body. Several commercial shoe insoles with these sensors are available for studying the pressure distribution under a person's foot (McPoil et al. 1995). There are many other force-measuring devices (e.g., load cell, strain gauge, isokinetic dynamometer) that help biomechanics scholars study the kinetics of movement.

6.3 Inertia Principle

Newton's first law of motion (**Law of Inertia**) describes the resistance of all objects to a change in their state of linear motion. In linear motion, the measure of inertia is an object's mass. Application of Newton's first law in biomechanics is termed the **Inertia Principle**. This section will discuss how teachers, coaches, and therapists adjust movement inertia to accommodate a task, improve movement, or mediate risk of injury. Our focus will be on the linear inertia (mass) of movement, so the inertial resistance to rotation will be summarized in ▶ Chap. 7.

The first example of application of the inertia principle is to reduce mass in order to increase the ability to rapidly accelerate. Obvious examples of this principle in track are the racing flats/shoes used in competition versus the heavier shoes sometimes used in training. The heavier shoes used in training provide greater protection for the foot and a small inertial overload. When race day arrives, the smaller mass of the shoes makes the athlete's feet feel light and quick. We will see in ▶ Chap. 7 that this very small change in mass, because of its position, makes a much larger difference in resistance to rotation (angular inertia). Let's add a little psychology and conditioning to the application of lowering inertia.

◘ **Fig. 6.5** A major consequence of Newton's third law is that the forces we exert on an object with larger inertia often create motion in the direction opposite of those forces. In running, the downward backward push of the foot on the ground (**F$_A$**) late in the stance **a** creates a ground reaction force which acts forward and upward, propelling the runner through the air. A defensive player trying to make a tackle (**F$_T$**) from a poor position **b** may experience reaction forces (**F$_R$**) that create eccentric muscle actions and injurious loads

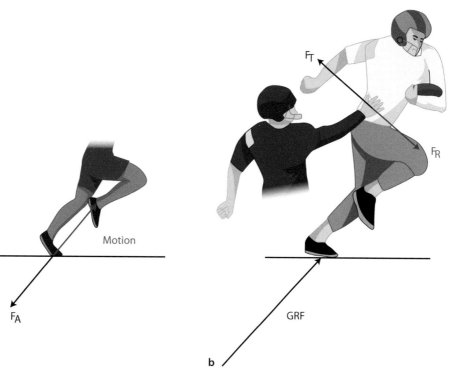

6

Warm-up for many sports involves a gradual increase in intensity of movements, often with larger inertia. In baseball or golf, warm-up swings are often taken with extra weights added (Fig. 6.6)., which when taken off make the "stick" feel very light and fast when it comes time to swing the club in competition.

In movements where stability is desired over mobility, the Inertia Principle indicates that mass should be increased. Linemen in football and centers in basketball have tasks that may benefit more from increasing muscle mass to increase inertia, than from decreasing inertia to benefit quickness. Adding mass to a golf club or tennis racket will make for faster and longer shots if the implement can be swung with the same velocity at impact. If an exercise machine tends to slide around in the weight room, a short-term solution might be to store some extra weights on the base or legs of the machine. If these new weights are not a safety risk (in terms of height or potential for tripping people), the increased inertia of the station would likely make the machine safer.

Another advantage of increased inertia is that the added mass can be used to modify the motion of another body segment. The preparatory leg drives and weight shifts in many sporting activities have several benefits for performance, one being putting more body mass in motion toward a particular target. The forward motion of a good percentage of body mass can be transferred to the smaller body segments just prior to impact or release. We will be looking at this transfer of force later on in this chapter when we consider the **Segmental Interaction Principle**. The defensive moves of martial artists are often designed to take advantage of the inertia of an attacker. An opponent striking from the left has inertia that can be directed by a block to throw to the right.

An area where modifications in inertia are very important is strength and conditioning. Selecting masses and weights for training and rehabilitation is a complicated issue. Biomechanically, it is very important because the inertia and gravitational force of an external object has a major influence on amount of muscular force and how those forces can be applied (Zatsiorsky and Kraemer 2006). Baseball pitchers often train by throwing heavier or lighter than regulation baseballs (Escamilla et al. 2000). Think about the amount of force that can be applied in a bench press exercise versus a basketball chest pass. The very low inertia of the basket-

◘ Fig. 6.6 Mass added to sporting implements in warm-up swings makes the inertia of the regular implement (when the mass is removed) feel very light and quick. Do you think this common sporting ritual of manipulating inertia is beneficial? If so, is the effect more biomechanical or psychological?

ball allows it to accelerate quickly, so the peak force that can be applied to the basketball before it is accelerated away from the body is much lower than what can be applied to a barbell. The most appropriate load, movement, and movement speed in conditioning for a particular human movement is often difficult to define. The principle of specificity says the movement, speed, and load should be similar to the actual activity; therefore, the overload should only come from moderate changes in these variables so as to not adversely affect skill. In rehabilitative exercise, the inertia (masses) and movement speeds are purposefully low to protect healing tissues. These inertias and movement speeds only increase as muscular strength and coordination return from the diminished values after injury and disuse.

Suppose a high school track coach has shot put athletes in the weight room throwing medicine balls. As you discuss the program with the coach you find that they are using loads (inertia) substantially lower than the shot in order to enhance the speed of trunk and upper extremity motion. How might you apply the principle of inertia in this situation? Are the athletes fully using their lower extremities in a similar motion to shot putting? Can the athletes build up large enough forces before acceleration of the medicine ball, or will the force–velocity relationship limit muscle forces? How much lower is the mass of the medicine ball than that of the shot? All these questions, as well as technique, athlete reaction, and actual performance, can help you decide if training is appropriate. The biomechanical research on power output in multi-segment movements indicates that training loads should be higher than the 30 to 40% of maximum isometric force seen in vitro with individual muscles and muscle fibers (see the upcoming section on energy, work and power; Funato et al. 1996; Cronin et al. 2001a, b). Selecting the inertia for weight training has come a long way from "do three sets of 10 reps at 80% of your IRM."

6.4 Muscle Angle of Pull: Qualitative and Quantitative Analysis of Vectors

In our study of linear kinetics, it is a good time to review the special mathematics required to handle vector quantities like force and acceleration. The linear kinetics of a biomechanical issue called *muscle angle of pull* will be explored in this section. While a qualitative understanding of adding force vectors is enough for most kinesiology professionals, quantifying forces and their components due to their angle of action (pull for muscles) provides a deeper level of explanation and understanding of the causes of human movement. We will see that the linear kinetics of the pull of a muscle often changes dramatically because of geometric changes in its angle of pull when joints are rotated.

6.4.1 Qualitative Vector Analysis of Muscle Angle of Pull

While the attachments of a muscle do not change, the angle of the muscle's pull on bones changes with changes in joint angle. The angle of pull is critical to the linear and angular effects of that force. Recall that a force can be broken into parts or components. These pulls of a muscle's force on a bone in two dimensions are conveniently resolved into longitudinal and rotational components. This local (anatomical or relative) frame of reference helps us study how muscle forces affect the body, but do not tell us about the orientation of the body to the world like absolute frames of reference do. ◘ Figure 6.7 illustrates typical angles of pull and these components for the biceps muscle at two points in the range of motion. The linear kinetic effects of the biceps brachii on the forearm can be illustrated with arrows that represent force vectors.

◘ **Fig. 6.7** Typical angles of pull of the biceps brachii muscle in an arm curl. The angular positions of the shoulder and elbow affect the angle of pull of the muscle, which determines the size of the components of the muscle force. Muscle forces (**F**) are usually resolved along the longitudinal axis of the distal segment (**F**$_L$ or stabilizing component) and at right angles to the distal segment to show the component that causes joint rotation (**F**$_R$ or rotary component)

6

The component acting along the longitudinal axis of the forearm ($\mathbf{F}_L$) does not create joint rotation, but provides a load that stabilizes or destabilizes the elbow joint. The component acting at right angles to the forearm is often called the rotary component ($\mathbf{F}_R$) because it creates a torque that contributes to potential rotation. Remember that vectors are drawn to scale to show their magnitude with an arrowhead to represent their direction. Note that in the extended position, the rotary component is similar to the stabilizing component. In the more flexed elbow position illustrated, the rotary component of biceps is larger than the smaller stabilizing component. In both positions illustrated, the biceps muscle tends to flex the elbow, but the ability to do so (the rotary component) varies widely.

This visual or qualitative understanding of vectors is quite useful in studying human movement. When a muscle pulls at a 45° angle, the two right-angle components are equal. A smaller angle of pull favors the longitudinal component, while the rotary component benefits from larger angles of pull. Somewhere in the midrange of the arm curl exercise the biceps has an angle of pull of 90°, so all the bicep's force can be used to rotate the elbow and there is no longitudinal component.

Vectors can also be qualitatively added together. The rules to remember are that the forces must be drawn accurately (size and direction), and they then can be added together in tip-to-tail fashion. This graphical method is often called drawing a *parallelogram of force* (□ Fig. 6.8). If the vastus lateralis and vastus medialis muscle forces on the right patella are added together, we get the resultant of these two muscle forces. The resultant force from these two muscles can be determined by drawing the two muscle forces from the tip of one to the tail of the other, being sure to maintain correct length and direction. Since these diagrams can look like parallelograms, they are called a parallelogram of force. Remember that there are many other muscles, ligaments, and joint forces not shown that affect knee function. It has been hypothesized that an imbalance of greater lateral forces in the quadriceps may contribute to patellofemoral pain syndrome (Callaghan and Oldham 1996). Does the resultant force ($\mathbf{F}_R$) in □ Fig. 6.8 appear to be directed lateral to the longitudinal axis of the femur?

6.4.2 Quantitative Vector Analysis of Muscle Angle of Pull

Quantitative or mathematical analysis provides precise answers to vector resolution (in essence subtraction to find components) or vector composition. Right-angle trigonometry provides the perfect tool for this process. A review of the major trigonometric relationships (sine, cosine, tangent) is provided in Appendix D. Suppose an

□ **Fig. 6.8** Any vectors acting on the same object, like the vastus medialis ($\mathbf{F}_{VM}$) and vastus lateralis ($\mathbf{F}_{VL}$) of the right knee, can be added together to find a resultant ($\mathbf{F}_R$). This graphical method of adding vectors is called a *parallelogram of forces*

athlete is training the isometric stabilization ability of their abdominals with leg raises in a Roman chair exercise station. □ Figure 6.9a illustrates a typical orienta-

◻ **Fig. 6.9** Right-angle trigonometry is used to find the components of a vector like the iliopsoas muscle force illustrated. Notice the muscle force is resolved into components along the longitudinal axis (stabilizing) of the femur and at right angles to the femur. The right-angle (rotary) component is the force that would tend to flex the hip or control extension (**F$_R$**)

tion and magnitude of the major hip flexors (the iliopsoas group) that hold their legs elevated. The magnitude of the weight of the legs and the hip flexor forces provide a large resistance for the abdominal muscles to stabilize. This exercise is not usually appropriate for untrained persons.

If an iliopsoas resultant muscle force of 400 N acts at a 55° angle to the femur, what are the rotary (**F$_R$**) and longitudinal (**F$_L$**)components of this force? To solve this problem, the rotating component is moved tip to tail to form a right triangle (◻ Fig. 6.9b). In this triangle, right-angle trigonometry says that the length of the adjacent side to the 55° angle (**F$_L$**) is equal to the resultant force times cos 55°. So the stabilizing component of the iliopsoas force is: **F$_L$** = 400(cos55°) = 229 N, which would tend to compress the hip joint along the longitudinal axis of the femur. Likewise, the rotary component of this force is the side opposite the 55° angle, so this side is equal to the resultant force times sin 55°. The component of the 400-N iliopsoas force that would tend to rotate the hip joint, or in this example isometrically hold the legs horizontally, is: **F$_R$** = 400 sin 55° = 327 N upward. It is often a good idea to check calculations with a qualitative assessment of the free body diagram. Does the rotary component look larger than the longitudinal component? If these components are the same at 45°, does it make sense that a higher angle would increase the vertical component and decrease the component of force in the horizontal direction?

When the angle of pull (θ) or push of a force can be expressed relative to a horizontal axis (2D analysis like above), the horizontal component is equal to the resultant times the cosine of the angle θ. The vertical component is equal to the resultant times the sine of the angle θ. Consequently, how the angle of force application affects the size of the components is equal to the shape of a sine or cosine wave. A qualitative understanding of these functions helps one understand where the largest changes occur and what angles of force application are best. Let's look at a horizontal component of a force in two dimensions. This is analogous to our iliopsoas example or how any force applied to an object will favor the horizontal over the vertical component? A cosine function is not a linear function like our spring example in ▸ Chap. 2. ◻ Figure 6.10 plots the size of the cosine function as a percentage of the resultant for angles of pull from 0 to 90°.

A 0° (horizontal) angle of pull has no vertical component, so all the force is in the horizontal direction. Note that, as the angle of pull begins to rise (0 to 30°), the cosine or horizontal component drops very slowly, so most of the resultant force is still directed horizontally. Now the cosine function begins to change more rapidly, and from 30 to 60° the horizontal component has dropped from 87 to 50% the size of the resultant force. For angles of pull greater than 60°, the cosine drops off fast, so there is a dramatic decrease in the horizontal component of the force, with the horizontal

6

component becoming 0 when the force is acting at 90° (vertical). We will see that the sine and cosine relationships are useful in angular kinetics as well. These curves allow for calculation of several variables related to angular kinetics from linear measurements. Right-angle trigonometry is also quite useful in studying the forces between two objects in contact or precise kinematic calculations.

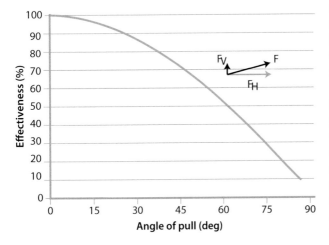

■ **Fig. 6.10** Graph of the cosine of angle 8 between 0 and 90° (measured from the right horizontal) shows the percentage effectiveness of a force (**F**) in the horizontal direction (**F**$_H$). This horizontal component is equal to **F** cos(8), and angle 8 determines the tradeoff between the size of the horizontal and vertical components. Note that the horizontal component stays large (high percentage of the resultant) for the first 30° but then rapidly decreases. The sine and cosine curves are the important nonlinear mathematical functions that map linear biomechanical variables to angular

6.5 Contact Forces

The linear kinetics of the interaction of two objects in contact is also analyzed by resolving the forces of contact into right-angle components. These components often use a local frame of reference like the two-dimensional muscle angle of pull above, because using horizontal and vertical components are not always convenient (■ Fig. 6.11b). The forces between two objects in contact are resolved into the **normal reaction** and **friction**. The normal reaction is the force at right angles to the surfaces in contact, while friction is the force acting in parallel to the surfaces. Friction is the force resisting the sliding of the surfaces past each other, while normal force presses the two surfaces together.

When the two surfaces are dry, the force of friction (**F**) is equal to the product of the coefficient of friction (μ) and the normal reaction (**F**$_N$), or **F** = μ• **F**$_N$. This is Coulomb's Law of Friction. The coefficient of friction depends on the texture and nature of the two surfaces and is determined by experimental testing. There are coefficients of static (non-moving) friction (μ$_S$) and kinetic (sliding) friction (μ$_K$). The coefficients of kinetic friction are typically 25% smaller than the maximum static friction. It is easier to keep an object sliding over a surface than to stop it and start the object sliding again. Conversely, if you want friction to stop motion, preventing sliding (like with anti-lock auto brakes) is a good strategy. ■ Figure 6.12 illustrates the friction force between an athletic shoe and a force platform as a horizontal force is increased. Please note that the friction grows in a linear fashion until the limiting friction is reached (μ$_S$ • **F**$_N$), at which point the shoe begins to slide across

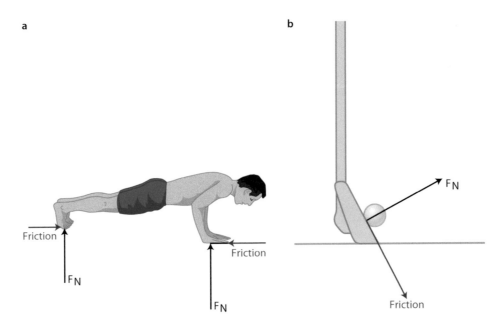

■ **Fig. 6.11** Forces of contact between objects are usually resolved into the right-angle components of *normal reaction* (**F**$_N$) and *friction*

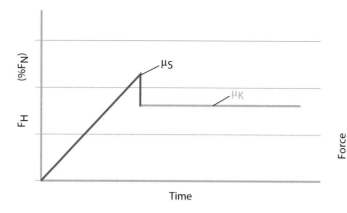

Fig. 6.12 The change in friction force between an athletic shoe and a force platform as a horizontal force is applied to the shoe. The ratio of the friction force ($\mathbf{F}_H$) on this graph to the normal force between the shoe and force platform determine the coefficient of friction for these two surfaces

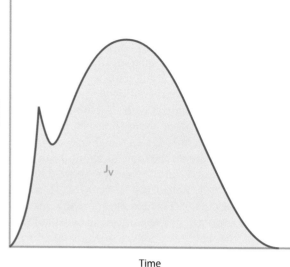

Fig. 6.13 The vertical impulse ($\mathbf{J}_v$) of the vertical ground reaction force for a foot strike in running is the area under the force–time graph

the force platform. If the weight on the shoe created a normal reaction of 300 N, what would you estimate the μ_S of this rubber/metal interface?

Typical coefficients of friction in human movement vary widely. Athletic shoes have coefficients of static friction that range from 0.4 to over 1.0 depending on the shoe and sport surface. In tennis, for example, the linear and angular coefficients of friction range from 0.4 to over 2.0, with shoes responding differently to various courts (Nigg et al. 1989). Epidemiological studies have shown that playing on lower-friction courts (clay) had a lower risk of injury (Nigg et al. 1989). Many teams that play on artificial turf use flat shoes rather than spikes because they believe the lower friction decreases the risk of severe injury. The sliding friction between ice and a speed skating blade has been measured, demonstrating coefficients of kinetic friction around 0.005 (van Ingen Schenau et al. 1989), and ski/snow kinetic friction is slightly larger (0.03-0.07).

6.6 Impulse–Momentum Relationship

Human movement occurs over time, so many biomechanical analyses are based on movement-relevant time intervals. For example, walking has a standardized gait cycle (Whittle 2001), and many sport movements are broken up into phases (usually, preparatory, action, and follow-through). The kinetic variables that are most relevant and used in these kinds of analyses are impulse (**J**) and momentum (**p**). These two variables are related to each other in the original language of Newton's second law: the change in momentum of an object is equal to the impulse of the resultant force in that direction. The impulse–momentum relationship is Newton's second law written over a time interval, rather than the instantaneous (**F** = m**a**) version.

Impulse is the effect of force acting over time. Impulse (**J**) is algebraically calculated as the product of force and time (**J** = **F** · t), so the typical units are N•s and lb•s. Impulse can be visualized as the area under a force–time graph (calculus is used to get this time-integral of force). The vertical ground reaction force during a foot strike in running can be measured using a force platform, and the area under the graph (integral with respect to time) represents the vertical impulse (Fig. 6.13). A person can increase the motion of an object by applying a greater impulse, and both the size of the force and duration of force application are equally important. Impulse is the mechanical variable discussed in the following section on the **Force–Time Principle**. In movement, the momentum a person can generate, or dissipate in another object, is dependent on how much force can be applied *and* the amount of time the force is applied.

Newton realized that the mass of an object affects its response to changes in motion. Momentum is the vector quantity that Newton said describes the quantity of motion of an object. Momentum (**p**) is calculated as the product of mass and velocity (**p** = m · **v**). The SI unit for momentum is kg•m/s. Who would you rather accidentally run into in a soccer game at a 5 m/s closing velocity: a 70- or 90-kg opponent? We will return to this question and mathematically apply the impulse–momentum relationship later on in this chapter once we learn about a similar kinetic variable called kinetic energy.

6

The association between impulse (force exerted over time) and change in momentum (quantity of motion) is quite useful in gaining a deeper understanding of all human movements (daily living, sports, exercise, rehabilitation). For example, many impacts create very large forces because the time interval of many elastic collisions is so short. For a golf ball to change from zero momentum to a considerable momentum over the 0.0005 seconds of impact with the club requires a peak force on the golf ball of about 10,000 N, or greater than 2200 pounds (Daish 1972). In a high-speed soccer kick, the ball is actually on the foot for about 0.016 seconds, so that peak forces on the foot are above 230 pounds (Tsaousidis and Zatsiorsky 1996; Tol et al. 2002; Nunome et al. 2006). Fortunately, for many catching activities in sport an athlete can spread out the force applied to the ball over longer periods of time. The Impulse–Momentum Relationship is the mechanical law that underlies the **force–time principle** introduced earlier in ▶ Chaps. 2 and 4. Let's revisit the application of the **force–time principle** with our better understanding of linear kinetics.

Interdisciplinary Issue: Acute and Overuse Injuries
An important area of research by many kinesiology and sports medicine scholars is related to musculoskeletal injuries. Injuries can be subclassified into acute injuries or overuse injuries. Acute injuries are single traumatic events, like a sprained ankle or breaking a bone in a fall. In an acute injury, the forces create tissue loads that exceed the ultimate strength of the biological tissues and cause severe physical disruption. The mechanical strength of human tissues is understood from testing of tissue samples. On rare occasions, knowledge of the causes of acute injury is enhanced when participants in lab or sport competition biomechanics studies are injured (e.g., Zernicke et al. 1977; Fong et al. 2009; Kong 2009).

Overuse injuries develop over time (thus, chronic) from a repetitive motion, loading, inadequate rest, or a combination of the three. Injuries from repetitive vocational movements or work-related musculoskeletal disorders (WMSDs) are examples of chronic injuries (Barr and Barbe 2002). Stress fractures and anterior tibial stress syndrome (shin splits) are classic examples of overuse injuries associated with running. Runners who overtrain, run on very hard surfaces, and are susceptible can gradually develop these conditions. If overuse injuries are untreated, they can develop into more serious disorders and injuries. For example, muscle overuse can sometimes cause inflammation of tendons (tendinitis), but if the condition is left untreated degenerative

---

changes begin to occur in the tissue that are called tendinoses (Khan et al. 2000). Severe overuse of the wrist extensors during one-handed backhands irritates the common extensor tendon attaching at the lateral epicondyle, often resulting in "tennis elbow."

The etiology (origin) of overuse injuries is a complex phenomenon that requires interdisciplinary research. The peak force or acceleration (shock) of movements is often studied in activities at risk of acute injury. It is less clear if peak forces or total impulse are more related to the development of overuse injuries. ◘ Figure 6.14 illustrates the typical vertical ground reaction forces measured with a force platform in running, step aerobics, and walking. Note that the vertical forces are normalized to units of bodyweight. Notice that step aerobics has peak forces near 1.8 BW because of the longer time of force application and the lower intensity of movement. Typical vertical ground reaction forces in step aerobics look very much like the forces in walking (peak forces of 1.2 BW and lower in double support) but tend to be a bit larger because of the greater vertical motion. The peak forces in running typically are about 3 BW because of the short amount of time the foot is on the ground. Do you think the vertical impulses of running and step aerobics are similar? Landing from large heights and the speed involved in gymnastics are very close to injury-producing loads. Note the high peak force and rate of force development (slope of the F–t curve) in the running ground reaction force. Gymnastic coaches should limit the number of landings during practice and utilize thick mats or landing pits filled with foam rubber to reduce the risk of injury in training because the rate of loading and peak forces are much higher (8 BW or more) than running.

6.7 Force–Time Principle

The applied manifestation of Newton's Second Law of Motion as the Impulse–Momentum Relationship is the **force–time principle**. The **force–time principle** states that the time available for force application is as important as the size of the forces used to create or modify movement. If a person can apply force over a longer period of time (large impulse), they will be able to achieve a greater speed (change in momentum) than if they used similar forces in a shorter time interval. Unfortunately, in many human movements there is not an unlimited amount of time to apply forces, and there are several muscle mechanical characteristics that complicate application of this principle. Force–time principle is concerned with

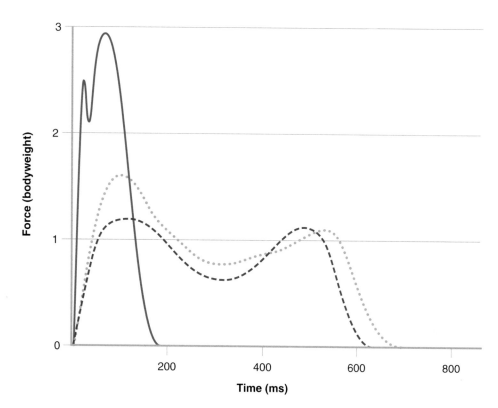

■ **Fig. 6.14** Typical vertical ground reaction forces (in units of bodyweight) for running (solid), walking (dashed), and step aerobic exercise (dotted)

the temporal strategy of force application in movements, while the force–time relationship (electromechanical delay) discussed in ► Chap. 4 noted how muscle tension build-up takes time. The electromechanical delay is clearly related to how a person selects the appropriate timing of force application. The force–time principle will be illustrated in using forces to slow down an external object, as well as to move, project, or strike an object effectively. The movement of interest, muscle characteristics, and the mechanical strengths of tissues all affect optimal application of forces to create motion.

Movers can apply forces in the opposite direction of the motion of an object to gradually slow down the object. Movements like catching a ball or landing from a jump (■ Fig. 6.15) employ primarily eccentric muscle actions to gradually slow down a mass over some period of time. Positioning the body to intercept the object early allows the mover to maximize the time the object can be slowed down. How does a gymnast maximize the time of force application to cushion the landing from a dismount from a high apparatus? Near complete extension of the lower extremities at touchdown on the mat allows near maximal joint range of motion to flex the joints and more time to bring the body to a stop.

The primary biomechanical advantage of this longer time of force application is safety, because the peak force experienced by the body (and consequently the stress in tissues) will be lower than during a short application of force. Moving the body and reaching with the extremities to maximize the time of catching also has

■ **Fig. 6.15** The extension of the limbs before contact and increasing flexion in landing increases the time that forces can be applied to slow down the body. In this chapter we will see that this decreases the peak force on the body and decreases the risk of injury

6

strategic advantages in many sports. A team handball player intercepting the ball early by extending their arms out from the body in a catch not only has a higher chance of a successful catch, but they may prevent an opponent from intercepting the ball.

Imagine you are a track coach whose observations of a discus thrower indicate they are rushing their motion across the ring. The force–motion principle and the force–velocity relationship make you think that slowing the increase in speed on the turns and motion across the circle might allow for longer throws. There is likely a limit to the benefit of increasing the time to apply because maximizing discus speed in an appropriate angle at release is the objective of the event. Are there timing data for elite discus throwers available to help with this athlete, or is there a little art in the application of this principle? Are you aware of other sports or activities where coaches focus on a controlled build-up in speed or unrushed rhythm?

There are sometimes limits to the benefit of increasing the time of force application. In movements with high demands on timing accuracy (baseball batting or a tennis forehand), the athlete should not maximize the time of force application because extra speed is of lower importance than temporal accuracy. If a tennis player preferred a large loop backswing where they used a large amount of time and the force of gravity to create racket head speed, the player will be vulnerable to fast and unpredictable strokes from an opponent. The wise opponent would mix up shot placements, spin, and increase time pressure to make it difficult for the player to get their long stroke in.

Suppose a patient rehabilitating from surgery is having difficulty using even the smallest weights in the clinic. How could a therapist use the force–time principle to provide a therapeutic muscular overload? If bodyweight or assistive devices were available, could the therapist have the patient progressively increase the time they isometrically activate muscles rather than lift/move weights? While this approach would tend to benefit muscular endurance more so than muscular strength, these two variables are related and tend to improve the other. Increasing the time of muscle activity in isometric actions or by modifying the cadence of dynamic exercises is a common training adaptation used in rehabilitation and strength training.

There are a few movements that do allow movers to nearly maximize the time of force application to safely slow down an object. In landing from a jump, the legs are extended at contact with the ground, so there is near maximal joint range of motion to flex the joints and absorb impact forces. A softball infielder is taught to lean forward and extend her glove hand to field a ground ball so that she can absorb the force of the ball over a longer

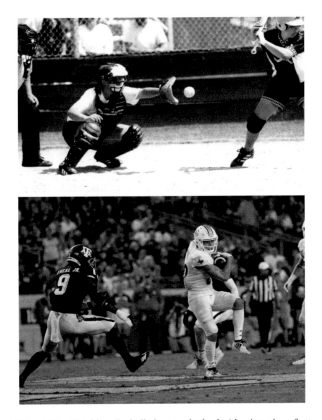

■ **Fig. 6.16**   Catching the ball close to the body (the American football example) is a poor application of the force–time principle because there is minimal time or range of motion to slow down the ball. Photo used by permission of Texas State University Department of Athletics. The softball catcher has increased the time and range of motion that can be used to slow down the ball

time interval. ■ Figure 6.16 illustrates two people catching balls: which athlete is using a technique that is correctly applying the force–time principle? Young children often catch by trapping the object against the body and even turn their heads in fear. Even skilled football players (■ Fig. 6.16) occasionally rely on their talent or sense of self-preservation more than coaching and use a similar catching technique. For how much time can forces be applied to slow the balls in these cases? The momentum of the ball in these situations is often so great that the force between the person's body and the ball builds up so fast that the ball bounces out of their grasp. If these people extended their arms and hands to the ball, the time the force is applied to slow down the ball could be more than ten times longer. Not only does this increase the chance of catching the ball, but it decreases the peak force and potential discomfort involved in catching.

Athletes taught to reach for the ground and "give" with ankle, hip, and knee flexion dramatically increase the time of force application in landing and decrease the peak ground reaction forces. Exactly how the muscles are positioned and pre-tensed prior to landing affects which muscle groups are used to cushion landing (DeVita and Skelly 1992; Kovacs et al. 1999; Zhang et al. 2000). How to teach the important skill of landing has only recently become well researched with the linking of landing an cutting movements with ACL injuries.

The sound of an impact often tells an athlete about the severity of a collision, so this has been used as a teaching point in catching and landing. It has also been shown that focusing attention on decreasing the sound of landing is an effective strategy to decrease peak forces during landing (McNair et al. 2000). Increasing the "give" of the cushioning limbs increases the time of force application and decreases the tone and intensity of the sound created by the collision.

> **Activity: Impulse–Momentum Relationship**
> Fill a few small balloons with water to roughly softball size. Throw the water balloon vertically and catch it. Throw the balloon several times trying to maximize the vertical height thrown. Imagine that the water balloon represents your body falling and the catching motions represent your leg actions in landing. What catching technique points modify the force and time of force application to the balloon to create a vertical impulse to safely (without breaking) reduce the momentum of the balloon to zero?

In some movements there are other biomechanical factors involved in the activity that limit the amount of time that force can be applied. In these activities, increasing time of force application would decrease performance, so the only way to increase the impulse is to rapidly create force during the limited time available. A good example of this is long jumping. Recall that in the kinematics chapter we learned that long jumpers have low takeoff angles (approximately 20°). The takeoff foot is usually on the board for only 100 ms, so there is little time to create vertical velocity. Skilled long jumpers train their neuromuscular system to strongly activate the leg muscles prior to foot strike. This allows the jumper to rapidly increase ground reaction forces so they can generate vertical velocity without losing too much horizontal velocity. Similar temporal limitations are at work in running or throwing. In many sports where players must throw the ball quickly to score or prevent an opponent scoring, the player may make a quicker throw than they would during maximal effort without time restrictions. A quick delivery may not use maximal throwing speed or the extra time it takes to create that speed, but it meets the objective of that situation. Kinesiology professionals need to instruct movers as to when using more time of force application will result in safer and more effective movement and when the use of longer force application is not the best movement strategy.

> **Application: Force–Time and Range-of-Motion Principle Interaction**
> Human movements are quite complex, and several biomechanical principles often apply. This is a challenge for the kinesiology professional, who must determine how the task, performer characteristics, and biomechanical principles interact. In the sport of weight lifting, coaches know that the initial pull on the bar in snatch and clean-and-jerk lifts cannot be maximal until the bar reaches about knee height. This appears counter to the force–time principle, where maximizing initial force over the time of the lift seems to be advantageous. It turns out that there are ranges of motion (postural) and muscle mechanical issues that are more important than a rigid application of the force–time principle. The whole-body muscular strength curve for this lift is maximal near knee level, so a fast bar speed at the strongest body position compromises the lift because of the decrease in muscle forces in increasing speed of concentric shortening (Garhammer 1989; Zatsiorsky 1998). In other words, there is a force–motion advantage of a strategy emphasizing athlete focus on maximal effort as the bar reaches knee level. Olympic lifts require a great deal of practice and skill. The combination of speed and force, as well as the motor skills involved in Olympic lifting, makes these whole-body movements popular conditioning exercises for dynamic sports requiring combinations of high external force and speed (Garhammer 1989).

6.8 Work–Energy Relationship

The final approach to studying the kinetics of motion involves laws from a branch of physics dealing with the scalar concepts of work, energy, and power. Since much of the energy in the human body, machines, and on the earth are in the form of heat, these laws are used in thermodynamics to study the flow of heat energy. Biomechanists are interested in how average mechanical energy flows inform the production of steady state movements or transfer of energy between bodies. These methods are inferior and incomplete ways to study the kinetics of human motion compared to the vector and causative expressions of Newton's Laws (Vigotsky et al. 2019); nevertheless, they are sometimes useful in analyzing steady-state motion.

6.8.1 Mechanical Energy

In mechanics, **energy** is the capacity to do work. In the movement of everyday objects, energy can be viewed as the mover of stuff (matter), even though at the atomic level matter and energy are more closely related. Energy is measured in Joules (J) and is a scalar quantity. One Joule of energy equals 0.74 ft·lbs. Energy is a scalar because it represents an ability to do work that can be transferred in any direction. Energy can take many forms (e.g., heat, chemical, nuclear, motion, or position). There are three mechanical energies that are due to an object's motion or position in a gravitational field.

The energies of motion are linear and angular **kinetic energy**. Linear or translational kinetic energy can be calculated using the following formula: $KE_T = \frac{1}{2}mv^2$. There are several important features of this formula. First, note that squaring velocity makes the energy of motion primarily dependent on the velocity of the object. The energy of motion varies with the square of the velocity, so doubling velocity increases the kinetic energy by a factor of 4 (2^2). Squaring velocity also eliminates the effect of the sign (+ or −) or vector nature of the velocity term. Angular or rotational kinetic energy can be calculated with a similar formula: $KE_R = \frac{1}{2}I\omega^2$. We will learn more about angular kinetics in ▶ Chap. 7.

The mathematics of kinetic energy ($\frac{1}{2}mv^2$) looks surprisingly similar to momentum (mv). However, there are major differences in these two quantities. First, momentum is a vector quantity describing the quantity of motion in a particular direction. Second, kinetic energy is a scalar that describes how much work an object in motion could perform on another object. The variable momentum is used to document the current state of motion, while kinetic energy describes the potential for future interactions. Let's consider a numerical example from American football. Imagine you are a small (80 kg) halfback spinning off a tackle with one yard to go for a touchdown. Who would you rather run into just before the goal line: a quickly moving defensive back or a very large lineman not moving as fast? ◻ Figure 6.17 illustrates the differences between kinetic energy and momentum in an inelastic collision. Applying the impulse–momentum relationship is interesting because this will tell us about the state of motion or whether a touchdown will be scored. Notice that both defenders (small and big) have the same amount of momentum (−560 kg•m/s), but because the big defender has greater mass you will not fly backwards as fast as in the collision with the defensive back. The impulse–momentum relationship shows that you do not score either way (negative velocity after impact: V_2), but the defensive back collision looks very dramatic because you reverse direc-

tions with a faster negative velocity. The work–energy relationship tells us that the total mechanical energy of the collision will be equal to the work the defender can do on you. Some of this energy is transferred into sound and heat, but most of it will be transferred into deformation of your pads and body! Note that the sum of the energies of the two athletes and the strong dependence of kinetic energy on velocity results in nearly twice (2240 versus 1280 J) as much energy in the collision with the defensive back. In short, the defensive back hurts the most because it is a very high-energy collision, potentially adding injury to the insult of not scoring.

There are two types of mechanical energy that objects have because of their position or shape. One is gravitational potential energy and the other is strain energy. Gravitational **potential energy** is the energy of the mass of an object by virtue of its position relative to the surface of the earth. Potential energy can be easily calculated with the formula: $PE = mgh$. Potential energy depends on the mass of the object, the acceleration due to gravity, and the height of the object. Raising an object with a mass of 35 kg a meter above the ground stores 343 J of energy in it ($PE = 35 • 9.81 • 1 = 343$). If this object were to be released, the potential energy would gradually be converted to kinetic energy as gravity accelerated the object toward the earth. This simple example of transfer of mechanical energies is an example of one of the most important laws of physics: the **Law of Conservation of Energy**.

The **Law of Conservation of Energy** states that energy cannot be created or destroyed; it is just transferred from one form to another. The kinetic energy of a tossed ball will be converted to potential energy or possibly strain energy when it collides with another object. A tumbler taking off from a mat has kinetic energy in the vertical direction that is converted into potential energy on the way up and back into kinetic energy on the way down. A bowler who increases the potential energy of the ball during the approach can convert this energy to kinetic energy prior to release (◻ Fig. 6.18). In a similar manner, in golf or tennis a forward swing can convert the potential energy from preparatory movement into kinetic energy. A major application area of conservation of energy is the study of heat or thermodynamics.

The First Law of Thermodynamics is the law of conservation of energy. This is the good news: when energy is added into a machine, we get an equal amount of other forms of energy out. Unlike these examples, examination of the next mechanical energy (strain energy) will illustrate the bad news of the Second Law of Thermodynamics: that it is impossible to create a machine that converts all input energy into some useful

□ **Fig. 6.17** Comparison of the kinetic energy (scalar) and momentum (vector) in a football collision. If you were the running back, you would not score a touchdown against either defender, but the work done on your body would be greater in colliding with the smaller defender because of their greater kinetic energy than the more massive defender

Kinetic Energy

$$KE_{Back} = \frac{1}{2}\,80(2)^2 = 160\ J$$

$$KE_{Small} = \frac{1}{2}\,70(-8)^2 = 2240\ J$$

$$KE_{Big} = \frac{1}{2}\,140(-4)^2 = 1120\ J$$

$$Total_{Small} = 2400\ J$$
$$Total_{Big} = 1280\ J$$

Conservation of Momentum

$$m_1V_1 + m_2V_2 = m_1V_2 + m_2V_2$$

$$Small\ 80(2) - 70(8) = (70 + 80)\,V_2$$

$$V_2 = -2.7\ m/s$$

$$Big\ 80(2) - 140(4) = (80 + 140)\,V_2$$

$$V_2 = -1.8\ m/s$$

output energy. In other words, man-made devices will always lose energy in some non-useful form and never achieve 100% efficiency. This is similar to the energy losses (hysteresis) in strain energy stored in deformed biological tissues studied in ▶ Chap. 4.

Strain energy is the energy stored in an object when an external force deforms that object. Strain energy can be viewed as a form of potential energy. A pole vaulter stores strain energy in the pole when loading the pole by planting it in the box. Much of the kinetic energy stored in the vaulter's body during the run up is converted into strain energy and back into kinetic energy in the vertical direction. Unfortunately, again, not all the strain energy stored in objects is recovered as useful energy. Often large percentages of energy are converted to other kinds of energy that are not effective in terms of producing movement. Some strain energy stored in many objects is essentially lost because it is converted into sound waves or heat. Some machines employ heat production to do work, but in human movement, heat is a by-product of

many energy transformations that must be dissipated. Heat is often even more costly than the mechanical energy in human movement because the cardiovascular system must expend more chemical energy to dissipate the heat created by vigorous muscular action.

The mechanical properties of an object determine how much of any strain energy is recovered in restitution as useful work. Recall that many biomechanical tissues are viscoelastic and that the variable hysteresis (area between the loading and unloading force–displacement curves) determines the amount of energy lost to unproductive energies like heat. The elasticity of a material is defined as its stiffness. In many sports involving elastic collisions, a simpler variable can be used to get an estimate of the elasticity or energy losses of an object *relative to another object*. This variable is called the **coefficient of restitution** (COR and e are common abbreviations). The COR is a dimensionless number usually ranging from 0 (perfectly plastic collision: mud on your mother's kitchen floor) to near 1 (very elastic

6

◘ **Fig. 6.18** Raising a bowling ball in the approach stores more potential energy in the ball than the kinetic energy from the approach. The potential energy of the ball can be converted to kinetic energy in the downswing

3 m/s

$$KE_B = \tfrac{1}{2}(6.4)(3)^2 = 28.8 \text{ J}$$

$$PE_B = 6.4(9.81)(1.5) = 92.4 \text{ J}$$

1.5 m

pairs of materials). The coefficient of restitution cannot be equal to or greater than 1 because of the second law of thermodynamics. High COR represent elastic collisions with little wasted energy, while lower coefficients of restitution do not recover useful work from the strain energy stored in an object.

The COR can be calculated as the relative velocity of separation divided by the relative velocity of approach of the two objects during a collision (Hatze 1993). The most common use of the COR is in defining the relative elasticities of balls used in sports. Most sports have strict rules governing the dimensions, size, and specifications, including the ball and playing surfaces. Officials in basketball or tennis drop balls from a standard height and expect the ball to rebound to within a small specified range allowed by the rules. In these uniformly accelerated flight and impact conditions where the ground essentially doesn't move, e can be calculated with this formula: e = (bounce/drop)$^{1/2}$. If a tennis ball were dropped from a 1-meter height and it rebounded to 58 cm from a concrete surface, the coefficient of restitution would be $(58/100)^{1/2} = 0.76$. Dropping the same tennis ball on a short pile carpet might result in a 45-cm rebound, for an e = 0.67. The COR for a sport ball varies depending on the nature of the other object or surface it interacts with (Cross 2000), the velocity of the collision, and other factors like temperature. Squash players know that it takes a few rallies to warm up the ball and increase its COR. Also, putting softball or baseballs in a refrigerator will take some of the slugging percentage out of a

strong hitting team. COR is a useful variable for sports balls and implements because their relative elasticity is easier to test than mounting it into an expensive materials testing machine at a university.

Most research on the COR of sport balls has focused on the elasticity of a ball in the vertical direction, although there is a COR in the horizontal direction that affects friction and the change in horizontal ball velocity for oblique impacts (Cross 2002). The horizontal COR strongly affects the spin created on the ball following impact. This is a complicated phenomenon because balls deform and can slide or rotate on a surface during impact. How spin, in general, affects the bounce of sport balls will be briefly discussed in the section on the spin principle in ▶ Chap. 8.

6.8.2 Mechanical Work

All along we have been defining mechanical energies as the ability to do mechanical work. Now we must define **work** and understand that this mechanical variable is not the same as most people's common perception of work as some kind of effort. The mechanical work done on an object is defined as the product of the resultant force and displacement in the direction of the force (W = **F** • **d**). Joules are the units of work: one joule of work is equal to one Nm. In the English system, the units of work are usually written as foot-pounds (ft•lb) to avoid confusion with the angular kinetic variable torque, whose unit is the lb•ft. Note that the algebraic formula assumes the force applied to the load is constant over the duration of the movement. The person in ◘ Fig. 6.19 does mechanical work to increase the potential energy of the weight stack. If a force is applied to an object and there is no movement (**F** below the object's inertia or inadequate to overcome friction, etc.), then there is no work done. Work is a very interesting variable because it requires motion (and consequently time), but the magnitude of this scalar quantity is not related to the timing of the force application! Calculus is necessary to calculate the work of the true time-varying forces applied to objects they move.

Since mechanical work can only be done on an object when it is moved relative to the line of action of a resultant force. A more complete algebraic definition of mechanical work in the horizontal (x) direction that takes into account the component of motion in the direction of the force on an object would be $W = (F \cos \theta) \cdot d_x$. For example, a person pulling a load horizontally on a dolly given the data in ◘ Fig. 6.20 would do

435 Nm or Joules of work. Only the horizontal component of the force times the displacement of object determines the work done. Note also that the angle of pull in this example is like the muscle angle of pull analyzed earlier. The smaller the angle of pull, the greater the horizontal component of the force that does work to move the load. The vertical component of pull does not do any mechanical work, although it may decrease the weight of the dolly or load and, thereby decrease the rolling friction to be overcome. What is the best angle to pull in this situation depends on many factors. Factoring in rolling friction and the strength (force) ability in various pulling postures might indicate that a higher angle of pull that doesn't maximize the horizontal force component may be "biomechanically" effective for this person. The inertia of the load, the friction under the person's feet, and the biomechanical factors of pulling from different postures all interact to determine the optimal angle for pulling an object. In fact, in some closed kinematic chain movements (like cycling) the optimal direction of force application does not always maximize the

◘ **Fig. 6.19** The mechanical work done on a weight in this rowing exercise is the product of the resultant force and the displacement

70 N

0.5 m

◘ **Fig. 6.20** Mechanical work is calculated as displacement of the object in the direction of the force (resultant assuming no rolling friction). This calculation is accurate if the 80-N force is constant during horizontal displacement of the dolly. If you were pulling this dolly, what angle of pull would you use?

F = 80 N

$W = 80 (\cos 25°) 6 = 435$ Nm

$\theta = 25°$

F cos 25°

d = 6.0 m

6

effectiveness or the component of force in the direction of motion (Doorenbosch et al. 1997).

Despite energy and work being a scalar quantities (no need for sign designating direction), the vectors used in their calculation make for tempting use of these signs. This is common in many kinesiology/exercise science journals so it will be discussed now, but remember this addition of signs/directions to these variables while intuitively appealing for application are not strictly correct and can lead to confusion (Knudson 2009; Winter et al. 2016; Vigotsky et al. 2019). A patient performing rowing exercises (◘ Fig. 6.19) could be considered to perform "positive work" (W = 70 • 0.5 = +35 Nm or Joules) on the weight against gravity. In essence, energy flows from the patient to the weights (increasing their potential energy) in the concentric phase of the exercise. In the eccentric phase of the exercise, the work is apparently "negative," meaning that potential energy is being transferred from the load to the patient's body. This example also assumes that the energy losses in the pulleys are negligible as they change the direction of the force created by the patient. We know from ▶ Chap. 4 that this is really controlled by the forces created by eccentric muscle actions, not gravity ramping up resistance to overcome the patient.

Mechanical work does not directly correspond to people's sense of muscular effort. Isometric actions, while many times taking considerable effort or intensity, do not perform mechanical work. This dependence on the object's displacement of mechanical work makes the work–energy relationship useful in biomechanical studies where the motion of an object may be of more interest than temporal factors.

This brings us to the **Work–Energy Relationship**, which states that the mechanical work done on an object is equal to the change in mechanical energy of that object. Biomechanical studies have used the work–energy relationship to study the average kinetics of movements over time. One approach calculates the changes in mechanical energies of the segments to calculate work, while the other calculates mechanical power and integrates these data with respect to time to calculate changes in work and energy. The next section will discuss the concept of mechanical power flows.

6.8.3   Mechanical Power

Mechanical power is an important kinetic variable for analyzing many human movements because it incorporates time into work and energy calculations. Power is defined as the rate of doing work, so mechanical power is the time derivative of mechanical work or work divided by time (P = W/t). Note that a capital "P" is used because lower-case "p" is the SI symbol for momentum. Typical units of power are Watts (one J/s) and horsepower. One horsepower is equal to 746 W. Maximal mechanical power flows are achieved by the right combination of force and velocity that maximizes the mechanical work done on an object. This is clear from the other formula derived from W/t for calculating power: P = **F** • **v**. Prove to yourself that the two equations for power are the same by substituting the formula for work W and do some rearranging that will allow you to substitute **v** for its mechanical definition.

If the concentric lift illustrated in ◘ Fig. 6.19 was performed within 1.5 seconds, we could calculate the average power flow to the weights. The "positive" work done on the weights was equal to 35 J, so P = W/t = 35/1.5 = 23.3 W. Recall that these algebraic definitions of work and power calculate a mean value over a time interval for constant forces. The peak instantaneous power flow to the weight in ◘ Fig. 6.19 would be higher than the average power calculated over the whole concentric phase of the lift. The force–motion principle would say that the patient increased the vertical force on the resistance to more than the weight of the stack to positively accelerate it and would reduce this force to below the weight of the stack to gradually stop the weight at the end of the concentric phase. Instantaneous power flow to the weights also follows a complex pattern based on a combination of the force applied and the motion of the object in the direction of the force that drives the motion.

What movements do you think require greater peak mechanical power delivered to a barbell: the lifts in the sport of Olympic weight lifting or power lifting? Don't let the names fool you. Since power is the rate of doing work, the movements with the greatest mechanical power must have high forces and high movement speeds in the same direction. Olympic lifting has mechanical power outputs much higher than power lifting and power lifting is clearly a misnomer given the true definition of power. The dead lift, squat, and bench press in power lifting are high-strength movements with large external resistances but very slow velocities. The faster movements of Olympic lifting, along with somewhat smaller weights, clearly create a greater power flow to the bar than power lifting. Peak power flows to the bar in power lifting are between 370 and 900 W (0.5–1.2 hp), while the peak power flow to a bar in Olympic lifts is often as great as 4000 W or 5.4 hp (see Garhammer 1989). Olympic lifts are often used to train for dynamic sport movements and Olympic weight lifters can create significantly more whole-body average and peak mechanical external power flows than other athletes (McBride et al. 1999).

Many people have been interested in the peak and average mechanical power flows of whole-body and multi-segment movements. It is believed that higher external power flows are critical for quick, maximal anaerobic movements. In the coaching and kinesiology literature, these movements have been described as "explosive." This terminology may communicate the point of high rates of force development, intensity, and high levels of both speed and force, but a literal interpretation of this jargon is not too appealing! Remember that the mechanical power output calculated for a human movement will strongly depend on the model (point mass, linked segment, etc.) used and the time interval used in the calculation. The average or instantaneous power flows within the body, and the external flows from the body to external objects are quite different. In addition, other biomechanical factors affect how much mechanical power is developed during movements. Having a high peak power is useless, if the rest of the time needed in the movement, there is little mechanical work being done (low power flow).

The development of maximal power output in human movement depends on the direction of the movement, the number of segments used, and the inertia of the object. If we're talking about a simple movement with a large resistance, the right mix of force and velocity may be close to 30 to 45% of maximal isometric strength because of the force–velocity relationship (Kaneko et al. 1983; Wilson et al. 1993; Izquierdo et al. 1999). ◻ Figure 6.21 shows the in vitro concentric power output of skeletal muscle derived from the product of force and velocity in the force–velocity relationship. In movements requiring multi-joint movements, specialized dynamometer measurements indicate that the best resistances for peak power production and training are likely to be higher than the 30–45% and differ between the upper and lower extremities (Funato et al. 1996, 2000; Newton et al. 1996). The best conditioning for "explosive" movements may be the use of moderate resistances (just less than strength levels that are usually >70% 1RM), which are moved as quickly as possible. Oftentimes these exercises use special equipment like the Plyometric Power System, which allows for the resistance to be thrown (Wilson et al. 1993). The disadvantage of high-speed exercise is that it focuses training on the early concentric phase, leaving much of the range of motion submaximally trained. Even slow, heavy weight training exercises have large submaximal percentages (24–52%) of range of motion due to negative acceleration of the bar at the end of the concentric phase (Elliott et al. 1989).

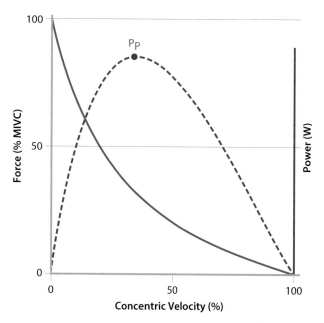

◻ **Fig. 6.21** Schematic of the in vitro mechanical power output of isolated skeletal muscle. Note that peak power (P_p) in concentric actions does not occur at either the extremes of force or velocity (about 1/3 of maximum shortening velocity and maximim isometric voluntary contraction [MIVC]).

Interdisciplinary Issue: Efficiency

One area of great potential for interdisciplinary cooperation is in determining the efficiency of movement. This efficiency of human movement is conceptually different from the classical definition of efficiency in physics. Physics defines efficiency as the mechanical work output divided by the mechanical work input in a system, a calculation that helps engineers evaluate machines and engines. For endurance sports like distance running, adjusting a formula to find the ratio of mechanical energy created to the metabolic cost of energy to create the movement appears to be an attractive way to study human "efficiency" (van Ingen Schenau and Cavanagh 1999). Progress in this area has been hampered by the wide variability of individual performance and confusion about the various factors that contribute to this movement efficiency (Cavanagh and Kram 1985). Cavanagh and Kram argued that the efficiency of running, for example, could be viewed as the sum of several efficiencies (e.g., biochemical, biomechanical, physiological, psychomotor) and other factors. Examples of the complexity of this area are the difficulty in defining baseline metabolic energy expenditure and calculating the true mechanical work because more work is done than is measured by ergometers. For instance, in cycle ergometry the mechanical work used to move the limbs is not measured. Biomechanists are

6

also struggling to deal with the zero-work paradox in movements where there no net mechanical work is done, like in cyclic activities, co-contracting muscles, or forces applied to the pedal in an ineffective direction. ◘ Figure 6.22 illustrates the typical forces applied to a bicycle pedal at 90° (from vertical). The normal component of the pedal force does mechanical work in rotating the pedal (F_N), while the other component does no work that is transferred to the bike's flywheel. Movement efficiency is an area where cooperative and interdisciplinary research may be of interest to many scientists and may be an effective tool for improving human movement.

There are several field tests to estimate short-term "explosive" leg power, but the utility and accuracy of these tests are controversial. The Margaria test (Margaria et al. 1966) estimates power flow to overcome gravity from running upstairs, and various vertical jump equations (see Johnson and Bahamonde 1996; Sayers et al. 1999) have been proposed that are based on the original Sargent (1921) vertical jump test. Companies now sell mats and apps that estimate the height and power of a vertical jump (from time and projectile equa-

tions). Although peak and average mechanical power output in such jumps is higher than high-force strength movements, these tests and devices are limited because the resistance is limited to body mass, the many factors that affect jump height, and the assumptions used in the calculation. There has been a long history of criticism of the assumptions and logic of using vertical jump height to estimate overall muscular power in short, impulsive activities like jumping (Adamson and Whitney 1971; Barlow 1971; Winter 2005; Knudson 2009; Winter et al. 2016). One has to question the logic of putting focus on an average or peak power estimate from jump height and athlete body mass for Newton's Laws (Impulse-Momentum Relationship) that directly determines jump performance. Instantaneous measurements of power from force platforms or kinematic analysis are more accurate but are expensive and time-consuming. Future studies will help determine the role of mechanical power in various movements, how to best train for these movements, and what field tests help coaches monitor athletes.

6.9 Segmental Interaction Principle

Human movement can be performed in a wide variety of ways because of the many kinematic **degrees of freedom** our linked segments provide. In ► Chap. 5 we saw that coordination of these kinematic chains ranges along a continuum from simultaneous to sequential joint rotations. Kinetics provides several ways in which to examine the potential causes of these coordination patterns. The two expressions of Newton's second law and the work–energy relationship have been employed in the study of the coordination of movement. This section proposes a Principle of Segmental Interaction that can be used to understand the origins of movement so that professionals can modify movement to improve performance and reduce risk of injury.

The **Segmental Interaction Principle** says that forces acting between the segments of a body can transfer force or energy between segments. The biomechanics literature has referred to this phenomenon in several ways (Putnam 1993). The contribution of body segments to movement has been called coordination of temporal impulses (Hochmuth and Marhold 1978), the kinetic link principle (Kreighbaum and Barthels 1996), summation of speed (Bunn 1972), summation or continuity of joint torques (Norman 1975), the sequential or proximal-to-distal sequencing of movement (Marshall and Elliott 2000), and the transfer of energy or transfer of momentum (Miller 1980; Lees and Barton 1996). The many names for this phenomenon and the three ways to document the kinematics and kinetics of motion are a good

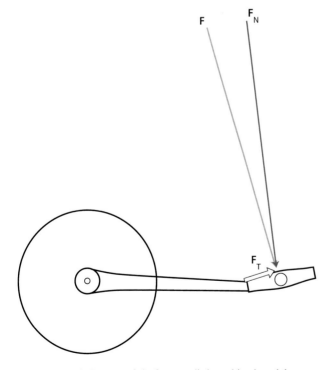

◘ **Fig. 6.22** Only some of the force applied to a bicycle pedal creates work and mechanical power. Note how the angle of the pedal illustrated means that a small component of F_T actually resists the normal force (F_N) creating rotation

indication of the difficulty of the problem and the controversial nature of the causes of human motion.

Application: Strength Versus Power

The force–velocity relationship and domains of strength discussed in ▶ Chap. 4, as well as this chapter's discussion of mechanical power should make it clear that muscular strength and power are not the same thing. Like the previous discussion on power lifting, the common/colloquial use of the term *power* is often inappropriate or inaccurate. Muscular strength is the expression of maximal tension in isometric or slow velocities of shortening. We have seen that peak power is the right combination of force and velocity that maximizes mechanical work. In cycling, the gears are adjusted to find this peak power point. If cadence (pedal cycles and, consequently, muscle velocity of shortening) is too high, muscular forces are low and peak power is not achieved. Similarly, power output can be submaximal if cadence is too slow and muscle forces high. The right mix of force and velocity seems to be between 30 and 70% of maximal isometric force and depends on the movement and environment. The best gearing of a bicycle (best mix of force and velocity) depends on the athlete, grade, and wind. There is no one universal peak external power condition for any movement, nor is training that maximizes some external power flow does not necessarily mean this is the best training force and velocity for a criterion sport movement of interest. Kinesiology professionals need to keep up with the growing research on the biomechanics of conditioning and sport movements. Future research will help refine our understanding of the nature of specific movements and the most appropriate exercise resistances and training programs.

Currently it is not possible to have definitive answers on the linear and angular kinetic causes for various coordination strategies because of complex multi-segment effects and limitations of models of biomechanical systems. This text has chosen to emphasize the forces transferred between segments as the primary kinetic mechanism for coordination of movement. Most electromyographic (EMG) research has shown that in sequential movements muscles are activated in short bursts that are timed to take advantage of the forces and geometry between adjacent segments (Feldman et al. 1998; Roberts 1991). This coordination of muscular kinetics to take advantage of "passive dynamics" or "motion-dependent" forces (gravitational, inertial forces) has been observed in the swing limb during walking (Mena et al. 1981), running (Phillips et al. 1983), kicking (Roberts 1991; Koike et al. 2019), throwing (Feltner 1989; Hirashima et al. 2002, 2008; Naito and Maruyama 2008), tennis (Koike and Harada 2014) limb motions toward targets (Galloway and Koshland 2002; Liu et al. 2013), and limb adjustments to unexpected obstacles (Eng et al. 1997). Passive and multi-segment dynamics is also a part of coordination of common movement like walking, reaching, and standing (Mena et al. 1981; Fox and Delp 2010; Liu et al. 2013; Caruthers et al. 2016) so researchers trying to design robots that effectively use this biomechanical control strategy (Fox and Delp 2010).

Some biomechanists have theorized that the segmental interaction that drives the sequential strategy is a transfer of force or energy from the proximal segment to the distal segment. This theory originated from observations of the close association between the negative acceleration of the proximal segment (see the activity on Segmental Interaction below) with the positive acceleration of the distal segment (Plagenhoef 1971; Roberts 1991). This mechanism is logically appealing because the larger force and energy potential of large muscle groups can be transferred distally and is consistent with the large forces and accelerations of small segments (that a not normally possible acting alone) late in baseball pitching (Feltner and Dapena 1986; Roberts 1991; Fleisig et al. 1995). ◘ Figure 6.23 illustrates a schematic

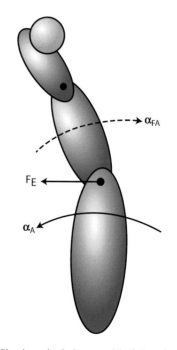

◘ **Fig. 6.23** Simple sagittal plane model of throwing illustrates the Segmental Interaction Principle. Joint forces ($\mathbf{F}_E$) from a slowing proximal segment create a segmental interaction to angularly accelerate the more distal segments (α_{FA}). This contributes to greater forearm, hand, and ball velocity to the right.

6

of throwing where the negative angular acceleration of the arm (α_A) creates a backward elbow joint force (F_E) that accelerates the forearm (α_{FA}). This view of the Segment Interaction Principle states that slowing the larger proximal segment will transfer force or energy to the distal segment. It is clear that this movement strategy is highly effective in creating high-speed movements of distal segments, but the exact mechanism of the segmental interaction principle is not clear.

When you get down to this level of kinetics, you often end up with a chicken-or-egg dilemma. In other words, which force/torque was created first and which is the reaction force/torque (Newton's third law)? There are some scholars who have derived equations that support the proximal-to-distal transfer of energy (Roberts 1991; Hong et al. 2000), while others conclude that the acceleration of the distal segment causes slowing of the proximal segment (Putnam 1991, 1993; Nunome et al. 2002, 2006; Sorensen et al. 1996). Whatever the ultimate, underlying mechanism or direction of force transfer, fast human movements utilize a sequential (proximal-to-distal) coordination that relies on the transfer of forces and energy between segments. We are truly fortunate to have so many muscles and degrees of freedom to create a wide variety and speeds of motion.

A good example of the controversy related to the Segmental Interaction Principle is the role of the hand and wrists in the golf swing. Skilled golf shots can be accurately modeled as a two-segment (arm and club) system with motion occurring in a diagonal plane. Golf pros call this the swing plane. Some pros say the golfer should actively drive the club with wrist action on the downswing, while others teach a relaxed or more passive wrist release. One simulation study found that correctly timed wrist torques could increase club head speed by 9% (Sprigings and Neal 2000), but the small percentage and timing of these active contributions suggests that proximal joint forces are the primary accelerator of the club. Jorgensen (1994) has provided simple qualitative demonstrations and convincing kinetic data that support the more relaxed use of wrist action and explain how weight shifts can be timed to accelerate the golf club. The delayed "uncocking" of grip angular kinematics of highly skilled golfers tends to support the conclusion that most of the force for high-speed golf downswings comes from proximal segments and passive dynamic, not active wrist action late in the swing.

It is clear that forces are transferred between segments to contribute to the motion of the kinematic chain (Zajac and Gordon 1989). The exact nature of that segmental interaction remains elusive and people vary in strength and skill, so kinesiology professionals can expect performers to have subtle variations (sequential to simultaneous) of combinations of joint motion.

It would be unwise to speculate too much on the muscular origins of that transfer. This view is consistent with the EMG and biomechanical modeling research reviewed in ► Chap. 2. So how can kinesiology professionals prescribe conditioning exercises and learning progressions so as to maximize the segmental interaction effect? Currently, there are few answers, but we can make a few tentative generalizations about conditioning and learning motor skills.

Physical conditioning for any human movement should clearly follow the training principle of *specificity*. Biomechanically, this means that the muscular actions and movements should emulate the movement as much as possible. Since the exact kinetic mechanism of segmental interaction is not clear, kinesiology professionals should select exercises that train all the muscles involved in a movement. In soccer kicking, it is not 100% clear whether it is the activity of the quadriceps or hip flexors that predominantly contribute to late angular acceleration of the lower leg. Selecting exercises that train both muscles is clearly indicated, although the dominate size of proximal joint moments over distal joint moments argues for greater emphasis there. More recent trends in rehabilitation and sport performance conditioning have focused on training with "functional" movements that emulate the movement, rather than isolating specific muscle groups. The resistance, body motion, speed, and balance aspects of "functional" exercises may be more specific forms of training; unfortunately, there has been limited research on this topic. What outcome or performance metric is best is not always clear and training studies with control groups are time-consuming and expensive.

> **Activity: Segmental Interaction**
> Segmental interaction or the transfer of force from a proximal to a distal segment can be easily simulated using a two-segment model. Suspend a rigid stick (ruler, yardstick, racket) between the tips of your index finger and thumb. Using your hand/forearm as the proximal segment and the stick as the distal segment, simulate a kick. You can make the stick extend or kick without any extensor muscles by using intersegmental reaction forces (passive dynamics and Newton's Third Law). Accelerate your arm in the direction of the kick (positive). When you reach peak speed, rapidly slow (negatively accelerate) your arm and observe the positive acceleration of the stick below your hand. Positive acceleration of your arm creates an inertial lag in the stick, while negative acceleration of your arm creates a backward force at the joint, which creates a torque that positively accelerates the stick. This also works in "overarm" swing motions, but it is easy to smack yourself with the stick using this as a demonstration.

Interdisciplinary Issue: Kinematic Chain

A *kinematic chain* is an engineering term that refers to a series of linked rigid bodies. The concept of kinematic chains was developed to simplify the mathematics of the kinematics and kinetics of quantitative analysis linked mechanical systems. An early pre-biomechanics kinesiology textbook (Steindler 1955) adapted this terminology to refer to the linked segments of the human body as a "kinetic chain" and to classify movements as primarily "open" or "closed" kinetic chains. A closed kinetic chain is a movement where the motion of the distal segment is restrained by "considerable external resistance." Steinder was an MD without training in mechanics when he proposed these terms in this pre-biomechanics book. Over the years, unfortunately, the rehabilitation and conditioning professions have adopted this terminology, referring to open kinetic chain exercises (knee extension) and closed kinetic chain exercises (leg press or squat). Considerable research has focused on the forces and muscle activation involved in various exercises classified as open or closed kinetic chains. This research has shown both similarities and differences in muscular function between similar open and closed kinetic chain movements. There are, however, problems in uniquely defining a closed chain or what constitutes "considerable resistance." The vague nature of the classification of many exercises has prompted calls to avoid this terminology (Dillman et al. 1994; Blackard et al. 1999; di Fabio 1999).

Learning the sequential coordination of a large kinematic chain is a most difficult task. Unfortunately, there have been relatively few studies on changes in joint kinetics accompanied by learning. Assuming that proximal force or energy was transferred distally in a sequential movement (like our immature volleyball spike in the previous chapter), it would not be desirable to practice the skill in parts because there would be little force to learn to transfer. Recent studies have reinforced the idea that sequential skills should be learned in whole at submaximal speed, rather than in disconnected parts (see Sorensen et al. 2000). Most modeling and EMG studies of the vertical jump have also shown the interaction of muscle activation and coordination (Bobbert and van Zandwijk 1999; Bobbert and van Soest 1994; van Zandwijk et al. 2000), while some other studies have shown that strength parameters do not affect coordination (Tomioka et al. 2001). Improvements in computers, software, and biomechanical models may allow more extensive studies of the changes in kinetics as skills are learned. Currently, application of the **Segmental**

Interaction Principle involves corrections in body positioning and timing. Practice should focus on complete repetitions of the whole skill performed at submaximal speeds. Improvement should occur with many practice repetitions, while gradually increasing speed. This perspective is consistent with more recent motor learning interest in a dynamical systems theory understanding of coordination, rather than centralized motor program (Schmidt and Wrisberg 2007). Interdisciplinary research involving motor learning and biomechanics scholars may one day provide strong evidence on what are the best motor skill teaching and learning strategies.

Application: Arm Swing Transfer of Force

Many movements incorporate an arm swing that is believed to contribute to performance. How much does arm swing contribute to vertical jump performance? Several studies have shown that the height of a jump increases by about 10% with compared to those without arm swing (Feltner et al. 1999). There are several possible mechanisms involving multiple transfers of force or momentum between the arms and body (Lees et al. 2004). Logically, vigorous positive (upward) acceleration of the arms creates a downward reaction force on the body that increases the vertical ground reaction force. It has also been hypothesized that this downward force creates a pre-loading effect on the lower extremities that limits the speed of knee extension, allowing greater quadriceps forces because of the force–velocity relationship. A detailed kinetic study (Feltner et al. 1999) found that augmenting knee torques early in a jump with arm swings combined with slowing of trunk extension may be the mechanisms involved in a good arm swing during a vertical jump. Late in the jump, the arms are negatively accelerated, creating an upward reaction force at the shoulder and increasing the height of the center of mass at take-off. While the arms do not weigh a lot, the vigor of these movements does create large forces, which can be easily seen by performing this arm swing pattern standing on a force platform.

What segmental interactions create and transfer this force? This answer is less clear and depends on the model and kinetic variable used during analysis. The muscular and segmental contributions to a vertical jump have been analyzed using force platforms (Luthanen and Komi 1978a, b), computer modeling (Bobbert and van Soest 1994; Pandy et al. 1990), joint mechanical power calculations (Hubley and Wells 1983; Fukashiro and Komi 1987; Nagano et al. 1998), angular momentum (Lees and Barton 1996), and net joint torque contributions to vertical motion (Hay

6

et al. 1981; Feltner et al. 1999, 2004). While the jumping technique may look quite similar, there is considerable between-subject variation in the kinetics of the vertical jump (Hubley and Wells 1983). The problems involved in partitioning contributions include defining scalar energy transfer, energy transfer of biarticular muscles, muscle co-activation, and bilateral differences between limbs. While there is much yet to learn, it appears that the hip extensors contribute the most force/energy, closely followed by the knee extensors, with smaller contributions by the ankle plantar flexors. Conditioning for vertical jumping should utilize a variety of jumps and jump-like exercises. If specific muscle groups are going to be isolated for extra training, the hip and knee extensors appear to be the groups with the greatest contribution to the movement.

6.10 Summary

Linear kinetics is the study of the causes of linear motion. There are several laws of mechanics that can be applied to a study of the causes of linear motion: Newton's laws, the impulse–momentum relationship, and the work–energy relationship. The most common approach involves Newton's Laws of Motion, called the laws of *Inertia*, *Momentum/Acceleration*, and *Reaction*. Inertia is the tendency of all objects to resist changes in their state of motion. The Inertia Principle suggests that reducing mass will make objects easier to accelerate, while increasing mass will make objects more stable and harder to accelerate. Applying the Inertia Principle might also mean using more mass in activities where there is time to overcome the inertia, so that it can be used later in the movement. When two objects are in contact, the forces of interaction between the bodies are resolved into right-angle directions: normal reaction and friction. The Impulse–Momentum Relationship says that the change in momentum of an object is equal to the impulse of the resultant forces acting on the object. This is Newton's second law when applied over a time interval. The real-world application of this relationship is the Force–Time Principle. Energy is the capacity to do mechanical work; mechanical energies include strain, potential, and kinetic energy. The Work–Energy Relationship indicates that mechanical work equals the change in mechanical energy. Mechanical power is the rate of doing work and can also be calculated by the product of force and velocity. The Segmental Interaction Principle says that force can be transferred between segments. While the exact nature of these transfers has been difficult to determine,

both simultaneous and sequentially coordinated movements take advantage of the energy transferred through the many degrees of freedom of the linked segment system of the body.

> **Interdisciplinary Issue: Power in Vertical Jumping**
> One of the contentious uses of the word "power" occurs in the strength and conditioning literature, specifically as it relates to the use of the vertical jump as a measure of lower extremity muscular function. Soon after the Sargent (1921) jump test that was published, many authors have tried to use the standing vertical jump as a measure of the external power or "explosive" anaerobic muscular power. There is a correlation between measures of external power flow to a force platform and jump height, so many regression equations can be used to estimate average or peak power from jump height and body mass. Despite eloquent arguments, Newton's second law, and experiments showing net impulse is really the mechanical variable that determines jump height (Adamson and Whitney 1971; Barlow 1971; Winter 2005), the coaching and conditioning literature continues to use the terms "muscular" or "muscle power" in misleading ways related to vertical jump tests (Knudson 2009; Winter et al. 2016; Vigotsky et al. 2019). Students can help the field progress by correct use of terminology and contributing to interdisciplinary research in this area. When measurement, biomechanics, strength and conditioning, and exercise physiology scholars collaborate and consistently use terminology, real progress can be made in understanding muscular performance.

❓ Review Questions

1. Which has more inertia, a 6-kg bowling ball sitting on the floor or one rolling down the lane? Why?
2. What are the two ways to mathematically express Newton's second law?
3. When might it be advantageous for a person to increase the inertia used in a movement?
4. Do smaller or larger muscle angles of pull on a distal segment tend to create more joint rotation? Why?
5. What are strategies to increase the friction between a subject's feet and the floor?
6. What two things can be changed to increase the impulse applied to an object? What kinds of human movement favor one over the other?
7. If the force from the tibia on the femur illustrated below was 1000 N acting at 30° to the femur, what are the longitudinal (causing knee compression)

and normal (knee shear) components of this force? Hint: move one component to form a right triangle and solve.

8. Give human movement examples of the three mechanical energies.

9. Compare and contrast muscular strength and muscular power.

10. How is momentum different from kinetic energy?

11. A rock climber weighing 800 N has fallen and is about to be belayed (caught with a safety rope) by a 1500-N vertical force. Ignoring the weight of the rope and safety harness, what is the vertical acceleration of the climber? Hint: remember to sum forces with correct signs (related to direction).

12. Draw a free-body diagram of a proximal segment of the body showing all forces from adjacent segments. Draw a free body diagram of an adjacent segment using Newton's third law to determine the size and direction at the joint.

13. What are the potential kinetic mechanisms that make a sequential motion of segments in high-speed movements the optimal coordination?

14. Do the angles of pull (relative to the body) of free weights change during an exercises? Why?

15. An Olympic lifter exerts a 4000-N upward (vertical) force to a 30-kg barbell. What direction will the bar tend to move, and what is its vertical acceleration?

Key Terms

Conservation of energy (Law of Conservation of Energy) · Degrees of freedom · Direct dynamics · Energy · Force platform · Force–time principle · Friction · Impulse · Impulse–momentum relationship · Inertia · Inertia principle · Inverse dynamics · Kinetic energy · Law of Acceleration · Law of Inertia · Law of Reaction · Momentum · Normal reaction · Potential energy · Power (mechanical) · Strain energy · Work (mechanical) · Work–energy relationship

Suggested Reading

Abernethy, P., et al. (1995). Strength and power assessment: Issues, controversies and challenges. *Sports Medicine, 19*, 401–417.

Caruthers, E. J., et al. (2016). Muscle force and their contributions to vertical and horizontal acceleration of the center of mass during sit-to-stand transfer in young, healthy adults. *Journal of Applied Biomechanics, 32*, 487–503.

Dowling, J. J., & Vamos, L. (1993). Identification of kinetic and temporal factors related to vertical jump performance. *Journal of Applied Biomechanics, 9*, 95–110.

Greenemeier, L. (2016). Blade runners: Do high-tech prostheses give runners an unfair advantage? *Scientific American*, scientificamerican.com/article/blade-runners-do-high-tech-prostheses-give-runners-an-unfair-advantage/.

Jorgensen, T. P. (1994). *The physics of golf*. New York: American Institute of Physics.

Knudson, D. (2009). Correcting the use of the term "power" in the strength and conditioning literature. *Journal of Strength and Conditioning Research, 23*, 1902–1908.

Lees, A., & Barton, G. (1996). The interpretation of relative momentum data to assess the contribution of the free limbs to the generation of vertical velocity in sports activities. *Journal of Sports Sciences, 14*, 503–511.

Linthorne, N. P. (2001). Analysis of standing vertical jumps using a force platform. *American Journal of Physics, 69*, 1198–1204.

McPoil, T. G., et al. (1995). A comparison of two in-shoe plantar pressure measurement systems. *The Lower Extremity, 2*, 95–103.

Zajac, F. E. (2002). Understanding muscle co-ordination of the human leg with dynamical simulations. *Journal of Biomechanics, 35*, 1011–1018.

Zajac, F. E., & Gordon, M. E. (1989). Determining muscle's force and action in multi-articular movement. *Exercise and Sport Sciences Reviews, 17*, 187–230.

Zatsiorsky, V. M. (2002). *Kinetics of human motion*. Champaign: Human Kinetics.

Internet Resources

▶ Crash Course Physics. PBS lessons related to linear kinetics are: Newton's Laws, Friction, and Work, Energy, and Power. *klru.pbslearningmedia.org/collection/crash-course-physics/*

▶ Khan Academy Newton's Laws. *khanacademy.org/science/physics/forces-newtons-laws*

▶ Manchester Metropolitan University Forces in Locomotion. *youtube.com/watch?v=q8kZxwk_mGQ*

▶ Manchester Metropolitan University Force Platform Forces. *youtube.com/watch?v=-1CqNEMj_iU*

▶ NBC News Learn. Video lessons on the Science of Golf, Science of NFL Football, and Science of the Winter Olympics. *nbclearn.com/portal/site/learn/resources*

▶ SMU Locomotor Lab Ground Reaction Forces in Running. *youtube.com/watch?v=PfHNOwmmik4*

Angular Kinetics

Key Questions

➤ **What do you need to know to begin?**
Knowledge of musculoskeletal anatomy, algebra, kinematics, and linear kinetics

➤ **Why do you need to know this material?**
Understanding the causes of angular motion matches the major joint rotations of the body and informs general recommendations for technique and training.

➤ **What is the key idea?**
Biokinetic variables and principles for analysis of angular motion are essential tools in establishing the causes of rotations in human movement.

Angular kinetics explains the causes of rotary motion and employs many variables similar to the ones discussed in the previous chapter on linear kinetics. In fact, Newton's laws have angular analogs that explain how moments/torques tend to create rotation. The net moment acting on an object creates an angular acceleration inversely proportional to the angular inertia called the moment of inertia. Angular kinetics is quite useful because it explains the causes of the joint rotations of the human body and provides a quantitative way to determine the center of gravity of the human body. The application of angular kinetics is illustrated with the principles of **Inertia** and **Balance**.

7.1 Moment of Force/Torque

The rotating effect of a force is called a **moment of force** or **torque**. A force acting off-center creates a moment that can tend to rotate a body or create bending in a deformable body (■ Fig. 7.1). In rigid body models, bending is ignored (negligible), so an off-center force create both linear acceleration and an angular acceleration from the moment of force.

Recall that a moment of force or torque is a vector quantity, and the usual two-dimensional convention is that counterclockwise rotations are positive. Moment is calculated as the product of force (**F**) and the **moment arm**. The moment arm or leverage is the perpendicular displacement ($\mathbf{d}_\perp$) from the line of action of the force to the axis of rotation (■ Fig. 7.2). The biceps femoris pictured in ■ Fig. 7.2 has mo-

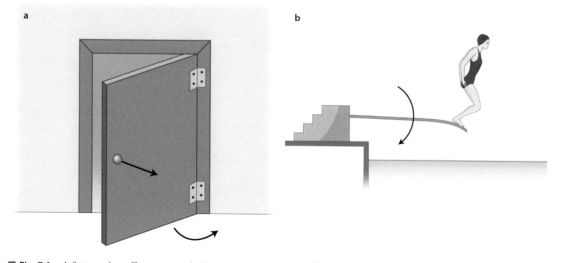

■ Fig. 7.1 A force acting off-center on a body can create a rotation effect and a bending effect. (With permission from Ozkaya et al. (2017)

© Springer Nature Switzerland AG 2021
D. Knudson, *Fundamentals of Biomechanics*, https://doi.org/10.1007/978-3-030-51838-7_7

7

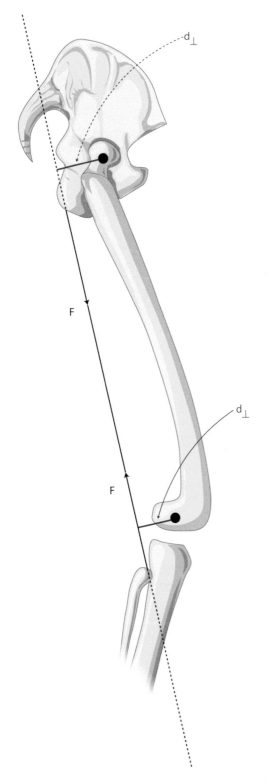

■ Fig. 7.2 The moment arms (**d**$_\perp$) for the biceps femoris muscle. A moment arm is the right-angle distance from the line of action of the force to the axis of rotation

ment arms that create hip extension and knee flexion torques. An important point is that the moment arm is always the shortest (right angle) displacement between the force line of action and axis of rotation, not the length of the rigid body that is commonly assumed to be the leverage. This text will use the term torque synonymously with moment of force, even though there is a more specific mechanics-of-materials meaning for torque.

In algebraic terms, the formula for torque is **T = F · d**$_\perp$, so that typical units of torque are N•m and lb•ft. Like angular kinematics, the usual convention is to define counterclockwise (ccw) torques positive and clockwise ones negative. Note that the size of the force and the moment arm are *equally* important in determining the size of the torque created. This has important implications for maximizing performance in many activities. A person wanting to create more torque can increase the applied force or increase their effective moment arm. Increasing the moment arm is often easier and faster than months of conditioning! ■ Fig. 7.3 illustrates two positions where a therapist can provide resistance with a hand dynamometer to manually test the isometric strength of the elbow extensors. By positioning their arm more distal (position 2), the therapist increases the moment arm and decreases the force they must create to balance the torque created by the patient and gravity (**T**$_p$).

Let's look at another example of applying forces in an optimal direction to maximize the torque created. A biomechanics student takes a break from her studies to bring a niece to the playground. Let's calculate the torque the student creates on the merry-go-round in the horizontal plane by the force **F**$_1$ illustrated in ■ Fig. 7.4. Thirty pounds of force times the moment arm of 4 feet is equal to 120 lb•ft of torque. This torque can be considered positive because it acts counterclockwise. If on the second spin the student pushes with the same magnitude of force (**F**$_2$) in a different direction, the torque and angular motion created would be smaller because of the smaller moment arm (**d**$_B$). Use the conversion factor in Appendix B to see how many N•m are equal to 120 lb•ft of torque.

Good examples of torque measurements in exercise science are the joint torques measured by isokinetic dynamometers. The typical maximum isometric torques of several muscle groups for apparently healthy males are listed in ■ Table 7.1. These torques should give you a good idea of some "approximate" maximal

Fig. 7.3 Increasing the moment arm for the therapist's (position 2) manual resistance makes it easier to perform a manual muscle test that balances the extensor torque created by the patient (T_p)

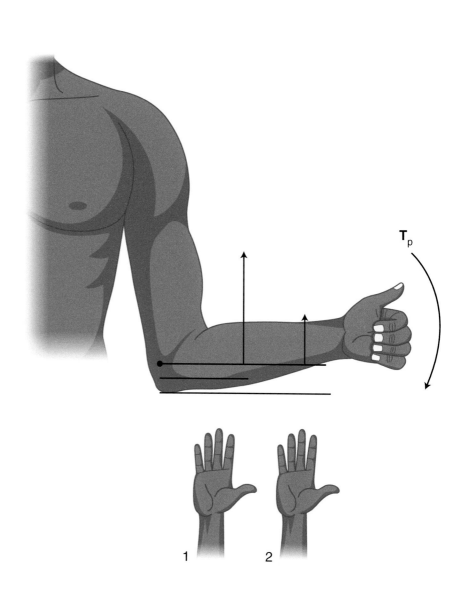

values for many major joints. Peak torques from **inverse dynamics** (see ▶ Sect. 7.4) in sporting movements can be larger than those measured in isokinetic testing because of antagonist activity in isokinetics testing, segment interaction in dynamic movements, the stretch-shortening cycle, and eccentric muscle actions. Most isokinetic norms for the strength of muscle groups are normalized to body mass or weight (e.g., lb•ft/lb) and categorized by gender and age. Recall that the shape of the torque–angle graphs from isokinetic testing reflects the integration of many muscle mechanical variables. The angle of the joints affects the torque that the muscle group is capable of producing because of variations in moment arm, muscle angle of

Fig. 7.4 Calculating the torque created by a person pushing on a merry-go-round involves multiplying the force times its moment arm. This torque can be converted to other units of torque with conversion factors (Appendix B). Qualitatively estimate the lower torque created by the same magnitude by less effectively applied second push (**F₂**)

7

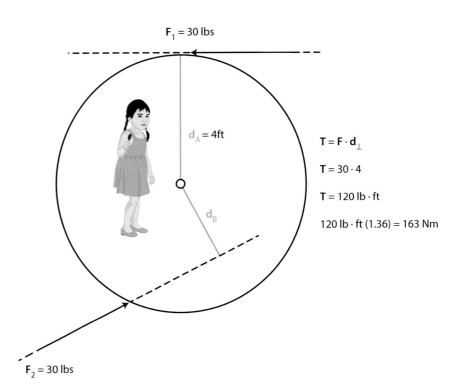

$F_1 = 30$ lbs

$d_A = 4ft$

d_B

$F_2 = 30$ lbs

$T = F \cdot d_\perp$

$T = 30 \cdot 4$

$T = 120$ lb · ft

120 lb · ft $(1.36) = 163$ Nm

Table 7.1 Typical isometric joint torques measured by isokinetic dynamometers

	Peak torques	
	N•m	lb•ft
Trunk extension	258	190
Trunk flexion	177	130
Knee extension	204	150
Knee flexion	109	80
Hip extension	150	204
Ankle plantar flexion	74	102
Elbow flexion	20	45
Wrist flexion	8	11
Wrist extension	4	5

pull, and the force–length relationship of the muscle. There are several shapes of torque–angle diagrams of muscle groups, but they most often look like an inverted "U" because of the combined effect of changes in muscle moment arm and force–length relationship (**Fig. 7.5**).

Torque is a good variable to use for expressing muscular strength because it is not dependent on the point of application of force on the limb like force sensors (clinical hand dynamometers). The torque an isokinetic machine (**T**) measures will be the same for either of the two resistance pad locations illustrated in **Fig. 7.6** if the person's effort and joint torque are the same. Sliding the pad toward the subject's knee will decrease the moment arm for the force applied by the subject, increasing the force on the leg (**F₂**) at that point to create the same torque. Using torque instead of force created by the subject allows for easier comparison of measurements between different dynamometers.

Application: Muscle Balance and Strength Curves
Recall that testing with an isokinetic dynamometer documents the strength curves (joint torque–angle graphs) of muscle groups. Normative torques from isokinetic testing also provide valuable information on the ratio of strength between opposing muscle groups. Many dynamometers have computerized reports that list test data normalized to body mass or weight and can be expressed as a ratio of the peak torque of opposing muscle groups. For example, peak torques created by the hip flexors tend to be 60 to 75%

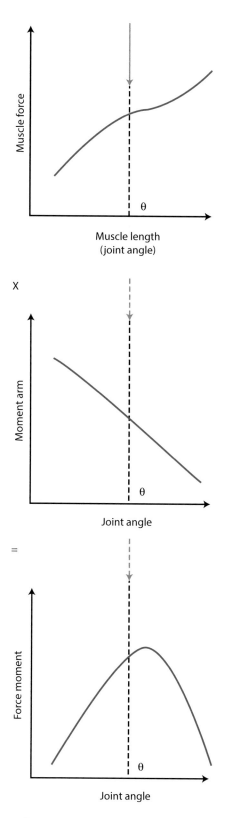

Fig. 7.5 Joint torque–angle diagrams represent the muscular strength curves of muscle groups. The shapes of joints vary based primarily upon the combined effect of changes in muscle length properties and muscle moment arms. (Reprinted by permission from Zatsiorsky (1995))

of peak hip extensor torques (Perrin 1993). Another common strength ratio of interest is the ratio of the quadriceps to the hamstrings. This ratio depends on the speed and muscle action tested, but peak concentric hamstring torque is typically between 40 and 50% of peak concentric quadriceps torque (Perrin 1993), which is close to the physiological cross-sectional area difference between these muscle groups. Greater emphasis has been placed on more functional ratios (see Aagaard et al. 1998), like hamstring eccentric to quadriceps concentric strength (H_{ecc}:Q_{con}), because hamstrings are often injured ("pulled" in common parlance) when they slow the vigorous knee extension and hip flexion before foot strike in sprinting. In conditioning and rehabilitation, opposing muscle group strength ratios are often referred to as muscle balance. Isokinetic (see Perrin 1993) and hand dynamometer (see Phillips et al. 2000) testing are the usual clinical measures of strength, while strength and conditioning professionals usually use one-repetition maxima (1RM) for various lifts. These forms of strength testing to evaluate muscle balance are believed to provide important sources of information on the training status, performance, and potential for injury of athletes. In rehabilitation and conditioning settings, isokinetic and other forms of strength testing are useful in monitoring progress during recovery. Athletes are cleared to return to practice when measurements return to some criterion/standard, a percentage of pre-injury levels or a percentage of the uninvolved side of their body. It is important for kinesiology professionals to remember that the strength (torque capability) of a muscle group is strongly dependent on many factors: testing equipment, protocol, and body position, among others, affect the results of strength testing (Schlumberger et al. 2006). If standards in testing are being used to qualify people for jobs or athletic participation, there needs to be clear evidence correlating the criterion test and standard with safe job performance.

7.2 Summing Torques

The state of an object's rotation depends on the balance of torques created by all the forces acting on the object. Remember that summing or adding torques acting on an object must take into account the vector nature of torques. All the muscles of a muscle group sum together to create a joint torque in a particular direction. These muscle group torques must also be summed with torques from antagonist muscles, ligaments, and external forces to determine the net torque at a joint. **$\Box$** Figure 7.7

Fig. 7.6 Isokinetic dynamometers usually measure torque because torque does not vary with variation in pad placement. Positioning the pad distally decreases the force the leg applies to the pad for a given torque because the moment arm for the leg is larger. The dynamometer is designed to create an opposing torque (**T**) to keep the match the torque crated by the person, maintaining near isokinetic (no or constant rotation) conditions

Fig. 7.7 The shoulder flexion torques of anterior deltoid and long head of the biceps can be summed to obtain the resultant flexion torque acting to oppose the gravitational torque from the weight of the arm

illustrates the forces of the anterior deltoid and long head of the biceps in flexing the shoulder in the sagittal plane. If ccw torques are positive, the torques created by these muscles would be positive. The net torque of these two muscles is the sum of their individual torques or 6.3 N•m (60 • 0.06 + 90 • 0.03 = 6.3 N•m). If the weight of this person's arm multiplied by its moment arm created a gravitational torque of −16 N•m, what is the net torque acting at the shoulder? Assuming there are no other shoulder flexors or extensors active to make forces, we can sum the gravitational torque (−16 N•m) and the

net muscle torque (6.3 N•m) to find the resultant torque of −9.7 N•m. This means that there is a resultant-rotating effect acting at the shoulder that is an extension torque, where the shoulder flexors are acting eccentrically to lower the arm. Torques can be summed about any axis, but be sure to multiply the force by the moment arm and then assign the correct sign to represent the direction of rotation before they are summed. Following the sign (+ and −) conventions of linear kinematics and kinetics in a plane automatically creates the correct ccw positive angular kinetic rotational conventions.

Recall the isometric joint torques reported in **Table 7.1**. Peak joint torques during vigorous movement calculated from inverse dynamics are often larger than those measured on isokinetic dynamometers (Veloso and Abrantes 2000). There are several reasons for this phenomenon, including transfer of energy from biarticular muscles, differences in muscle action, and co-activation. Coactivation of antagonist muscles is a good example of summing opposing torques. EMG research has shown that isokinetic joint torques underestimate net agonist muscle torque because of coactivation of antagonist muscles in maximal efforts (Kellis and Baltzopoulos 1997, 1998; Aagaard et al. 2000).

Summing moments and understanding moment arms are crucial to the angular kinetics of the simple machines called levers. A **lever** is a rigid segment that is rotated about an axis in order to either increase force or movement. There are three classes of levers based on the location of the axis of rotation (fulcrum or pivot) and two opposing forces acting on the lever (**Fig. 7.8**).

One force we will call the effort ($\mathbf{F}_E$) since it can be considered the person or machine force driving the lever system and the resistance ($\mathbf{F}_R$) that must be moved. A first-class lever (like a playground see/saw or scissors with forces on opposite sides) can magnify force or motion depending on the ratio of the moment arms for the two forces (mechanical advantage). Remember the moment arm is not always the length of the lever but is the perpendicular distance from the line of action of a force and the axis of rotation. A second-class lever (like a wheelbarrow) magnifies force because the moment arm of the effort is larger than the moment arm of the resistance. Most body segments are third-class levers because the moment arms for the effort (muscles) are smaller than the moment arms of the resistance (body segment mass and/or external forces at the end of the segment). Notice that third class levers magnify motion/speed (large distance traveled compared to distance attachment travels) so it is a good thing muscles create large forces to drive these anatomical levers. A biceps brachii must make a large force to make a torque larger than the torque created by the forearm/hand/dumbbell, but a small amount of shortening of the muscle creates greater rotation and speed at the hand. Early biomechanical research was interested in using anatomical leverage principles for a theory of high-speed movements, but this turned out to be a dead end because of the discovery of sequential coordination and the influence of passive dynamics (Segmental Interaction principle) of these movements (Roberts 1991).

Activity: Torque and Levers
Take a desk ruler (~12 inch but any length will do), and balance it on a sturdy small cylinder like a highlighter or symmetric drinking glass. Place a coin (quarter, etc.) at the 11 inch position, and note the behavior of the ruler. Tap the 1 inch position on the ruler with your index finger, and note the motion of the dime. Which torque was larger: the torque created by the dime or your finger? Why? Tap the ruler with the same effort on

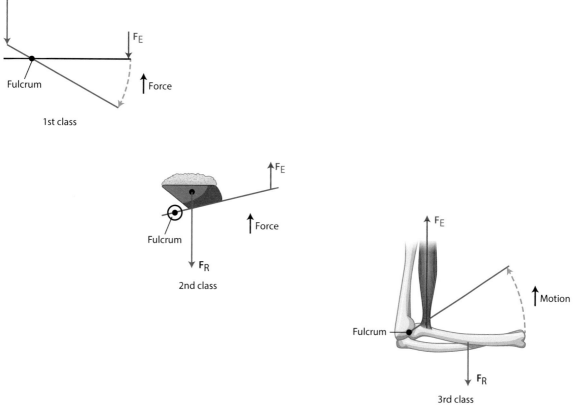

☐ **Fig. 7.8** The three classes of levers. Notice the moment arms for the effort force ($\mathbf{F}_E$) and resistance force ($\mathbf{F}_R$), their relative size, and the distance (arcs of motion) these levers create. First class levers can magnify force (like car jack illustrated) or motion. Second-class levers magnify force (wheelbarrow). Most human joints (elbow) are third class levers that magnify force different positions on the ruler with the dime at 11 inches, and note the motion of the dime. Modify the position (axis of rotation) of the highlighter to maximize the moment arm for the dime, and note how much force your finger must exert to balance the lever in a horizontal position. How much motion in the dime can you create if you tap the ruler? Can you create all classes of levers and magnify motion or the force created?

7.3 Angular Inertia (Moment of Inertia)

A *moment of force* or *torque* is the mechanical effect that creates rotation, but what is the resistance to angular motion? In linear kinetics we learned that mass was the mechanical measure of inertia. In angular kinetics, inertia is measured by the **moment of inertia**, a term pretty easy to remember because it uses the terms *inertia* and *moment* from moment of force. Like the mass (linear inertia), **moment of inertia** is the resistance to angular acceleration. While an object only has one mass, the object has an infinite number of moments of inertia! This is because the object can be rotated about an infinite number of axes and different planes of motion. We will see that rotating the human body is even more interesting because the links allow the configuration of a multi-segment body to move mass around relative to various axes of rotation.

The symbol for the moment of inertia is I. Subscripts are often used to denote the axis of rotation associated with a moment of inertia. The smallest moment of inertia of an object in a particular plane of motion is always about its center of gravity (I_0). Biomechanical studies also commonly use moments of inertia about the proximal (I_P) and distal (I_D) ends of body segments. The formula for a rigid body moment of inertia about an axis (A) is $I_A = \Sigma mr^2$. To determine the moment of inertia of a ski in the transverse plane about an anatomically longitudinal axis ($\blacksquare$ Fig. 7.9), the ski is cut into eight small masses (m) of known radial distances (**r**) from the axis. The sum of the product of these masses and the squared radius is the moment of inertia of the ski about that axis. Note that the SI units of moment of inertia are $kg \cdot m^2$.

The formula for moment of inertia shows that an object's resistance to rotation depends more on distribution of mass (r^2) than just object mass (m). This large increase in moment of inertia from changes in location of mass relative to the axis of rotation (because **r** is squared) is very important in human movement. Modifications in the position of joints changing the moment of inertia of body segments can help or hinder ro-

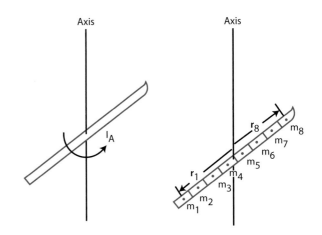

$\blacksquare$ **Fig. 7.9**   The moment of inertia of a ski about a specific axis can be calculated by summing the products of the masses of small elements (m) and the square of the distance from the axis (**r**)

tary movement. The use of or change in the moment of inertia of implements or tools (bat, prosthetics, etc.) can dramatically affect their feel and influence a person's coordination and skill in movement.

Activity: Moment of Inertia
Take a long object like a baseball bat, tennis racket, or golf club, and hold it in one hand. Slowly swing the object back and forth in a horizontal plane to eliminate gravitational torque from the plane of motion. Try to sense how difficult it is to initiate or reverse the object's rotation. You are trying to subjectively evaluate the moment of inertia of the object. Grab the object in several locations, and note how the moment of inertia changes. Add mass to the object (e.g., put a small book in the racket cover) at several locations. Does the moment of inertia of the object seem to be more related to mass or the location of the mass?

Most all persons go through adolescence with some short-term clumsiness. Much of this phenomenon is related to motor control problems from large changes in the moment of inertia from rapid growth. Imagine the balance and motor control problems from a major shift in leg moment of inertia if a young person grows two shoe sizes and 4 inches in a 3-month period without equal changes in strength. How much larger is the moment of inertia of this teenager's leg about the hip in the sagittal plane if this growth (dimension and mass) was about 8%? Would the increase in the moment of inertia of the leg be 8% or larger? Why?

When we want to rotate our bodies, we can skillfully manipulate the moment of inertia by changing the configuration of our body segments, moving their mass relative to the axis/axes of rotation. Bending the joints of the upper and lower extremities brings segmental masses close to the proximal joint axis of rotation (hip or shoulder), dramaticallydecreasing the limb's moment of inertia. This bending allows for easier angular acceleration and motion about a proximal joint axis. For example, the faster a person runs, the greater the knee flexion in the swing limb, which makes the leg easy to rotate and to get into position for another footstrike. Diving and skilled gymnastic tumbling both rely on decreasing the moment of inertia of the human body to allow for more rotations and then increasing the length of the body to slow rotation down in preparation for water entry or landing.

■ Figure 7.10 shows the differences in the moment of inertia for a human body in the sagittal plane for different body segment configurations relative to the axis of rotation. Notice the dramatically larger I_H for the extended body position rotating about the hands (c) and the smaller I_{CoM} for the body rotating about the body center of mass (a & b).

Variations in the moment of inertia of external objects or tools are also very important to performance. Imagine you are designing a new unicycle wheel. You design two prototypes with the same mass but with different distributions of mass. Which wheel design (see ■ Fig. 7.11) do you think would help a cyclist maintain balance on such a small area: wheel A or wheel B? Think about the movement of the wheel when a person balances on a unicycle. Does agility (low inertia) or consistency of rotation (high inertia) benefit the cyclist? If, on the other hand, you are developing an exercise bike that would provide slow and smooth changes in resistance,

which wheel would you use? What kind of wheel would you design for wheelchair sports in order to maximize quick motions (chair/athlete agility)? A heavy ski boot and ski dramatically affect the moments of inertia of your legs about the hip joint.

The moment of inertia of many sport implements (golf clubs, tennis rackets, skis, bats) is commonly called the "swing weight." A longer implement can have a similar swing weight to a shorter implement by keeping mass proximal and making sure the added length has low mass. It is important to realize that the three-dimensional nature of sports equipment means that there are moments of inertia about the three principal or dimensional axes of the equipment. Tennis

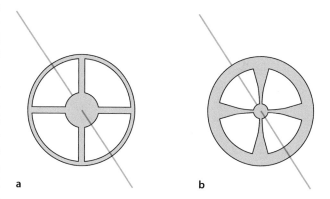

■ Fig. 7.11 The distribution of mass most strongly affects moment of inertia, so wheel **a** with mass close to the axle would have much less resistance to rotation than wheel **b**. Wheel **a** would make it easier for a cyclist to make quick adjustments of the wheel back and forth to balance a unicycle

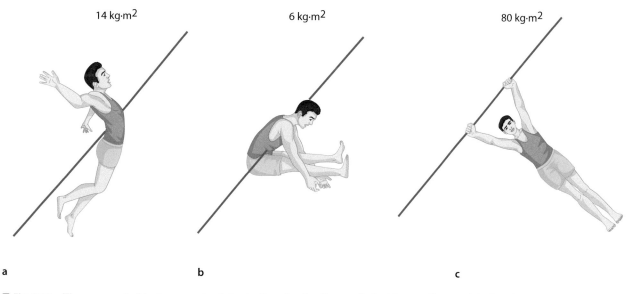

14 kg·m² 　　 6 kg·m² 　　 80 kg·m²

a 　　　 b 　　　 c

■ Fig. 7.10 The movement of body segments relative to the axis of rotation makes for large variations in the moment of inertia of the body. Typical sagittal plane moments of inertia and axes of rotation for a typical athlete are illustrated for long jump **a**, **b** and high bar **c** body positions

7

players often add lead tape to their rackets so as to increase shot speed and racket stability. Tape is often added to the perimeter of the frame for stability (by increasing the polar moment of inertia) against off-center impacts in the lateral directions (reduced angular acceleration of the frame from off-center ball impacts on the string bed). Weight at the top of the frame would not affect this lateral stability but would increase the moment of inertia about the grip increasing the inertial resistance of the racket to ball impact. The large radius of this mass (from his grip to the tip of the racket), however, would make the racket more difficult to swing. Recent baseball/softball bat designs allow for variations in where bat mass is located, making for wide variation in the moment of inertia for a swing. It turns out that an individual batting style affects the likely optimal bat mass (Bahill and Freitas 1995) and moment of inertia (Watts and Bahill 2000) for a particular hitter.

You can now see that the principle of inertia can be extended to angular motion of biomechanical systems. This application of the concepts related to moment of inertia is a bit more complex than mass in linear kinetics. For example, a person putting on snowshoes will experience a dramatic increase (larger than the small mass of the shoes implies) in the moment of inertia of the leg about the hip in the sagittal plane because of the long radius for this extra mass. An injured person wearing a prosthetic lower leg will have to learn to accommodate not only the loss of foot and shank muscular forces but the different moments of inertia of the artificial foot/leg. Angular inertia is most strongly related to the distribution of mass, so an effective strategy to decrease this inertia in multi-segment systems (leg or arm) is to bring segment masses close to the desired axis of rotation. Coaches can get players to "compact" their extremities or body to make it easier to initiate rotation.

7.4   Newton's Angular Analogs

Newton's laws of motion also apply to angular motion, so each may be rephrased using angular variables. The angular analog of Newton's third law says that for every torque there is an equal and opposite torque. The angular acceleration of an object is proportional to the resultant torque, is in the same direction, and is inversely proportional to the moment of inertia. This is the angular expression of Newton's second law. Likewise, Newton's first law demonstrates that objects tend to stay in their state of angular motion unless acted upon by an unbalanced torque. Biomechanists often use rigid body, linked-segment models of the human body and apply Newton's

laws to calculate the net forces and torques acting on body segments that created movements of interest.

Recall that this working backward from video measurements of acceleration (second derivatives) using both the linear and angular versions of Newton's second law is called **inverse dynamics**. Such analyses to understand the resultant forces and torques that create movement were first done using laborious hand calculations and graphing (Elftman 1939; Bressler and Frankel 1950), but they are now done with the assistance of computers and mathematical computation programming languages. The resultant or net joint torques calculated by inverse dynamics do not account for co-activation of opposing muscles and represent the sum of many muscles, ligaments, joint contact, and other anatomical forces (Winter 1990; Baltzopoulos 2021).

Despite the imperfect nature of these net torques (Winter 1990; Hatze 2000), inverse dynamics provides good estimates of the net motor control signals to create human movement (Winter and Eng 1995) and can detect changes with fatigue (Apriantono et al. 2006) or practice/learning (Schneider et al. 1989; Yoshida et al. 2004). ◘ Figure 7.12 illustrates the net joint torques at the hip and knee in a soccer toe kick. These torques are similar to the torques reported in three-dimensional studies of high-speed soccer kicks (Nunome et al. 2002; Nunome et al. 2006). The kick is initiated by a large hip flexor torque that rapidly decreases before impact with

◘ **Fig. 7.12**   The net joint hip (thick line) and knee (thin line) joint torques in a soccer kick calculated from inverse dynamics. The backswing (BS), range of deepest knee flexion (DKF), forward swing (FS), and impact (IMP) are illustrated. (Adapted with permission from Zernicke and Roberts (1976))

the soccer ball. The knee extensor torque follows the hip flexor torque and also decreases to near zero at impact. This near-zero knee extensor torque could be expected because the foot would be near peak speed at impact, with the body protecting the knee from hyperextension. If the movement were a punt, there would usually be another rise and peak in hip flexor torque following the decline in knee torque (Putnam 1983). It is pretty clear from this planar (2D) example of inverse dynamics that the hip flexor musculature likely makes a larger contribution to kicking than the knee extensors.

It is not as easy to calculate or interpret 3D kinetics since a large joint torque might have a small resistance arm and not make a large contribution to a desired motion or a torque might be critical to positioning a segment for another torque to be able to accelerate the segment (Sprigings et al. 1994; Bahamonde 2000). For example, sometimes large joint torques act in directions where there is no anatomical joint rotation and create stress in ligaments that resist that motion. In fast throwing and striking sports, the forceful shoulder horizontal adduction and internal rotation in the forward swing create a large valgus torque at the elbow. The varus torques (30–60 Nm) calculated in inverse dynamics studies (Feltner and Dapena 1986; Fleisig et al. 1995; Bahamonde and Knudson 2003) indicates significant stress on the ulnar collateral ligament and wrist flexor muscles that stabilize the elbow medially. For these and other reasons, some biomechanists use Newton's laws with more detailed modeling and simulations to input muscle and external forces/moments to predict subsequent motion (**direct dynamics**).

The resultant joint torques calculated in inverse and direct dynamics are sometimes multiplied by the joint angular velocity to derive net joint power flows. When the product of a net joint torque and joint angular velocity are positive (in the same direction), the muscle action is *hypothesized* to be primarily concentric and generating positive work. Negative joint powers are *hypothesized* to represent eccentric actions of muscle groups slowing down an adjacent segment. These joint powers can be integrated with respect to time to calculate the net work done at joints. Other studies first calculated mechanical energies (kinetic and potential energies) and summed them to estimate work and eventually calculate power. Unfortunately, this summing of mechanical energy analyses does not agree well with direct calculation of joint power from torques because of difficulties in modeling the transfer of mechanical energies between external forces and body segments (Aleshinsky 1986a, b; Wells 1988) and coactivation of muscles (Neptune and van den Bogert 1998). Like ▶ Chap. 6 noted, problems of accurate interpretation result from

discarding the vector and causative nature of forces and torques, replacing/converting them to scalar measures like energy, work, and power (Vigotsky et al. 2019).

7.5 Equilibrium

An important concept that grows out of Newton's first and second laws is *equilibrium*. **Mechanical equilibrium** occurs when the forces and torques acting on an object sum to zero. Newton's second law accounts for both linear and angular conditions of **static equilibrium** ($\Sigma F = 0$, $\Sigma T = 0$), where an object is motionless or moving at a constant velocity. **Dynamic equilibrium** is used to refer to the kinetics of accelerated bodies using Newton's second law ($\Sigma F = m \cdot a$, $\Sigma T = I \cdot \alpha$).

In a sense, dynamic equilibrium fits the definition of equilibrium if you rearrange the equations (i.e., $\Sigma F - m \cdot a = 0$). The $m \cdot a$ term in the previous equation is sometimes referred to as the inertial force. This inertial force is not a real force and can cause confusion in understanding the kinetics of motion, so it is best to avoid this interpretation and term.

This text will focus on static equilibrium examples because of their simplicity and because summation of forces and torques is identical to dynamic equilibrium. Biomechanics studies often use static or quasi-static analyses (and thus employ static equilibrium equations and avoid difficulties in calculating accurate accelerations) in order to study slow movements with small accelerations. The initial occupational lifting standards set by the National Institute for Occupational Safety and Health (NIOSH) were based in large part on static biomechanical models and analyses of lifting. Static equilibrium will also be used in the following section to calculate the center of gravity of the human body.

Equilibrium and angular kinetics are the mechanical tools most often used in the study of balance. We will see in the next two sections that the center of gravity of the human body can be calculated by summing moments in a static equilibrium form, and these kinds of data are useful in examining the state of mobility and stability of the body. This control of stability and ability to move is commonly called *balance*. What mechanics tells us about balance is summarized in the Principle of Balance.

7.6 Center of Mass/Gravity

A natural application of angular kinetics and anthropometrics is the determination of the center of gravity of multi-segment systems like the human body. The **center of gravity** is the location in space where the weight (gravitational force) of an object can be considered to act. An object's center of mass (mass centroid) is the same as its center of gravity. The center of gravity of small rigid objects (pencil, pen, bat) can be easily found by trying to balance the object on your finger. The point where the object balances is in fact the center of gravity, which is the theoretical point in space where you could replace the weight of the whole object with one downward force. There is *no requirement* for this location to be in a high-mass area or even within or on the object itself. Think about where the center of gravity of a basketball or any hollow uniform sphere/ball would be.

The center of gravity of the human body can move around, because joints allow the masses of body segments to move. In the anatomical position, the typical location of a body's center of gravity in the sagittal plane is at a point equivalent to about 57 and 55% of the height for males and females, respectively (Hay and Reid 1982). Can you name some structural and weight distribution differences between the sexes that account for this general difference? Knowing where the force of gravity acts in various postures of the human body allows biomechanists to study the kinetics and stability of these body positions.

Interdisciplinary Issue: The Spine and Low Back Pain
One of the most common musculoskeletal complaints is low back pain (LBP). The medical literature would say that most often the etiology (origin) LBP is often related to soft tissue (muscle, tendons, ligaments) and idiopathic (of unknown origin). The diagnostic accuracy of advanced imaging techniques like magnetic resonance imaging (MRI) for identifying spinal abnormalities (e.g., disk herniation) that correlate with function and symptoms of LBP is poor (Beattie and Meyers 1998). The causes of LBP are complicated and elusive because there is often no clear sign of specific soft tissue injury. Biomechanics can contribute clues that may help in understanding LBP and potentially effective treatments. Mechanically, the spine is like a stack of blocks separated by small cushions (McGill 2016). Stability of the spine is primarily a function of the ligaments and muscles, which act like the guy wires that stabilize a tower or the mast of a boat. These muscles are short and long and often must simultaneously stabilize and move the spine. Total spine motion is a summation of the small motions at each intervertebral level (Ashton-Miller and Schultz 1988). Biomechanical studies of animal and cadaver spines usually examine loading and rotation between two spinal levels in what is called a motion segment. Individuals even exhibit

different strategies for rotation of motion segments in simple trunk flexion movements (Nussbaum and Chaffin 1997; Gatton and Pearcy 1999), so that neuromuscular control likely plays an important role in injury and rehabilitation (Ebenbichler et al. 2001; McGill 2016). Occasionally a person is unfortunate and gets injured in a biomechanical study. Cholewicki and McGill (1992) reported X-ray measurements of the "buckling" of a single spinal segment that occurred during a heavy deadlift. Biomechanics research using computer models and EMG are trying to understand how muscles and loads affect the spine and the nature of this motion segment buckling (Preuss and Fung 2005). This information must be combined with occupational, epidemiological, neurologic, and rehabilitative research to understand the development and treatment of LBP.

There are two main methods used to calculate the center of gravity of the human body, and both methods employ the equations of static equilibrium. One lab method, which requires a person to hold a certain body position fixed for a short length of time, is called the **reaction change** or *reaction board* method. The other method used in research is called the *segmental method*. The **segmental method** uses anthropometric data (population specific averages) and mathematically breaks up the body into segments to calculate the center of gravity.

The reaction board method requires a rigid board with special feet and a scale (1D) or scales (2D) to measure the ground reaction force under the feet of the board.

The "feet" of a reaction board are knife-like edges or small points similar to the point of a nail. A 1D reaction board, a free-body diagram, and static equilibrium equations to calculate the center of gravity in the sagittal plane are illustrated in ◘ Fig. 7.13. Note that the weight force of the board itself is not included. This force can be easily added to the computation, but an efficient biomechanist "zeros" the scale with the board in place to exclude extra terms from the calculations. The person in ◘ Fig. 7.13 weighs 185 pounds, the distance between the edges is 7 feet, and the scale reading is 72.7 pounds. With only three forces acting on this system and everything known but the location of the center of gravity, it is rather simple to apply the static equilibrium equation for torque and solve for the center of gravity ($d_\perp$). Note how the sign of the torque created by the subject's body is negative according to convention, so a negative $d_\perp$ (to the left) of the reaction board edge fits this standard, and horizontal displacement to the left is negative. In

Fig. 7.13 Application of static equilibrium and a reaction board to calculate whole body center of gravity. Summing torques about the reaction board edge at the feet and solving for the moment arm ($\mathbf{d}_\perp$) for gravity locates the center of gravity/center of mass

$$\Sigma T = 0$$
$$-185(-d_\perp) - 72.7(-7) = 0$$
$$d_\perp = 2.75 \text{ ft}$$

72.7 lbs

-185 lbs

this case, the subject's center of gravity is 2.75 feet up from the edge of the reaction board. If the subject were 5.8 feet in height, his center of gravity in this position would be 47% of his or her height.

In the segmental method, the body is mathematically broken up into segments. The weight of each segment is then estimated from mean anthropometric data. For example, according to Plagenhoef et al. (1983), the weight of the forearm and hand is 2.52 and 2.07% for a man and a woman, respectively. Mean anthropometric data are also used to locate the segmental centers of gravity (percentages of segment length) from either the proximal or distal point of the segment. Figure 7.14 depicts calculation of the center of gravity of a high jumper clearing the bar using a three-segment biomechanical model. This simple model (head+arms+trunk, thighs, legs+feet) illustrates the segmental method of calculating the center of gravity of a linked biomechanical system. Points on the feet, knee, hip, and shoulder are located and combined with anthropometric data to calculate the positions of the centers of gravity of the various segments of the model. Most biomechanical studies use rigid body models with more (20 or more) segments to more accurately calculate the whole-body center of gravity and other biomechanical variables. If a biomechanist were studying a high jump with high-speed video (120 Hz), a center of gravity calculation much like this would be made for every video image of the movement.

The segmental method is also based on the equations of static equilibrium. The size and location (moment arm) of the segmental forces are used to calculate and sum the torques created by each segment. If this body posture in the snapshot were to be balanced by a torque in the opposite direction (product of the whole body-

Segment Centers of Gravity (X, Y)
Shank/feet (3,2, 6.6)
Thighs (5.0, 10.5)
Head/Arms/Trunk (11.0, 9.0)

$$\Sigma T_0 = 0$$

$$182\,(d_\perp) - 22(3.2) - 36(5) - 124(11) = 0$$
$$d_\perp = 8.9 \text{ in}$$

-36 lbs

-22 lbs

-124 lbs

182 lbs

Fig. 7.14 Calculating the horizontal position of the whole body center of gravity of a high jumper using the segmental method and a three-segment model of the body. Most sport biomechanical models use more segments, but the principle for calculating the center of gravity is the same. Filled circle represents the high jump bar

weight acting in the opposite direction times the center of gravity location: $182 \cdot \mathbf{d}_\perp$), the total torque would be zero. By applying the law of statics and summing torques about the origin of our frame of reference, we calculate

7

that the person's bodyweight acts 8.9 inches from the origin. These distances are small because the numbers represent measurements on an image. In a 2D biomechanical analysis, the image-size measurements are scaled to real-life size by careful set-up procedures and imaging a control object of known dimensions for scaling the image data to life-size dimensions.

Finding the height of the center of gravity in the ◻ Fig. 7.14 example is identical, except that the y coordinates of the segmental centers of gravity are used as the moment arms. Students can then imagine the segment weight forces acting to the left, and the height of the center of gravity is the y coordinate that, multiplied by the whole bodyweight acting to the right, would cancel out the segmental torques toward the left. Based on the subject's body position and the weights of the three segments, guess the height in centimeters of the center of gravity. Did the center of gravity pass over the bar? Finish the calculation using the Y coordinate data in ◻ Fig. 7.14 to check your guess. The segmental method can be applied using any number of segments and in all three dimensions during 3D kinematic analysis. There are errors associated with the segmental method, and more complex calculations are done in situations where errors (e.g., trunk flexion/extension, abdominal obesity) are likely (Kingma et al. 1995).

Activity: Center of Gravity and Moment of Inertia
Take a 12 inch or other rigid ruler, and balance it on your finger to locate the center of gravity. Lightly pinch the ruler between your index finger and thumb at the 1 inch point, and allow the ruler to hang vertically below your hand. Swing the ruler in a vertical plane, and sense the resistance of the ruler to rotation. Tape a coin (nickel or quarter) to various positions on the ruler, and note how the center of gravity shifts and how the resistance to rotation changes. Which changes more: center of gravity or moment of inertia? Why? What factors make it difficult to sense changes in ruler moment of inertia (hint: an angular kinetic variable from gravity would not complicate things if the ruler were rotated in a horizontal plane)?

7.7 Balance Principle

We have seen than angular kinetics provides mathematical tools for understanding rotation, center of gravity, and rotational equilibrium. The movement concept of balance is closely related to these angular kinetic variables. **Balance** in the field of kinesiology is a person's ability to control their body position relative to some base of support (◻ Fig. 7.15). This ability is needed in both static equilibrium conditions (e.g., handstand on a balance beam) and during more dynamic movement (e.g., shifting the center of gravity from the rear foot to the forward foot). Balance can be enhanced by improving body segment positioning or posture. These adjustments should be based on biomechanical principles. There are also many sensory organs and cognitive processes involved in the control of movement (balance), but this section focuses on the mechanical or technique factors affecting balance and outlines application of the Principle of Balance.

Before we apply this principle to several human movements, it is important to reexamine the mechanical paradox of stability and mobility (recall the stability–mobility paradox in joint stability/mobility in ▶ Chap. 4). It turns out that optimal posture depends on the right mix of stability and mobility for the movement of interest. This is not always an easy task, because stability and mobility are inversely related mechanically. Highly stable postures allow a person to resist changes in position, while the initiation of movement (mobility) is facilitated by the adoption of a less stable posture. The skilled mover learns to control the position of their body for the right mix of stability and mobility for a task.

The biomechanical factors that can be changed to modify stability/mobility are the base of support and the position and motion of the center of gravity relative to the base of support. The base of support is the two-dimensional area formed by the supporting segments or areas of the body (◻ Fig. 7.16). A large base of support provides greater stability because there is greater area over which to keep the bodyweight under control. Much of the difficulty in many gymnastic balancing skills (e.g., handstand or scale) comes from the small base of support on which to center bodyweight.

The posture of the body in stance or during motion determines the position of the center of gravity relative to the base of support. Since gravity is a major external force our body moves against, the horizontal and vertical positions of the center of gravity relative to the base of support are crucial in determining the stability/mobility of that posture. The horizontal distance from the edge of the base of support to the center of gravity (line of action of gravity) determines how far the weight must be shifted to destabilize a person (◻ Fig. 7.17a). If the line of gravity falls outside the base of support, the gravitational torque tends to tip the body over the edge of the base of support. The vertical distance or height of the center of gravity affects the geometric stability of the body. When the position of the center of gravity is higher, it is easier to move beyond the base of support than in postures with a lower center of gravity. Positioning the line of gravity outside the base of support can facilitate the rotation of the body by the force of gravity (◻ Fig. 7.17b).

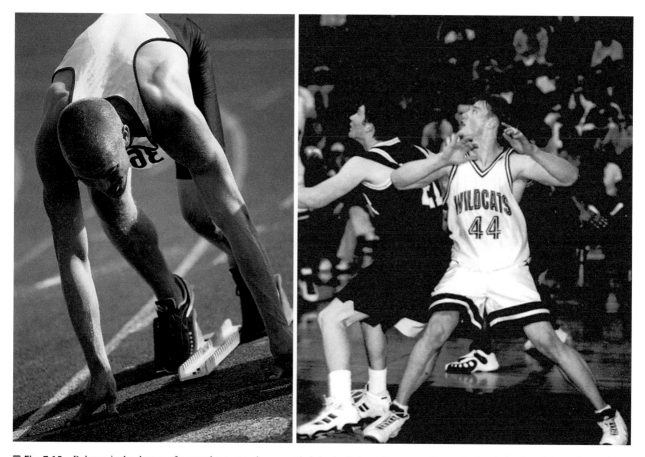

Fig. 7.15 Balance is the degree of control a person has over their body. Balance is expressed in static (track start) and dynamic conditions (basketball player boxing out an opponent). (Track image used with permission from Getty Images)

Fig. 7.16 The base of support is the two-dimensional area within all supporting or suspending points of the biomechanical system

7

☐ **Fig. 7.17**   The position of the line of gravity relative to the limits of the base of support determines how far the weight must be shifted for gravity to tend to topple the body **a** or the size of the gravitational torque helps create desired rotation **b**

Biomechanical studies of balance often document the motion of the two important forces of interest, bodyweight and the reaction force under the base of support. Video measurements using the segmental method measure the motion of the center of gravity over the base of support. Imagine where the center of gravity would be and how it would move in the base of supports illustrated in ☐ Fig. 7.16. Force platforms allow the measurement of the misnomer **center of pressure**, the location of the resultant reaction force relative to the base of support. In quiet standing, the center of gravity sways around near the center of the base of support, while the center of pressure moves even faster to push the weight force back to the center of the base of support. The total movement and velocities of these two variables are potent measures of a person's balance.

Recall that the inertia (mass and moment of inertia) and other external forces like friction between the base and supporting surface all affect the equilibrium of an object. There are also biomechanical factors (muscle mechanics, muscle moment arms, angles of pull, and so on) that affect the forces and torques a person can create to resist forces that would tend to disrupt their balance. The general base of support and body posture technique guidelines in many sports and exercises must be based on integration of the biological and mechanical bases of movement. For example, many sports use the "shoulder width apart" cue for the width of stances because this base of support is a good compromise between stability and mobility. Wider bases of support would increase po-

tential stability but put the limbs in a poor position to create torques and expend energy, creating opposing friction forces to maintain the base of support.

The Principle of Balance is based on the mechanical trade-off between stability and mobility. The Principle of Balance is similar to the Coordination Continuum because the support technique can be envisioned as a continuum between high stability and high mobility. The most appropriate technique for controlling your body depends on where the goal of the movement falls on the stability–mobility continuum. Coaches, therapists, and teachers can easily improve the ease of maintaining stability or initiating movement (mobility) in many movements by modifying the base of support and the positions of the segments of the body. It is important to note that good mechanical posture is not always required for good balance. High levels of skill and muscular properties allow people to have excellent balance in adverse, inherently unstable situations. A skater gliding on one skate and a basketball player caroming off defenders and still making a lay-up are examples of good balance in less than ideal conditions. A person walking and running dynamically control the mass of the body over a narrow (side-to-side) base of support. Running is essentially bouncing from one foot area base of support to another, sometimes at high speeds (sometimes over four steps per second and a meter apart).

Imagine that a physical therapist is helping a patient recover from hip joint replacement surgery. The patient has regained enough strength to stand for short lengths of time but must overcome some discomfort and instability when transitioning weight between both limbs in walking. The patient can walk safely between parallel bars in the clinic, so the therapist has the patient use a cane. This effectively increases the base of support, because the therapist thinks increasing stability (and safety) is more important.

If we combine angular kinetics with the Principle of Balance, it is possible to determine on what side of the body an assistive device like a cane should be held. If the cane was held on the same (affected hip) side, the base of support would be larger, but there would be little reduction in the pain of the hip implant because the gravitational torque of the upper body about the stance hip would not be reduced much. If the patient held the cane in the hand on the opposite (unaffected hip) side, the base of support would also be larger, and the arm could now support the weight of the upper body, which would reduce the need for hip abductor activity by the affected/recovering hip. Diagram the increase in area of the base of support from a single-leg stance in walking to a single-leg stance with a cane in each hand. Estimate the percentage increase in base of support area using the cane in each hand.

Classic examples of postures that would maximize mobility are the starting positions during a (track or swimming) race where the direction of motion is known. The track athlete in ◘ Fig. 7.18 has elongated his stance in the direction of his start and in the "set" position moves his center of gravity near the edge of his base of support. The blocks are not extended too far backward because this interacts with the athlete's ability to shift weight forward and generate forces against the ground. For a summary of the research on the effect of various start postures on sprint time, see Hay (1993). Hay's seminal book on the biomechanics of sports techniques also provides a good summary of early research on basic footwork and movement technique factors in many sports.

In many sports, athletes must take on defensive roles that require quick movement in many directions. The Principle of Balance suggests that postures that foster mobility over stability have smaller bases of support, with the center of gravity of the body not too close to the base of support. When athletes have to be ready to move in all directions, most coaches recommend a slightly staggered (one foot slightly forward) stance with feet about shoulder width apart. Compare the stance and posture of the volleyball and basketball players illustrated in ◘ Fig. 7.19. Compare the size of the base of support, and estimate the location of the center of gravity in both body positions. What posture differences are apparent, and are these related to the

◘ **Fig. 7.18** The starting position of a sprinter in the blocks shifts the line of gravity toward the front of the stance and the intended direction of motion. This stance favors mobility forward over stability

predominant motion required in that sport? Bases of support need only be enlarged in directions where stability is needed or the direction of motion is known.

There are movement exceptions to strict application of the Principle of Balance because of high skill levels or the interaction of other biomechanical factors. In well-learned skills like walking, balance is easily maintained without conscious attention over a very narrow base of support. Gymnasts can maintain balance on very small bases of support as the result of considerable skill and training. A platform diver doing a handstand prior to a dive keeps their base of support smaller than one shoulder width because extra side-to-side stability is not needed and the greater shoulder muscle activity that would be required if the arms were not directly underneath the body. Another example might be the jump shot in basketball. Many coaches encourage shooters to "square up" or face the basket with the body when shooting. Ironically, the stance most basketball players spontaneously adopt is staggered, with the shooting side foot slightly forward. This added base of support in the forward–backward direction allows the player to transition from pre-shot motion to the primarily vertical motion of the jump. It has also been hypothesized that this stagger in the stance and trunk (not squaring up) helps the player keep the shooting arm aligned with the eyes and basket, facilitating side-to-side accuracy (Knudson 1993).

Interdisciplinary Issue: Sex Differences
It is generally considered that the lower center of gravity in women gives them better balance than men. What is the biomechanical significance of the structural and physiological differences between men and women?

While there is substantial research on the physiological differences between the sexes, there is less comparative research on the biomechanical differences. Motor control and ergonomic studies have observed significant differences in joint angles during reaching (Thomas et al. 1998) and lifting (Lindbeck and Kjellberg 2000). Greater interest in sex differences seems to focus on issues related to risk of injury, for example, to like the anterior collateral ligament (ACL) (Malinzak et al. 2001; Charlton et al. 2002), although research on females and sex differences are still rare compared to research on males (Mujika and Taipale 2019). More research is needed specifically on female athletes. Normalizing kinematics (to stature) and kinetics (body mass) will likely show females with similar performance to males matched for skill and training; however greater public interest in male sports may contribute to the majority of biomechanics research on male participants. Biomechanics and other kinesiology sciences can help the world resolve rules for equitable competition between sexes or a more nuanced understanding of a continuum of sex differences for athletes with different gender expressions.

Application: Estimating Joint Moments of Walking
The ground reaction forces measured by force platforms in walking are used in clinical biomechanics labs to calculate net forces and torques in joints (inverse dynamics). These analyses calculate net forces and torques at the foot, then the lower leg, thigh, and so on. For the sagittal and frontal planes illustrated (❏ Fig. 7.20), can you visualize how the typical ground reaction force typ-

❏ **Fig. 7.19** Comparison of the ready positions of a basketball player and a volleyball player. How are the mechanical features of their stance adapted to the movement they are preparing for?

ically creates a knee flexor and adductor torques in stance? Can you draw the moment arms relative to the knee joint axis for these forces? The stance limb activates muscles to create a net knee extensor torque to support bodyweight in the sagittal plane (A), and a knee abductor torque to stabilize the knee in the frontal plane (B). These estimated moments calculated from projection of the ground reaction force to joints further up a linked-segment system will be close to the loading the body experiences. They are not as accurate as a full inverse dynamics analysis that calculates forces/torques in sequence up each segment of the system (Wells 1981). Wells' study is further confirmation of the segmental interaction principle, where professionals should speculate with care about muscle and external force actions when distributed to a linked segment system.

Balance is a key component of most motor skills. While there are many factors that affect the ability to control body mobility and stability, biomechanics focuses on the base of support and position of the center of gravity. Mechanically, stability and mobility are inversely related. Coaches can apply the Principle of Balance to select the base of support and postures that will provide just the right mix of stability/mobility for a particular movement. Angular kinetics is the ideal quantitative tool for calculating center of gravity and for examining the torques created by gravity that the neuromuscular system must balance.

7.8 Angular Impulse/Angular Momentum

Recall that Newton's second law expressed the insight that forces applied over time (impulse) are equal to the change in momentum (product of mass and velocity) of an object. This second law also applies to rotations, so the resultant angular impulse (torque•time) will equal the change in angular momentum of the object about the axis of interest.

Projectiles spinning rapidly, be they athletes or balls, conserve angular momentum because gravity acts through the center of mass and fluid forces are often not large enough to slow rotation. A fortunate by-product of this conservation of angular momentum is the inertial tendency to stabilize the axis of rotation of the spinning object. If one was to spin a top on the table or a basketball on the tip of our finger, the objects continue to spin in an upright orientation until friction gradually reduces the rotation and they fall over. Conservation of angular momentum is quite useful for the stabilization of the axis of rotation of objects (▶ Fig. 3.9). Skilled rotation in many sports/activities involves careful application of off-center forces (torque) over time and the manipulation of that angular momentum. Torques can modify the angular momentum or the person can move body mass with joint rotations, changing the mo-

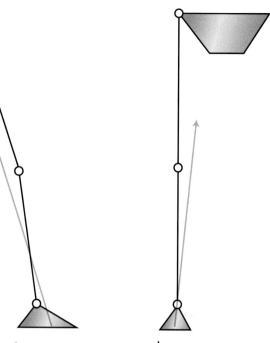

Fig. 7.20 Typical ground reaction force vectors in the stance phase of walking in the sagittal plane **a** and frontal plane **b**. What torques do these forces tend to make about the knee joint axes?

ment of inertia and the angular velocity. Any angular momentum has a stable axis of rotation, so rotation tends to stay in that plane.

7.9 Summary

The key mechanical variable in understanding the causes of rotary motion is the moment of force or torque. A force acting off-center on an object creates a torque. Torque is a tendency to rotate that is and is equal to the force times its moment arm (right-angle distance between axis of rotation and line of action of the force). Moments explain how levers either magnify force or motion. The moment of inertia is a variable expressing the angular inertia of an object about a specific axis of rotation. The moment of inertia most strongly depends on the distribution of mass relative to the axis of rotation of interest. The smallest moment of inertia in any plane for a body is about an axis at the center of mass/gravity. When all the torques acting on an object sum to zero, the object is said to be in static equilibrium. The equations of static equilibrium are often used to calculate the center of gravity/mass of objects. Biomechanics most often uses the reaction change and segmental methods to calculate the center of gravity of the human body. Balance is the ability of a person to control their body position relative to some base of support. The Balance Principle deals with the mechanical factors that affect balance, and the trade-off between stability and mobility in various body postures.

Torque applied over time creates angular impulse, determining the change in angular momentum of objects. An object's angular momentum in rotation contributes to stability about that axis of rotation.

❓ Review Questions

1. What are the two most important parameters that determine the size of a torque or moment of force?

2. What is the inertial resistance to angular acceleration object about an axis, and what factors affect its size?

3. Give examples of how the human body can position itself to increase or decrease its inertial resistance to rotation.

4. Calculate the shoulder flexion torque required to hold an 80 lb barbell just above your chest in a bench press (transverse plane). The horizontal distance from your shoulder axis to the barbell is 0.9 feet.

5. Restate Newton's three laws of motion in angular kinetic terms.

6. Explain how static equilibrium can be used to calculate the center of gravity of the human body.

7. Draw or trace a few freeze-frame images of the human body in several positions from sport or other human movements. Estimate the location of the center of gravity.

8. A mischievous little brother runs ahead of his sister and through a revolving door at a hotel. The little brother pushes in the opposite direction of his sister trying to exit. If the brother pushes with a maximum horizontal force of 40 pounds acting at a right angle and 1.5 feet from the axis of the revolving door, how much force will the sister need to create acting at 2.0 feet from the axis of rotation to spoil his fun?

9. What mechanical factors can be used to maximize stability? What does this do to a person's mobility?

10. What movement factors can a kinesiology professional qualitatively judge that show a person's balance in dynamic movements?

11. Say the force F_2 applied by the student in ▣ Fig. 7.4 acted 55° in from the tangent to the merry-go-round. Calculate the torque created by the student.

12. Draw a free-body diagram of a person standing on a reaction board (hint: the system is the body plus the board). Estimate the length of the board and the horizontal distance to the person's center of gravity. Calculate the reaction force on the board if you were the person on it.

13. If the rotary component of a brachialis force is 70 N and the muscle attaches 0.4 m from the axis of rotation, what is the flexor torque created by the muscle?

What other information do you need in order to calculate the resultant force created by the brachialis?

14. Since angular impulse and angular momentum are related, how long must a person apply a constant torque of 60 Nm in a horizontal plane to create a 100 kg·m^2/s increase in angular momentum of a bat?

15. Objects in rotation have stable axes of rotation because of what mechanical variable?

16. What are the three classes of levers and what mechanical effect do they improve?

❗ Key Terms

Angular impulse · Angular momentum · Balance principle · Base of support · Center of mass (center of gravity) · Moment arm · Moment of force (moment or torque) · Moment of inertia · Reaction change · Segmental method · Static equilibrium

⭐ Suggested Reading

Brown, L. E. (Ed.). (2000). *Isokinetics in human performance*. Champaign: Human Kinetics.

Chaffin, D. B., et al. (2006). *Occupational biomechanics* (4th ed.). New York: Wiley.

Huxham, F. E., et al. (2001). Theoretical considerations in balance assessment. *Australian Journal of Physiotherapy, 47*, 89–100.

Mann, R. V. (1981). A kinetic analysis of sprinting. *Medicine & Science in Sports & Exercise, 13*, 325–328.

McGill, S. M., & Norman, R. W. (1985). Dynamically and statically determined low back moments during lifting. *Journal of Biomechanics, 18*, 877–886.

Murray, M. P., et al. (1967). Center of gravity, center of pressure, and sup-portive forces during human activities. *Journal of Applied Physiology, 23*, 831–838.

Winter, D. A. (1984). Kinematic and kinetic patterns of human gait: Variability and compensating effects. *Human Movement Science, 3*, 51–76.

Winter, D. A. (1995). Human balance and posture control during standing and walking. *Gait and Posture, 3*, 193–214.

Winters, J. M., & Woo, S. L. Y. (Eds.). (1990). *Multiple muscle systems*. New York: Springer.

Zatsiorsky, V. M. (2002). *Kinetics of human motion*. Champaign: Human Kinetics.

🔗 Internet Resources

▸ Crash Course Physics. PBS lessons related to angular kinetics are: Rotational Motion, Statics, and Torque. *klru.pbslearningmedia.org/collection/crash-course-physics/*

▸ NASA Moment/Torque. *page–grc.nasa.gov/www/k-12/airplane/torque.html*

▸ NBC News Learn. Video lessons on the Science of Golf, Science of NFL Football, and Science of the Winter Olympics. *nbclearn.com/portal/site/learn/resources*

▸ Physiopedia page on sports. *biomechanics–physio-pedia.com/Biomechanics_In_Sport*

▸ Quintic Consultancy: Moment of Inertia in Diving. *quinticsports.com/moment-of-inertial*

▸ Torque Khan Academy. *khanacademy.org/science/physics/torque-angular-momentum/torque-tutorial/a/torque*

Fluid Mechanics

Key Questions

➤ **What do you need to know to begin?**
Knowledge of musculoskeletal anatomy, algebra, kinematics, linear and angular kinetics.

➤ **Why do you need to know this material?**
Understanding how fluid forces affect rigid bodies informs general recommendations for technique, equipment, and projectile motion.

➤ **What is the key idea?**
Fluids create significant forces that interact with and affect human motion and cannot normally be ignored when making recommendations on technique and equipment.

External forces that have a major effect on most human movements are related to immersion in or flow of fluids past a body. This chapter reviews the mechanical effect of moving through air and water, the two most common fluids encountered in human movement. Fluid forces usually result in considerable resistance to high-velocity movements through fluids, so many sport techniques and pieces of equipment are designed to minimize the fluid resistance force of drag. Fluid forces, however, can also be used to create desired movement, like the skillful use of lift forces to swim or the application of spin to projectiles. This chapter concludes with application of this use of fluid forces in the **Principle of Spin**.

8.1 Fluids

You may have studied the various states of matter in physics or noticed that many substances are not easily classified as totally solid or liquid (fat, water, chocolate) depending on temperature. Mechanically, fluids are defined as substances that *flow or continuously deform when acted upon by shear forces*. A thorough review of all the nuances of fluid mechanics is not possible in this chapter, so key concepts related to the supporting force of immersion in fluids and the forces that arise from moving through fluids are reviewed. Several references are cited to guide students interested in digging deeper into the nuances of fluid mechanics.

8.2 Fluid Forces

For the purposes of this chapter, the major fluid forces that affect human motion are classified according to an object's position or velocity within a fluid. When an object is placed in a fluid, there is a resultant upward force or supporting fluid force called **buoyancy**. The fluid force related to how the fluid flows past the object is resolved into right-angle components called **lift** and **drag**. In most movements, people have considerable control over factors that affect these forces and can, therefore, change technique or equipment to improve performance. Let us see how these fluid forces affect human movement.

8.2.1 Buoyancy

The vertical, supporting force of a fluid is called buoyancy. When an inanimate object is put in a fluid (like water), the vector sum of gravity and the buoyant force determines whether or not the object will float (⬛ Fig. 8.1). The **Archimedes Principle** states that the size of the buoyant force is equal to the weight of the fluid displaced by the object. Folklore says that the famous Greek physicist/mathematician realized this important principle when noticing water level rise while getting into the water to take a bath.

A sailboat floats at a level where the weight of the boat and contents are equal in size to the weight of the volume of water displaced by the hull below water level. Flotation devices used for water exercise and safety increase the buoyancy of a person in two ways: having a lower density (mass/volume) than water and having a hollow construction. These flotation devices displace water that weighs substantially more than the device, increasing the buoyancy of the person, if you can get them and the floating devices in the water!

D. Knudson, *Fundamentals of Biomechanics*, https://doi.org/10.1007/978-3-030-51838-7_8

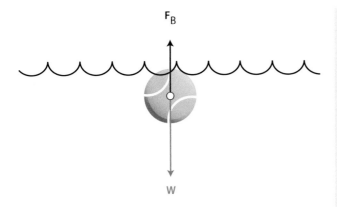

F_B

W

◘ Fig. 8.1   The resultant vector of gravity (**W**) and buoyancy (**FB**) will determine if an inanimate object floats. The ball illustrated will sink because the negative weight force is larger than the positive buoyant force

8

In a gravitational field, the mass of a fluid is attracted in a particular direction (vertically toward the center of the earth). The weight of water is typically 9800 N per cubic meter, but this figure gradually increases for water at greater depths. The deeper a scuba diver descends, the greater the fluid pressure around them (because of the greater mass of water essentially "on top" of them from the earth's gravitational perspective). This increased pressure in a particular volume of fluid means that the volume of water at lower depths weights more than a similar volume of water at the surface, so the buoyant force on objects tends to slightly increase as depth increases. A similar phenomenon occurs as we descend from a mountain, where the fluid pressure of the atmosphere on us increases. The buoyant force on the human body from the "sea" of atmospheric gases also depends on our depth (opposite of elevation), but is usually a fraction of a pound and can be ignored in vertical kinetic calculations of human movement.

The density of the human body is very close to that of water, largely due to the high water content of all tissues. Lean tissues (muscle and bone) have densities greater than water, while body fat tends to be less dense than water. The buoyant force on a swimmer varies with changes in body composition and when the person inhales or exhales. Taking a deep breath expands the chest, which increases the volume of the body and increases the buoyant force. If you have ever taught a swimming class you know that people typically fall into three groups based on their anthropometric somatotype and body composition: floaters, conditional floaters, and sinkers. The majority of your swim class can easily float when holding their breath (conditional floaters). There will be a few people who easily float (floaters) or cannot float (sinkers) without some form of propulsion or flotation device.

Activity
The next time you are at a pool, see if you can detect an increase in buoyant force with increasing depth. Hold a large sport ball (water polo, soccer, football) in one hand and gradually submerge it. Note the downward vertical force you exert to balance the buoyant force of the ball as it descends. Also note the horizontal forces you must exert to keep your hand forces balanced with the buoyant force and gravity! Another simple activity is to mark the water line on a floating ping-pong ball. Tape dimes to the ball and find the maximum buoyant force of a ping-pong ball. Does a forcibly submerged ball have potential energy?

The buoyant force in water acts upward at the center of buoyancy. The **center of buoyancy** is essentially the centroid of the volume of water displaced by an object. In the human body, the trunk makes up most of the volume, so the center of buoyancy is located 1–2 cm superior (McLean and Hinrichs 2000) to the center of gravity (◘ Fig. 8.2). Since so much body volume is in the upper

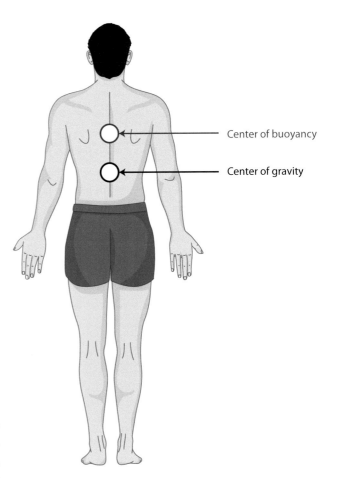

Center of buoyancy

Center of gravity

◘ Fig. 8.2   The center of buoyancy of the human body is superior to the center of gravity because of the large volume of the upper body trunk, moving the rest of the body makes smaller changes in the center of buoyancy than in the center of gravity. Note that the weight and buoyant forces can create a force couple that will tend to rotate the swimmer until these forces are colinear. Swimmers still scared of the water have great difficulty floating on their back because they tend to pike (flex the hip) and lift the head/upper trunk out of the water. The resulting loss of buoyant force (from less water displacement) tends to lower the swimmer deeper into the water, not keeping their head above water. If you are having difficulty getting a swimmer to relax and do a back float, how can you shift their limbs to shift the center of gravity and also maintain a large buoyant force? Are there assistive devices you could use and gradually remove to ease the nervous swimmer?

Application: Hydrotherapy

Therapeutic exercises in water utilize its buoyant force to unload the lower extremity. The amount of unloading of the body can be easily manipulated by the extent of submersion. This exercise modality differs from suspension systems that unload the body by pulleys lifting up the trunk up with counterweights or negative pressure chambers around treadmills. The flow of water also creates lift and drag forces that have been shown to create training resistances and differences in muscle activation in exercise (Poyhonen et al. 2001). Therapy pools that create currents for exercise likely exaggerate the neuromuscular differences between these movements and dry land movement.

We have seen that objects in a fluid experience a supporting force related to the position of the object in the fluid and the density of the object. The next section will deal with the interaction forces between an object and the fluid when there is relative motion between the two. These fluid motion forces can be quite large. The fluid forces between the air and your body are nearly identical if you are falling about a speed of 120 km/h while skydiving in a specific body position or if you are apparently still on top of a column of 120 km/h air flow in a simulator. In both these situations, the drag forces on the body are equal to your body weight. In the first case, the body is falling through essentially still air, while in the second case, the body is essentially stationary with air flowing upward over it.

8.2.2 Drag

The fluid force resisting motion between an object and a fluid is called **drag**. Drag acts in the same direction (parallel) as the relative flow of the fluid past an object and in the opposite direction of the object's motion in the fluid. Drag forces act on the fisherman (creek) and the fly (air) due to the relative motion of the fluid past the objects (see ◻ Fig. 8.3). If there are no propulsive forces acting on the object, like a projectile (▶ Chap. 5), the drag force tends to slow down the motion of the projectile through the fluid. Since the drag force acts parallel to the relative flow of the fluid, it is much like the contact force of friction studied in ▶ Chap. 6.

Research has shown that the size of the drag force (F_D) that must be overcome in a fluid can be calculated using the following formula: $F_D = \frac{1}{2} C_D \rho A_P V^2$. The coefficient of drag (C_D) is a dimensionless number much like the coefficient of friction or restitution. ◻ Table 8.1 lists the typical C_D values for air resistance for many common objects. We will see later that C_D depends on many object and fluid flow factors. Drag also depends on fluid density (ρ) and the projected frontal area (A_P) in the path of the fluid flow. The most important factor affecting drag is the relative velocity (V^2) of the fluid past the object.

Like the velocity term in kinetic energy, the force of drag varies with the square of the relative fluid velocity. This means that, all other things being equal, a cyclist that doubles and then triples his pace increases drag by 4 and 9 times compared to his initial speed! This explains why running faster or into a strong breeze feels much more difficult. The importance of the adjective "relative" can be easily appreciated by noting that it is easier to run with a strong breeze *behind* you. The dramatic effect of drag on sprint performance has forced the International Amateur Athletic Federation to not ratify sprint records if the wind assisting a runner exceeds 2.0 m/s. World records are always a controversial issue, but current weighting of records in many events does not take into account the effect of altitude (Mureika 2000) or latitude (Mizera and Horvath 2002). Remember that relative velocity means that we are talking about a local kinematic frame of reference—in other words, the speed and direction of fluid flow relative to the object of interest. These drag forces increase with the square of velocity and often dramatically affect performance.

The drag force on an object has several sources: *surface drag*, *pressure drag*, and *wave drag*. Understanding these drag forces is important for minimizing these resistances in many sports and activities.

Surface drag can be thought of as a fluid friction force, much like solid friction force studied in ▶ Chap. 6. Surface drag is also commonly called friction drag or skin friction drag. It results from the frictional force between fluid molecules moving past the surface of an object and the frictional force between the various layers of the fluid. Viscosity is the internal resistance of a fluid to flow. Air has a lower viscosity than water, which has a lower viscosity than maple syrup and most motor oils.

Fig. 8.3 The fluid force of drag (F_D) acts in a direction opposing the relative flow of fluid past the object

Table 8.1 Typical drag coefficients (C_D) for common objects in air

Streamlined human body	0.04
Airplane wing	0.05
Sports car	0.2–0.3
Sphere/ball	0.5
Bicycle	0.8–0.9
Human body standing	1.0–1.3
Motorcycle and rider	1.8

Data from: ▶ engineeringtoolbox.com/drag-coefficient-d_627.html

Assume a surfer is floating on their board waiting for the right wave (▢ Fig. 8.4). The fluid flow below the apparently stationary surfboard creates surface drag from the flow of the ocean under the board. Water mol-ecules immediately adjacent to the board are slowed by shear forces between them and the molecules of the board. So the fluid close to the board moves slower than the ocean water farther from the board. In fact, there is a region of water layers close to the board that moves more slowly because of viscous (fluid friction) forces between the fluid particles. This region of fluid affected by surface drag and viscosity near an object is called the **boundary layer**. Layers of fluid more distant from the object that are not affected by drag forces with the object represent the free stream velocity. Have you ever started driving your car and noticed a small insect on the hood or windshield? I am willing to wager that most of you noticed the considerable speed you had to drive to disrupt the boundary layer the insect stood in before it was swept away! We will see that this relative or free stream velocity is one of the most important factors affecting the drag and lift forces between objects and fluids.

Performers cannot change the viscosity of the fluid they move in, but they can modify the roughness of their body or equipment to decrease surface drag. Surfboards

Fig. 8.4 The water nearest a surfboard forms a boundary layer that flows more slowly (V_B) past the board than the free stream velocity (V_{FS}) because of friction with the board and fluid friction

Fig. 8.5 Form drag forces (F_D) result from a vacuum pressure formed in the pocket formed behind a submerged object **a**. Decreasing the pressure in this wake is how contouring the rear profile of an object (streamlining) decreases form drag **b**

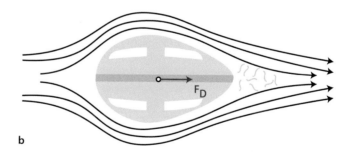

and skis are waxed, a swimmer may shave body hair, or very smooth body suits may be worn to decrease surface drag. Interestingly, some suits for high fluid force sports (swimming, cycling, ski jumping, etc.) actually introduce texture on portions of the fabric to modify both lift and drag forces (Benjanuvatra et al. 2002). While it is important to minimize surface drag, the largest fluid resistance in many sports tends to be from pressure drag.

The second kind of drag force that dominates the fluid resistance in many sports is **pressure drag**. Pressure drag is the resistance force to fluid flow that is created by a pressure differential when the fluid flows around a sub-

merged object. A simplified illustration of this phenomenon is presented in **Fig. 8.5**. The collision of the object and molecules of fluid creates a high pressure on the front of the object, while a lower-pressure region or wake is formed behind the object. The region of higher pressure "upstream" creates a resultant force backward on the object. We will see that the mechanics of this pressure differential is a bit more complicated and related to many factors. Fortunately, many of these factors can be modified to reduce the pressure drag in many movements by changing body shape. We will see shortly that some human movements use drag as a propulsive force.

8

To understand the variations in pressure drag, one must differentiate two different kinds of fluid flow in the boundary layer: *laminar* and *turbulent*. The air flow past a tennis ball can be highlighted by smoke introduced into a wind tunnel, depicted in ▣ Fig. 8.6, which shows both predominantly laminar and turbulent flow. Laminar flow typically occurs in low-velocity conditions with streamlined objects where the fluid particles can flow relatively undisturbed in parallel layers. Turbulent flow occurs when fluid molecules bounce off the object and each other, mixing in chaotic fashion.

The kind of fluid flow over an object also affects pressure drag. At low velocities, the boundary layer is laminar and cannot flow very far around a non-rotating sphere before peeling away from the surface (▣ Fig. 8.7a), creating a large pressure drag. At higher velocities, the boundary layer flow is turbulent and more resistant to the pressure gradient as it flows around the object. This results in a later point of separation (▣ Fig. 8.7b) and lower pressure drag than laminar flow. In most objects, there is not a distinct transition from laminar to turbulent flow, but a critical or transition region where flow is unstable can be either laminar

or turbulent. This transition region is important in the flight of spherical balls because the coefficient of drag can drop dramatically, creating a "drag crisis." Increasing the roughness of the ball (scuffing a baseball or putting dimples on a golf ball) can decrease the velocity where these lower drag forces occur. Scientists interested in fluid mechanics use a dimensionless ratio (the Reynolds number: Re) to combine the effects of object geometry on fluid flow to study fluid forces on objects. This chapter will not go into detail on Reynolds numbers, but interested students can see Mehta (1985) or Mehta and Pallis (2001a, b) for more information on Reynolds numbers related to sports balls.

Much of the variation in the flight characteristics of many sport balls is a result of differences in drag and lift forces that are directly related to variations in fluid flow in the transition region of Reynolds numbers. This provides a great opportunity for skill and coaching to modify the flight characteristics of many shots or throws in sports. Many of these important effects are counterintuitive. For example, slightly increasing the roughness of a sphere (golf or baseball) might decrease drag by promoting a more turbulent boundary layer, while increasing the lift forces generated. Another example is the nature of the felt on tennis balls. The felt has a major influence on the drag coefficient (Mehta and Pallis 2001a), so professional tennis players when serving select balls in part based on the amount of felt fluff and wear. In the next section, we will study how surface roughness of rotating balls can also be used to increase the fluid force of lift.

The two major techniques employed to decrease pressure drag in human movement are (a) decreasing the frontal area and (b) streamlining. The smaller the frontal area, the less the fluid must be accelerated to flow around the object. Extending the downstream lines of an object also decreases pressure drag by delaying separation and decreasing the turbulent wake behind the object. Swimming strokes often strike a balance between maintaining a streamlined body position and a one that maximizes propulsion. The high speeds and large surface areas in cycling make streamlined equipment and body positions critical (▣ Fig. 8.8).

▣ **Fig. 8.6**   The air flow past a tennis ball shows both laminar (L) and turbulent (T) fluid motion. The topspin on the ball deflects the air flow creating another fluid force called lift. (Photo courtesy of NASA Ames Research Center Fluid Mechanics Laboratory and Cislunar Aerospace, Inc)

a   b

▣ **Fig. 8.7**   Spheres like sport balls create different fluid flows and drag force depending on many factors. Primarily laminar flow **a** can result in large pressure drag because of early separation of the bound-

ary layer for a large wake, while turbulent flow **b** will often delay boundary separation and decrease pressure drag

◻ Fig. 8.8 High-speed sports like cycling (or skiing) use streamlining to decrease speed losses due to drag forces. Image used with permission from Getty Images

Application: Drafting

Sports with very high relative velocities of fluid flow are strongly affected by the resistance of the drag force. One strategy used to minimize drag forces in these sports (cycling, car racing) is drafting. Drafting means following closely behind another competitor, essentially following in their lower fluid pressure wake. The athlete in front will use more energy against greater pressure drag, while the drafting athlete experiences less fluid resistance and can use less energy while they draft. The strategy of the drafting athlete is often to outsprint the leader near the end of the race. In many team racing sports (cycling), it is the teammate who expends the extra energy to be in the front early in the race who makes it possible for other team members to finish in a higher final position. Drafting even has advantages in some lower-velocity events like swimming (Chatard and Wilson 2003). An athlete running about

1 m behind another runner can decrease air resistance, decreasing the metabolic cost of running by about 7% (Pugh 1971).

The third kind of drag is **wave drag**. At the surface of a fluid, it is possible that disturbances will create waves within the fluid that resist the motion of an object with area projecting at this surface. Wave drag can constitute a major resistance in swimming (Rushall et al. 1994). Triathletes swimming in the open water must overcome wave drag from both the wind and from their fellow competitors. Swimmers in enclosed pools are less affected by wave drag than those swimming in the open water because of lane makers and gutters designed to dampen waves. Small variations in lane placement, however, may affect the wave drag experienced by a swimmer.

8.2.3 Lift

The fluid force acting at right angles to the flow of fluid is called **lift** (◻ Fig. 8.9). Just like contact forces are resolved into right-angle components (friction and normal reaction), fluid forces are resolved into the right-angle forces of drag and lift. Since lift acts at right angles to the flow of the fluid, the direction of the lift force in space varies and depends on the shape, velocity, and rotation of the object. It is unwise to assume that the lift always acts upward. For example, the wings on race cars are designed to create a downward lift force to stabilize the car and keep it in contact with the ground.

The size of the lift force can also be modeled with a coefficient of lift (C_L) and a familiar equation: $F_L = \frac{1}{2} C_L \rho A_p V^2$. Just like drag, lift varies with the square of the relative velocity (V^2) of fluid. Earlier we characterized

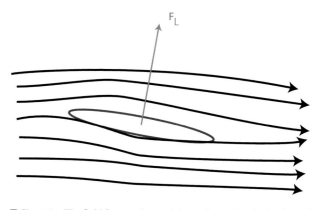

◻ Fig. 8.9 The fluid force acting at right angles to the relative flow of fluid is called lift. Lift acts in all directions, not just upward

drag as primarily a fluid resistance. Lift tends to be a fluid force and is often used for propulsion. One of the early leaders in swimming research, "Doc" Counsilman at Indiana University, used high-speed films of skilled swimmers to measure the complex patterns of arm and leg motions and was instrumental in demonstrating the importance of lift as a propulsive force in swimming (Counsilman 1971). Whether lift or drag is the primary propulsive force used in swimming is a controversial issue (Sanders 1998), and other theories like vortices (Arellano 1999) and axial fluid flow (Toussaint et al. 2002) are currently being examined. The important thing for swim coaches to realize is that precise arm and leg movements are required to use the hands and feet effectively, and that skilled swimmers learn to use both lift and drag forces for propulsion.

Synchronized swimming and competitive swimming tend to use small "sculling" hand movements to create lift forces for propulsion. A skilled swimmer precisely adjusts the pitch of their hands to maximize the down-the-pool resultant of the lift and drag forces (◘ Fig. 8.10). This is much like the high-tech propellers in modern aircraft that change the pitch of a blade based on flying conditions. The complexity of fluid flow over the human body has made it difficult to resolve the controversy over which fluid forces are most influential in propulsion. Another example of controversy and potential research is to understand why elite swimmers usually keep their fingers slightly spread. It is unknown if this improves performance from increased surface area for the hand, or that the flow through the fingers acts like a slotted airplane wing in modifying the lift created at lower speeds of fluid flow. Coaches should base their instruction on the kinematics of elite swimmers and allow scholars to sort out the basic science question whether lift, drag, or vortex (swirling eddies) modifying the flow of fluid is the primary propulsive mechanism for specific swimming strokes.

Activity

After trying the ball-submersion experiment, try out this little activity using freestyle swimming technique. Compare the number of arm pulls it takes to cross the pool using two extremes in arm pull technique. First, try an arm pull with a primarily paddling motion, straight (down the pool), circular (side view) downward under your shoulder. The next arm pull should be more like the traditional freestyle technique, sculling the hand/arm in a narrow "S" pattern (frontal plane view) down the body. The paddle stroke would use primarily drag for propulsion, while the sculling motion would combine lift and drag for propulsion. Attempt to match the speed/tempo of the pulls and employ a flotation assist (like a pull buoy), and no flutter kick for a true comparison. Which fluid force seems to be most effective in pulling your body through the water with the fewest strokes?

There are two common ways of explaining the cause of the lift force in fluids: Newton's Laws and Bernoulli's Principle. ◘ Figure 8.11 shows a side view of the air flow past a discus in flight. The lift force can be understood using Newton's second and third laws. The air molecules striking the undersurface of the discus are accelerated or deflected off its surface. Since the fluid is accelerated in the direction indicated, there

◘ Fig. 8.10 The inward sweep skill of a freestyle swimmer's hand may be selected to maximize the down-the-pool resultant of the lift and drag acting on the hand **a**. This angle of attack (8A) is critical to the lift and drag created **b**

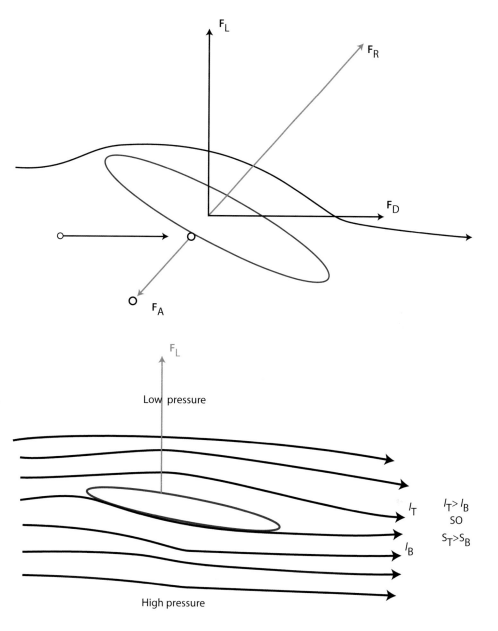

Fig. 8.11 The kinetics of the lift and drag forces can be explained by Newton's laws and the interaction of the fluid and the object. The air molecules (·) deflecting off the bottom of the discus creates the lift (**F**_L) and drag (**F**_D) acting on the discus

Fig. 8.12 The lift force (**F**_L) acting on a discus or airplane wing can be explained using Bernoulli's Principle. The greater distance (and faster speed of fluid flow) over the top of the wing (l_T) compared to the distance under the bottom (l_B) creates a pressure differential. The high pressure below and lower pressure above the wing lifts the airplane

must have been a resultant force (**F**_A) acting in that direction on the fluid. The reaction force (**F**_R) acting on the discus creates the lift and drag forces on the discus. **F**_R is large because there are many fluid molecules deflected downward.

The other explanation for lift forces is based on pressure differences in fluids with different velocities discovered by the Swiss mathematician Daniel Bernoulli. **Bernoulli's Principle** states that the pressure in a fluid is inversely proportional to the velocity of the fluid. In other words, the faster the fluid flow, the lower the pressure the fluid will exert on its environment. In many textbooks, this has been used to explain how lift forces are created on airplane wings. Airplane wings are designed to create lift forces from air flow pressure around the wing (◻ Fig. 8.12). Fluid molecules passing over the top of the wing cover a greater distance than molecules passing under the wing in the same amount of time and, therefore, have a greater average speed than the air flow under the wing. The lower pressure above the wing relative to below the wing creates a lift force toward the top of the wing. Unfortunately, this simplistic explanation is not technically correct. Rather, it is an oversimplification of a complex phenomenon. Bernoulli's equation only accounts for force changes due to fluid pressure (no work, heat, or friction) or a frictionless (inviscid) flow. This is not the case in most fluid dynamics situations (airplane wings or hands in the pool). Unfortunately, Bernoulli's Principle has also been commonly overgeneralized to lift forces on sport balls.

8.2.4 The Magnus Effect

Lift forces can also be created by the spin imparted to spherical balls. These lift forces arise because of pressure differences and fluid deflection resulting from ball spin. This phenomenon of lift force in spinning balls is called the **Magnus Effect**, after German engineer Gustav Magnus, though it may have been discovered a century earlier (Watts and Bahill 2000). Sport balls hit or thrown with topspin have trajectories that curve more downward than balls with minimal spin or backspin. This greater downward break comes from the vertical resultant force from gravity and the primarily downward lift force from the Magnus Effect.

> **Activity: Bernoulli's Principle**
> An easy way to demonstrate Bernoulli's Principle is to use a small (5 × 10 cm) piece of regular weight paper to simulate an airplane wing. If you hold the sides of the narrow end of the paper and softly blow air over the top with pursed lips of the sagging (from gravity) paper, the decrease in pressure above the paper (higher pressure below) will lift the paper, defeating gravity!

Recall that it was noted that Bernoulli's Principle is often overgeneralized to explain the lift force from the Magnus Effect. This oversimplification of a complex phenomenon essentially begins by noting that a rotating sphere affects motion in the boundary layer of air (◘ Fig. 8.13) because of the very small irregularities in the surface of the ball and the viscosity of the fluid molecules. Fluid flow past the ball is slowed where the boundary layer rotation opposes the flow, but the free stream fluid flow will be faster when moving in the same direction as the boundary layer. For the tennis ball with topspin illustrated in ◘ Fig. 8.13, Bernoulli's Principle would say that there is greater pressure above the ball than below it, creating a resultant downward lift force. As direct and appealing as this explanation is, it is incorrect because Bernoulli's Principle does not apply to the kinds of fluid flow past sport balls since the fluid flow has viscous properties that create a separation of the boundary layer (◘ Fig. 8.6). Bernoulli's Principle may only apply to pressure differences away from or outside the boundary layer of a spinning ball.

A better explanation of the lift force is based on how ball spin creates a deflection of the fluid, as evidenced by the shifted separation point of the boundary layer. At the spin rates that occur in sports, the boundary layers cannot "stick" to the ball all the way around because of an adverse pressure gradient behind the ball. Note how the topspin on the ball in ◘ Fig. 8.6 creates earlier separation of the boundary layer on the top of the ball, which results in upward deflection of the wake behind the ball (Mehta and Pallis 2001a). The backward motion of the boundary layer on the bottom of the ball increases the momentum of the boundary layer, allowing it to separate later (downstream), while the boundary layer separates sooner on top of the ball. This asymmetric separation of the boundary layer results in an upward deflection on the wake. An upward force on the fluid means that an equal and opposite downward lift (Newton's third law) is acting on the ball.

The lift force created by the Magnus Effect is apparent in the curved trajectory of many sport balls. Golf balls are hit with backspin to create lift forces in flight that resist gravity and alter the trajectory of shots (in-

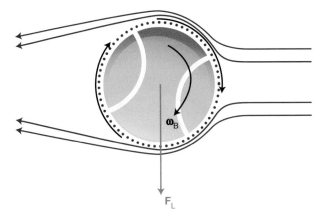

◘ Fig. 8.13 The lift forces created on spinning spheres is called the Magnus Effect. An overly simplified application of Bernoulli's Principle is often incorrectly used to explain the cause of this fluid force. Spin on the tennis ball drags the boundary layer of fluid in the direction of the spin. Fluid flow past the ball is slowed where the boundary layer opposes the free stream flow, increasing the fluid pressure. The topspin on this ball (like the ball in ◘ Fig. 8.6) creates a downward lift force that combines with gravity to make a steep downward curve in the trajectory crease distance of the shot). Golf balls given sidespin create lift forces that curve a ball's flight in the horizontal plane. ◨ Figure 8.14 illustrates a schematic of flight for various golf shots created using different sidespins in a horizontal plane. A tennis player imparting sidespin to a ball also creates a lateral lift force that makes the ball curve in flight, just like a soccer player can kick with a sidespin, creating a "banana" shot that curves around defenders to reach the goal. The flat trajectory of a fastball pitch in baseball results from an upward component of lift force that decreases the effect of gravity. Lift forces on sport balls are vectorially added to other forces like drag and weight to determine the resultant forces on the ball. The interaction of these forces creates the trajectory or flight path of the ball. The lift forces in a fastball are not larger than the weight of the baseball, so player perceptions of fastballs "rising" is an illusion based on their expectation that the ball will drop more before it crosses the plate. Fastballs do not rise; they just drop less than similar pitches with minimal backspin or topspin.

Much of the skill of baseball pitching relies on a pitcher's ability to vary the speed and spin of pitches. A curveball is pitched with an element of topspin that has a lift component in the same direction as gravity, so there is greater downward break as the ball nears the plate. The steepness of this break has resulted in hitters saying that a good curveball looks like it "drops off the table." Looking at the curveball in baseball (much like topspin shots in other sports, like volleyball and tennis) will provide a nice review of the kinetics of lift forces.

◨ Figure 8.15 shows a three-dimensional reconstruction of a major league curveball from two perspectives. The curveball has a gradual break that looks much steeper from the relatively poor vantage point of the hitter. Tracing high-speed ball trajectories has only been recently added to televised sports productions. Why does so much of the ball's break occur late in the trajectory when the hitter is swinging the bat and cannot change their swing? The major factors are the changing direction of the Magnus force in space and the slowing of the ball from drag.

Recall that the Magnus force acts perpendicular to the flow of fluid past the ball. This means that the Magnus force for a curveball primarily acts downward, adding to the drop created by gravity, but the horizontal

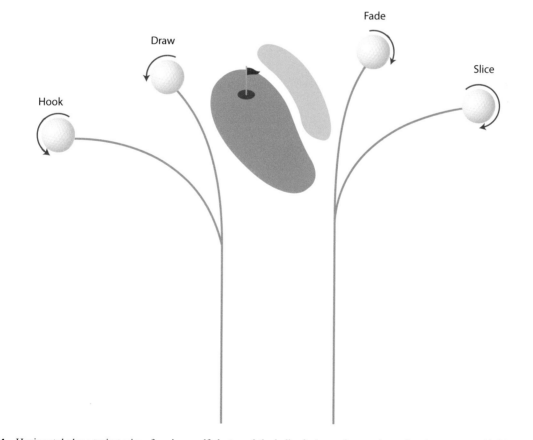

◨ **Fig. 8.14** Horizontal plane trajectories of various golf shots and the ball spin (curved arrows) creating these curves with Magnus forces

8

■ **Fig. 8.15**   Trajectory of the same curveball thrown by a major league player from two perspectives. Note the majority of the downward break occurs late in the trajectory, creating the illusion (to the hitter) of the ball "dropping off the table." (Adapted from Allman (1984))

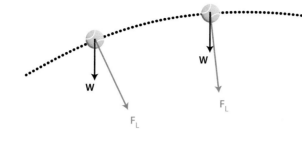

Fig. 8.16 The late "break" of a curveball can be explained by the changing direction of the Magnus force (**F**$_L$) on the ball and gravity (**W**). The downward deflection of the ball accelerates because the lift forces not only act downward with gravity, but backward (toward the pitcher). Slowing of the ball allows for more downward break

component of the lift force changes. As the direction of the pitch changes, so does the fluid flow and lift force (■ Fig. 8.16). As the ball breaks downward, the Magnus force has a backward component that slows the ball even more. This extra slowing and extra downward force contribute to the increasing "break" in the pitch late in its trajectory. Novice golfers can experience the same surprise if they consistently have trouble with "hooking" or "slicing" their drives. A "hooked" drive might initially look quite straight when the moderate sideward force is hard to detect due to the great initial speed of the ball. Unfortunately, as they watch their "nice" drive later in its trajectory, the ball seems to begin curving sideways late in flight. Diagram a transverse plane view of a "hooked" shot in golf. Draw the lift force acting on the ball and note its change in direction as the direction of the ball and relative flow of air changes.

The coefficient of lift (C_L) in spinning balls tends to be less sensitive to variations in Reynolds numbers than drag, so the size of the Magnus force depends mostly on spin and ball roughness (Alaways et al. 2001). Athletes can create more break on balls by increasing spin or increasing the surface roughness of the ball. In baseball, pitches can be thrown with four seams perpendicular to the throw, which increases C_L two to three times more than a two-seam rotation (Alaways et al. 2001).

Interesting exceptions to the dominant effect of lift forces on the flight of many sport balls are projections with minimal ball spin. A baseball "knuckleball" and a volleyball "floater" serve are examples of techniques where the ball is projected with virtually no spin. The erratic trajectory and break of these balls are due to unpredictable variations in air flow past the ball. As the ball gradually rotates, air flow can be diverted by a seam, valve stem, or an illegal scuff mark. These changes in surface irregularities make the ball take several small and unpredictable "breaks" during its trajectory. So spin, and the lack of it, on a sport ball has a major effect on trajectory.

Activity: Lift and Angle of Attack
When driving on an uncrowded road, roll down a window and put your hand out just into the flow of the air rushing past. Drive at a constant speed to standardize air speed and experiment with various hand shapes and angles of attack to the air. Think about the sport balls/objects your hand can simulate and note the drag you (and the simulated object) experience at that relative velocity of air flow. How much does the drag increase as you increase the frontal area of your hand? Can you make the lift force act downward? Find the angle of attack that seems to have the most lift and the least drag (maximum lift/drag ratio). See if your classmates have observed similar results.

8.3 Spin Principle

It is clear that fluid forces affect the motion of objects through a fluid. Lift is a key fluid force that can be modified by imparting spin on a projectile. The **Principle of Spin** is related to using the spin on a projectile to obtain an advantageous trajectory or bounce. Kinesiology professionals can use the principle of spin to understand the most successful techniques in many activities. The upward lift force created by backspin in a golf shot increases the distance of a drive (■ Fig. 8.17a), while the backspin on a basketball jump shot is primarily used to keep the ball close to the hoop when impacting the rim or backboard (■ Fig. 8.17b). The bottom of a basketball with backspin is moving faster than the center of the ball because the ball is rotating. This increases the friction force between the ball and the rim, decreasing the horizontal velocity of the ball, which makes the ball bounce higher off the rim or downward off the backboard. In applying the spin principle, professionals should weigh the trajectory and bounce effects of modifying projectile spin.

■ **Fig. 8.17** The principle of spin is used on a golf ball to create lift forces (**F**$_L$) that affect ball trajectory (**a**), while spin on a basketball is primarily used to modify ball rebound (**b**) to increase the chance of a made basket

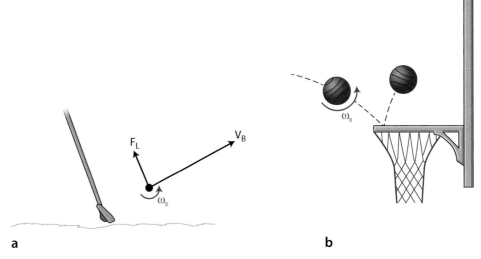

a

b

Applying spin to projectiles by throwing or striking have a key element in common that can be used to teach clients. The body or implement applies force to the ball off-center, creating a torque that produces spin on the projectile. The principles of torque production can be applied to the creation of spin, in that a larger force or a larger moment arm will increase the torque and spin produced. Coaching athletes to project or hit balls with minimal spin to create erratic trajectories requires that the object be in contact with a force in line with the ball's center of gravity. In volleyball, for example, the athlete is taught to strike through the center of the ball with minimal wrist snap. The flat impact through the ball's center of gravity and minimal torque from wrist rotation ensures that the ball will have minimal spin.

Unfortunately, the linear speed of the projectile is inversely proportional to the spin created. In other words, the more spin produced in the throw or hit comes at a cost of lower ball speed. This should be apparent from the vector nature of the force applied to a projectile. If the maximum force is applied by the player, angling that force off center (not aligned directly to the target) to create spin means the force component toward the target making linear speed must be smaller. In tennis, the lateral break of a ball with slice (sidespin) will not travel as fast as a flat serve (minimal spin) hit with the same effort. Much of the science of teaching and coaching is being able to evaluate a person's performance, diagnosing the factors related to spin and speed production that are appropriate for a specific situation.

There is one more advantage of imparting spin to a projectile that is not related to fluid or contact forces on a surface. This third advantage of projectile spin is related to Newton's laws and conservation of angular mo-

mentum. Any object in angular motion without external-acting torques (like a projectile) will conserve angular momentum. This inertia in a rotating object can be used to keep the projectile in a certain orientation. A pass in American football does not create significant lift force (it does not travel sideways across the spin), but the spin stabilizes the flight of the ball in a streamlined position. Divers and gymnasts (human body projectiles) can overcome this inertia and move body parts relative to an axis of rotation with internal muscle forces. In these situations, athletes can transfer angular momentum from one axis to another (e.g., add a twist in the middle of somersaults) by asymmetric motions of body parts. Coaches of these sports need to be familiar with this interesting application of the spin principle (see Yeadon 1991, 1997).

Knowing what advantage of spin in a particular situation is important and how to mechanically create it are critical, but this knowledge must be integrated with knowledge from other kinesiology disciplines. A physical educator could ask a junior high student to "use an eccentric force" or "increase the effective moment arm" and may be mechanically correct, but a good teacher selects an appropriate cue that communicates the essential correction without using such technical language. Think about what would be good cues for hitting a sport ball to create topspin, backspin, right sidespin, or left sidespin. Deciding whether cues about the ball (target) or body action (technique) are most relevant depends on the situation. This is another example of how a biomechanical principle must be integrated in an interdisciplinary fashion with other kinesiology disciplines (e.g., motor development, motor learning, psychology) when teaching clients or giving them feedback as intervention.

8.4 Summary

Fluid forces from air and water have a significant effect on human movement. The main fluid forces are buoyancy, lift, and drag. Buoyancy is the supporting or floating force that a fluid exerts on an object as it is submerged in the fluid. The size of the buoyant force can be determined by Archimedes Principle. The fluid force that acts in the same direction as the relative flow of fluid is drag, while the fluid force acting at right angles to the flow is lift. The size of the lift and drag forces depends on many factors, but they vary with the square of the relative velocity of the fluid. Lift forces can be created on spinning spherical balls through the Magnus Effect. Kinesiology professionals can apply the spin principle to help performers create spin on projectiles like sport balls. Imparting more spin to a projectile usually comes at the expense of a loss in some linear velocity, but the lift force can be used to gain an advantage from an altered flight or bounce relative to a no-spin projection.

? Review Questions

1. What are the major fluid forces and in what directions do they act?
2. What factors affect the fluid resistance acting on projectiles? What factor is most influential in creating fluid forces?
3. Compare and contrast the motion of the center of gravity and center of buoyancy with various body segment movements.
4. Explain why streamlining decreases fluid resistance.
5. How do fluid forces affect the optimal projection angles proposed earlier in ▶ Chap. 5?
6. Why do golf balls have dimples?
7. Draw or trace a person and estimate their center of buoyancy. Trace the person two more times with various exercise and swimming flotation devices and reestimate the likely center of buoyancy.
8. How is the density of water related to whether an object will float in water?
9. Explain why topspin serves in volleyball curve downward?
10. What are the benefits of imparting spin to round balls in sports?
11. How are the spin and the "break" of a ball in flight related?
12. Draw a free-body diagram of a golf ball in flight and explain how the resultant forces on the ball affect its flight.
13. Why do swimmers and cyclists shave their body, but baseball pitchers illegally roughen the surface of the ball?
14. What forces increase and decrease when exercising in watver?

! Key Terms

Archimedes Principle · Bernoulli's Principle · Boundary layer · Buoyancy · Center of buoyancy · Drag (surface, pressure, wave) · Free stream · Laminar flow · Lift · Magnus Effect · Spin principle · Turbulent flow

★ Suggested Reading

Alam, F. (2019). A review on aerodynamics and hydrodynamics in sports. *Energy Procedia, 160,* 798–805.

Alaways, L. W., et al. (2001). Identification of release conditions and aerodynamic forces in pitched-baseball trajectories. *Journal of Applied Biomechanics, 17,* 63–76.

Berger, M. A. M., et al. (1995). Hydrodynamic drag and lift force on human hand/arm models. *Journal of Biomechanics, 28,* 125–133.

McLean, S. P., & Hinrichs, R. N. (2000). Influence of arm position and lung volume on the center of buoyancy of competitive swimmers. *Research Quarterly for Exercise and Sport, 71,* 182–189.

McLean, S. P., & Hinrichs, R. N. (2000). Buoyancy, gender, and swimming performance. *Journal of Applied Biomechanics, 16,* 248–263.

Mehta, R. D., & Pallis, J. M. (2001). Sports ball aerodynamics: Effects of velocity, spin and surface roughness. In F. H. Froes & S. J. Haake (Eds.), *Materials and science in sports* (pp. 185–197). Warrendale, PA: The Minerals, Metals and Materials Society [TMS].

Mureika, J. R. (2000). The legality of wind and altitude assisted performances in the sprints. *New Studies in Athletics, 15*(3/4), 53–58.

Olds, T. (2001). Modelling of human locomotion: Applications to cycling. *Sports Medicine, 31,* 497–509.

Riewald, S., & Rodeo, S. (Eds.) (2015). *Science of swimming faster.* Champaign, IL: Human Kinetics.

Toussaint, H. M., et al. (2002). "Pumped-up propulsion" during front crawl swimming. *Medicine and Science in Sports and Exercise, 34,* 314–319.

Watts, R. G., & Bahill, A. T. (2000). *Keep your eye on the ball: The science and folklore of baseball* (2nd ed.). New York: W.H. Freeman.

Wei, T., et al. (2014). The fluid dynamics of competitive swimming. *Annual Review of Fluid Mechanics, 46,* 547–565.

Yeadon, M. R. (1997). The biomechanics of human flight. *American Journal of Sports Medicine, 25,* 575–580.

🔊 Internet Resources

▶ Aerodynamics–NASA educational pages on fluid mechanics. *grc.nasa.gov/WWW/K-12/airplane/short.html*
▶ BBC Human University–Uniformly Accelerated Motion. *youtube.com/watch?v=E43-CfukEgs&feature=youtu.be*
▶ Bernoulli vs. Newton–NASA webpage on the competing theories for lift forces in fluids. *grc.nasa.gov/WWW/K-12/airplane/bernnew.html*
▶ Bicycle Aerodynamics. *Biefluids.com/efluids/pages/bicycle.htm*
▶ Crash Course Physics–PBS lessons on fluids at rest and motion. *klru.pbslearningmedia.org/collection/crash-course-physics/*
▶ Cycling Aerodynamics and power calculation page. *exploratorium.edu/cycling/aerodynamics1.html*
▶ NBC News Learn–Video lessons on the Science of Golf and Science of the Winter Olympics. *nbclearn.com/portal/site/learn/resources*

Applications of Biomechanics in Qualitative Diagnosis of Movement

Kinesiology professionals in teaching/coaching and exercise/rehabilitation use biomechanics knowledge and principles to qualitatively diagnose the movement technique of their clients. Biomechanical principles must be integrated with knowledge from other kinesiology subdisciplines in the important professional skill of qualitative movement diagnosis (▶ Chap. 3). The chapters in Part III provide guided case studies/examples of applying biomechanics in qualitative movement diagnosis for two kinds of kinesiology professions: physical education teaching/coaching and exercise/rehabilitation common in strength and conditioning and allied health (AT/PT/OT). A variety of guided examples and questions for discussion are presented. The lab activities related to Part III provide students with opportunities to integrate biomechanical knowledge and principles with other subdisciplines of kinesiology. A sample table with the principles of biomechanics for technique evaluation in qualitative movement diagnosis can be found in Appendix E.

Photo by Nishant Ratnakar / CC BY-SA 4.0

U.S. Navy photo by Tom Watanabe

Applying Biomechanics in Teaching and Coaching

Key Questions

➤ **What do you need to know to begin?**
Knowledge of biomechanics (▶ Chaps. 1, 2, 3, 4, 5, 6, 7, and 8).

➤ **Why do you need to know this material?**
Kinesiology professionals know how to apply biomechanics concepts and principles to teaching and coaching motor skills and the qualitative diagnosis of technique.

➤ **What is the key idea?**
Movement technique can be taught and qualitatively diagnosed so clients can improve their performance and reduce risk of injury.

Professionals teaching motor skills (fitness instructors, physical educators, and coaches) teach a wide variety of human movement skills. Biomechanics concepts and principles provide a rationale for key teaching points (critical features) and for qualitative diagnosis of technique. Biomechanics also allows educators to identify exercises and physical activities that contribute to the physical development of various muscle groups and fitness components. Traditionally, careers in coaching have focused on working with the physically gifted in interscholastic athletics; however, there are many other levels of coaching: from parents who volunteer to coach their child's team, to the coach of a national team, and to a coach for an individual professional athlete. This chapter illustrates how biomechanical knowledge and the nine principles of biomechanics can be integrated with other sport sciences in qualitative diagnosis of six motor skills. Real movement performances have been illustrated, and typical teaching cues are used to show how biomechanics is applied to real-world teaching and diagnosis of movement technique. Biomechanical knowledge is also important to teachers and coaches when coordinating their efforts for clients with sports medicine professionals.

9.1 Qualitative Diagnosis of Kicking

A major professional skill of a physical educator or coach may be the qualitative diagnosis of movement technique to facilitate learning of motor skills. Biomechanics is the primary subdiscipline focusing on movement technique, so it is logical that teachers and coaches should use the principles of biomechanics in helping students move safely and effectively. Biomechanics provides knowledge relevant to all four tasks of qualitative diagnosis (▶ Fig. 3.11).

Imagine that you are an elementary physical educator planning a lesson on kicking as a lead-up to soccer, so you are involved in the preparatory task of qualitative diagnosis. In preparing to teach and qualitatively diagnose kicking technique, you list the critical features and teaching points of the movement (▢ Table 9.1). As students practice this skill, you are planning to evaluate these critical features and diagnose student performance using biomechanical principles. Which biomechanical principles seem most relevant to the critical features of high-speed place kicking?

Five of the critical features presented in ▢ Table 9.1 are strongly related to several of the biomechanical principles. The opposition and coordination involved in high-speed kicking are all strongly influenced by the principles of range of motion, coordination, and segmental interaction. In addition, the force–motion, force–time, and optimal projection principles are important in kicking as well. The teacher might plan to keep the principles of inertia, spin, and balance in the back of their mind, so they will not be a focus of observation. These three principles are not likely to play a significant role in the kicking executed by most primary school children.

A child making a full-effort kick toward a goal is observed to consistently have a technique like that illustrated in ◘ Fig. 9.1. Remember that good qualitative diagnosis requires the analyst to observe several performances so that clear trends of strengths and weaknesses can be identified, rather than jumping to conclusions or identifying unimportant "errors" (Knudson 2013). What critical features are strongly and weakly performed? These judgments are part of the evaluation process within the evaluation/diagnosis task of qualitative movement diagnosis.

◘ **Table 9.1** Critical features and teaching cues for fast place kicking

Critical feature	Possible teaching/intervention cues
Visual focus	Head down and focus on the ball
Opposition	Turn your hip to the ball
Foot plant	Plant your foot next to the ball
Coordination	Swing your hip and leg
Impact position	Kick the center of the ball
Follow-through	Follow-through toward the target

The child illustrated in ◘ Fig. 9.1 is clearly at a low developmental level of kicking. The teacher could praise the student's focus on the ball, strong approach, and balance during the kick. The list of biomechanical weaknesses is long at this beginning stage of learning. The biomechanical principles that are weakly incorporated into the kick are force–motion, optimal projection, inertia, range of motion, coordination continuum, and segmental interaction. The student applies a suboptimal force to the ball because they plant the support foot well behind the ball and impact the ball with their toe below the ball center of mass. The fastest kick results from the impact with the proximal instep (top of the shoelaces) of the foot near the ball center of mass. Low-trajectory shots are desirable, but this kick, rolling along the ground, will slow the ball down as it rolls, making it easier for opponents to intercept. Finally, the student could use considerable practice to increase the range of motion of the kick and to refine a well-timed sequential coordination with stretch-shortening cycles that transfer forces through segmental interactions. Highly skilled kickers will approach the ball at an angle to increase the contralateral hip range of motion that can be sequentially combined with the hip and knee motions of the kicking leg. Which of these weaknesses do you think is most important to kicking success? One effective intervention strategy would be to provide a cue to plant their foot next to the ball. This is a simple correction that might be related to other weaknesses and

◘ **Fig. 9.1** The technique of a young person making a high-speed soccer kick. The time between images is 0.08 s

Fig. 9.2 The technique of another soccer player kicking for maximum speed. Time between images is 0.08 s

might motivate the student with initial success and improvement.

Toward the end of the lesson, you notice another child consistently kicking as in the sequence illustrated in ◘ Fig. 9.2. What biomechanical principles are strongly or weakly performed in ◘ Fig. 9.2?

The student depicted in ◘ Fig. 9.2 is more skilled than the student from the previous example. Note the more vigorous approach to the ball. The intensity (inertia) of this approach is apparent in the length of the hurdle to the plant leg and the trunk lean used to maintain balance. It is hard to judge from the figure, but the ball is kicked at the desirable low trajectory. Some educators might conclude that all the biomechanical principles were well applied in this kick. The only two principles that might be slightly improved are range of motion and coordination. If the student were to approach the ball from a more oblique angle, the rotation of the pelvis on the left hip could be increased (range of motion) and combined (sequential coordination) with the good coordination of the kicking hip and knee.

Which of these small improvements, range of motion or coordination, do you think could be easily changed by this student in practice? Improvement in what principle would increase performance the most? These are the issues that are important for a physical educator or coach to examine in the diagnosis and intervention stages of qualitative diagnosis. After classes the teacher might review some research and review papers on kicking (Barfield 1998; Davids et al. 2000; Dorge et al. 2002).

9.2 Qualitative Diagnosis of Batting

Imagine you are a coach or physical educator working on batting with young boys or girls. Most primary school children receive some experience intercepting and striking objects from elementary physical education. The difficulty of the skill dramatically increases when these young people move from batting slow-moving or stationary (batting tee) objects to balls thrown with greater speed and spin. Use the technique points and cues in ◘ Table 9.2 to analyze the batting technique of the student illustrated in

◘ Fig. 9.3. Assume the technique illustrated is representative of most batting attempts off a batting tee by this child. What biomechanical principles seem to be well applied by this child, and what principles are poorly applied? More importantly, prioritize the weaknesses in an order that you think would result in the best batting performance if the weaknesses were improved.

Most all of the biomechanical principles are relevant to batting performance. The student in ◘ Fig. 9.3 strongly incorporates many biomechanical principles into batting. His strengths include balance, inertia, and coordination. He strides into the swing and gets the bat in line with the ball. The force–motion principle could be improved since the bat does not squarely collide with the ball (note the tipping batting tee). The principles of force–time and range of motion may be the weaknesses that could be im-

◘ **Table 9.2**   Critical features and teaching cues for batting

Critical feature	Possible teaching/intervention cues
Visual focus	Head down and focus on the ball
Opposition	Sideward stance
Readiness	Bat up and elbow back
Weight shift	Short stride toward the pitch
Coordination	Throw your hands through the ball
Follow-through	Follow-through around your body

◘ **Fig. 9.3**   A physical education student batting a ball from a tee. Time between images is 0.1 seconds

proved. The student exaggerates the stride and uses an abbreviated follow-through. The physical educator must diagnose the situation and decide if instruction should be focused on the larger than normal range of motion and time in the stride or on the less than expected time/range of motion in the follow-through. Weighing the importance of these principles so as to lead to potential improvement is very difficult. Remember, we noted that this student would soon be applying this skill in the more dynamic condition of impacting a moving ball.

Since the student has good balance, their long stride (which increases range of motion and time of force application) could generate more speed without adversely affecting accuracy. This is typical for a young person with limited upper body strength trying to clobber a ball off a batting tee. Maintaining a long (time and distance) stride in hitting pitched balls, however, is generally a bad tradeoff. Accuracy in contacting the ball becomes more important in dynamic hitting conditions.

It may even be possible to maintain a similar bat speed with a shorter stride if the student improves his/her follow-through. An abbreviated follow-through means that the hitter is slowing down the bat before impact. Skilled striking involves generating peak velocity at impact, delaying negative accelerations until that point (Knudson and Bahamonde 2001). The force–time and range of motion principles also imply that a short follow-through may increase the risk of injury since the peak forces slowing the movement must be larger. Since the follow-through is an important strategy for minimizing the risk of injury in many movements, the physical educator should rate this intervention ahead of adjusting the preparatory (stride) range of motion. Once the student gets comfortable swinging through the ball, they may have more bat speed at impact and might be more willing to reduce their stride and weight shift when hitting pitched balls. More recent research on baseball batting has focused on differences in various bats (Greenwald et al. 2001) and hitting from both sides of the plate (McLean and Reeder 2000).

9.3 Qualitative Diagnosis of Catching

Imagine that you are a junior high school physical educator teaching a basketball unit. You have been ingenious in getting the students to realize the rewards of moving without the ball and passing rather than "hogging" the ball by dribbling. There is one small problem that many of the students have poor catching skills. You previously taught students the critical features of catching (■ Table 9.3) using a variety of cues. In watching a passing drill, you notice a student receiving passes similar to what is illustrated in ■ Fig. 9.4. What biomechanical principles are well or poorly incorporated in catching the basketball? Diagnose the situation and prioritize

■ Table 9.3 Critical features and teaching cues for two-handed catching

Critical feature	Possible teaching/intervention cues
Readiness	Athletic stance
Visual focus	Watch the ball
Intercept	Move and reach toward the ball
Hand position	Thumbs in or thumbs out
Absorption	Give with your hands and arms

the importance of the biomechanical principles in successful catching for this player and think about what the best intervention would be. The player has good balance and uses simultaneous coordination in receiving the ball. The force–motion principle was well applied by predicting the location of the ball, intercepting the ball with the hands, and applying the force through the center of gravity of the ball. The two principles that could be improved are range of motion and force–time. Since you are a good physical educator, you also note the non-biomechanical factors relevant in this situation: the player appears to visually focus on the ball, is motivated, and is trying her best. Praise of these strengths might help motivate the student for some improvements you propose based on your diagnosis.

Diagnosis of this situation is not as difficult as many qualitative diagnoses because the two weaknesses demonstrated in this example are closely related. Increasing the range of motion in receiving the ball will generally increase the time of force application. You must decide if the player's catching and basketball performance would improve most if her attention were focused on reaching more to intercept the ball or emphasizing how the arms bring the ball in. Both biomechanical principles are important. Can you really say one is more important than the other? The player would clearly improve if she stepped and reached more to intercept the ball earlier and provide more body range of motion to slow the ball down. Increasing range of motion also has a secondary benefit by reducing the risk of a pass being intercepted. How the hand forces opposing ball motion are applied, however, has the most influence on whether a ball is caught or bounces out of a player's grasp. This is a case where some professionals might disagree on the most appropriate intervention. In class, you only have a few seconds, and you provide a cue to a student to focus on "giving" with her hands and arms as she receives the ball. You say, "See if you can give with your hands and arms as you catch the ball. Bring that ball in so you barely hear a sound." Subsequent observations and evaluations might allow you to refine this feedback if she still does not reach to intercept the ball.

Fig. 9.4   A junior high school basketball player catching a pass. Time between images is 0.1 seconds

9.4 Qualitative Diagnosis of Overarm Throwing

Imagine you are a youth baseball or softball coach scouting the throwing ability of potential players. You set the children up in the outfield to see how well they can throw the ball to home plate. The technique points and cues for overarm throwing are listed in ■ Table 9.4. One young person trying out for the team shows a throwing technique like illustrated in ■ Fig. 9.5. What are the strengths and weaknesses of their performance? Are there weaknesses you are confident they can overcome this season if they become part of your team?

The athlete in ■ Fig. 9.4 has a very immature throwing pattern, so he has weaknesses in several biomechanical principles. In fact, the straight arm sling this player uses likely places great stress on the throwing shoulder. The principle most in need of improvement is range of motion, which could improve with a more vigorous approach and a longer stride with the opposite leg. The

Table 9.4   Technique points and cues for overarm throwing

Technique points	Possible teaching/intervention cues
Approach/stride	Step with the opposite foot toward the target
Opposition and coordination	Turn your side to the target
Arm position	Align your arm with your shoulders
Shoulder internal rotation	Range of motion
Angle of release	Throw the ball low and flat
Relaxation	Be loose and relaxed

inertia of the throwing arm should be reduced in the propulsion phase by flexing the elbow to about 90°. The thrower does rotate their trunk away from and then into

■ **Fig. 9.5** A softball player throwing with maximum effort to home plate from the outfield

the throw, but sequential coordination that maximizes segmental interaction will require considerable practice. Like many young players, this person throws with a high initial trajectory, violating the optimal projection principle. Optimal throwing angles for maximum distance with baseballs and softballs are about 30° (Dowell 1978).

Some of these weaknesses can be corrected quickly, but some will likely take more than a full season. The athlete should be able to improve his approach, arm action, and angle of projection. Fine-tuning coordination of his throw will likely take longer than a few months. The biomechanics of coordination in overarm throwing is quite complex (Atwater 1979; Feltner and Dapena 1986; Fleisig et al. 1999). Consistent practice over a long period of time will gradually build the sequential rotation that optimizes segmental interactions to create a skilled overarm throw. To see if he listens and can easily change aspects of his throwing technique, ask him to step vigorously with his opposite foot and to throw the ball "lower." It is possible that a youth softball coach might select this player for his team based on other factors. Biomechanical technique in one skill may not be as important as motivational factors or the philosophy employed to help all players develop.

9.5 Qualitative Analysis of Dribbling Technique

Put yourself in the role of a youth soccer coach. After working on several dribbling drills, you begin a more game-like drill where one player consistently performs as in the illustration in ■ Fig. 9.6. Use the technique points and biomechanical principles in ■ Table 9.5 to help guide your observation and qualitative diagnosis of ■ Fig. 9.6. What biomechanical principles are strengths or weaknesses in this performance? Diagnose the performance and decide what would be a good intervention to help this player improve.

This young player shows good balance in this performance since he does not fall when stumbling over the ball. He has poor control of the ball, which likely contributed to him stepping on the ball. Despite a small stumble, he uses his trail leg to recover the ball. The player needs to adjust their application of the force–motion and range of motion principles to improve their dribbling. Providing a cue that improves one of these principles will likely also improve the angle of release or the optimal projection of the ball. Let's diagnose this situation by prioritizing these three weaknesses to provide the best intervention to help this player.

□ **Fig. 9.6**   A soccer player dribbling during a scrimmage

rect technique and the goal of ball control. More specific cues on effort or range of motion can follow if future observations of his dribbling yield similar results. Note that a young player is not cognitively ready for complex technique or strategic instruction. The biomechanical complexity of dribbling a soccer ball in the dynamic environment of a game must be appreciated by the coach, but not imposed on a young player too soon.

□ **Table 9.5**   Technique points and cues for soccer dribbling

Technique points	Possible teaching/intervention cues
Close to body	Keep the ball close to you
Kinesthetic awareness/ control	Feel the ball on your foot
Awareness of situation	Head up and watch the field
Arch of foot	Push the ball with the arch of the foot
Angle of release	Keep the ball close to the ground

9.6   Recruitment

As the golf coach for a university, you have many parents sending you videos of their children for potential scholarship consideration. These "daddy" or "momma" videos can be a nuisance, but you qualitatively diagnose the swings of the golfers on them for potential players you might have missed. This information combined with the player's performance in high school and tournaments will help you decide what athletes have potential for further investigation for a potential scholarship. The technique points and biomechanical principles of the full golf swing you use to analyze swings are presented in □ Table 9.6. For the player illustrated in □ Fig. 9.7, evaluate the strengths and weaknesses of their downswing. We will

Since this is a young player, you plan to praise his effort and a strong point before focusing attention on technique adjustments. Good intervention would be to praise his attention to the ball and recovery from the stumble. It is too early in this player's development to focus intervention on keeping his visual attention on the field. The best intervention may be a cue to "push the ball softly and keep it close to your body." This cue combines the force–motion principle and the range of motion principles and focuses the player's attention on cor-

◘ Table 9.6   Technique points and cues for the golf swing

Technique Points	Possible teaching/intervention cues
Weight shift	Push with rear and then the front foot
Swing plane	Swing forward and back on the same plane
Backswing	Slow and club not past horizontal
Tempo/ coordination	Delayed release/uncocking of the club
Impact/shot trajectory	Divot in front of ball
Follow-through	Long slow finish

now focus on how the relevant biomechanical principles would help you diagnose the weaknesses of this player and her potential as a golfer on your team. A coach would certainly want to know additional information about the swings timing that would be available in live replay, but also what environmental factors were also present (shot difficulty, competitive pressure/score, wind, etc.).

This player has an excellent full swing and control of the club. It is difficult to tell from this perspective, but it is likely this player keeps the club in a stable swing plane. The swing has an appropriate range of motion since the backswing terminates with the club virtually horizontal. The player has a good weight shift, hip and trunk twist, and a firm forward leg late in the swing. The follow-through is fine. The two technique points that are difficult to judge

◘ Fig. 9.7   The long iron swing of a prospective golf recruit from the video (and from the figure) are the coordination of the swing and the quality of the impact, short trajectory (optimal projection), and ball speed and spin for shot distance. In short, this particular player has several strengths that suggest she has an excellent golf swing. A good golf coach would be aware of the massive amount of research on the golf swing (Neal and Wilson 1985; Sprigings and Neal 2000; Williams & Sih, 2002). There are no obvious warning signs, but a complete diagnosis of this golf swing is difficult to obtain from a single video.

It is possible that the tape was edited to show only the best swings for many shots. To fully diagnose this golf swing, you clearly need to know about impact, shot trajectory relative to the intended target, and other environmental conditions of the swing. The sound of the impact might suggest that the ball is well hit; only observation of the ball's flight relative to the intended target will provide clues as to the player's potential and the many subtleties that set high-level golfers apart. A nearly perfect golf swing that strikes the ball with the club face angled away from the target or off-center can produce very poor golf shots. A good golf coach using video for qualitative diagnosis would get views from several vantage points and gather information on the flight of the ball. This distance and direction information can be written or in recorded form on the audio track of the video. Only an integrated qualitative diagnosis of all these factors over many strokes would allow the coach to correctly judge this player's potential.

Note how a diagnosis of possible strengths and weaknesses is severely limited when all we have is a single view of a golf swing. Remember that the biomechanical principles related to the golf swing also must be integrated with other kinesiology subdisciplines. This player might have a flawless swing in practice that turns rough and unpredictable under psychological pressure. If this player's tournament results are good, the coach might invest time talking to their high school coach and plan a trip to see them in competition.

9.7 Summary

Teachers and coaches apply biomechanics concepts in the teaching and qualitative diagnosis of motor skills. Educators and coaches often use cue words or phrases to communicate technique critical features and intervention to their clients. Several real-world movement technique examples were illustrated of the integration of biomechanical principles with experience and other kinesiology subdisciplines (Knudson 2013). Teachers and coaches must take into account psychological, conditioning, learning, and other factors in selecting the best intervention to help their clients improve performance or reduce risk of injury.

❓ Discussion Questions

1. What biomechanical principles are more important in kicking versus trapping a soccer ball?

2. What are the typical teaching points or cues for baseball/softball batting? What biomechanical principles are relevant in these teaching points?

3. Which biomechanical principles are relevant to the pushup exercise? How does changing hand position from a wide base of support to a narrow base of support modify the importance of these principles?

4. What biomechanical principles are most relevant to catching a softball? Catching a medicine ball?

5. What are typical teaching points in jumping to rebound a basketball? What points are most important based on the principles of biomechanics?

6. What biomechanical principles are important in throwing a pass in American football?

7. Are certain biomechanical principles more important to the advanced athlete? Which and why?

8. Athletics coaches often have the opportunity of working closely with a smaller number of performers over a greater length of time than other kinesiology professionals. Does this concern for long-term performance increase or decrease the importance of biomechanical principles?

9. Which biomechanical principles are relevant to athlete quickness? Can biomechanics be used to coach an athlete to be quicker? If so, how does this improvement compare to improvement from conditioning?

10. While the "daddy" or "momma" videos discussed above might give the coach a general indication of the swings of players, what important aspects of golf competition may not show up on these videos? What important biomechanical issues might be difficult to determine from inadequate camera views?

11. Prioritize the following factors based on their importance in coaching beginning, intermediate, and advanced athletes for a specific sport: biomechanics, maturation, physiology, and psychology.

12. What is your position on child specialization in one sport, practiced year-round? Is this "professionalization" of youth sport best for the young person from a performance and a health perspective?

⭐ Suggested Reading

Dyson, G. (1986). *Mechanics of athletics* (8th ed.). New York: Holmes & Meier.

Ecker, T. (1996). *Basic track and field biomechanics* (2nd ed.). Los Altos: Tafnews Press.

Hay, J. G. (1993). *The biomechanics of sports techniques* (4th. ed.). Englewood Cliffs: Prentice-Hall.

Hughes, M. (2019). *Essentials of performance analysis in sport* (3rd ed.). New York: Routledge.

Knudson, D. (2013). *Qualitative diagnosis of human movement* (3rd ed.). Champaign: Human Kinetics.

Knudson, D., & Morrison, C. (1996). An integrated qualitative analysis of overarm throwing. *JOPERD, 67*(6), 31–36.

Applying Biomechanics in Exercise and Rehabilitation

Key Questions

➤ **What do you need to know to begin?**
Knowledge of biomechanics (▶ Chaps. 1, 2, 3, 4, 5, 6, 7, and 8)

➤ **Why do you need to know this material?**
Allied health and conditioning professionals know how to apply biomechanics concepts and principles to teaching and the qualitative diagnosis of exercise technique.

➤ **What is the key idea?**
Exercise technique can be taught and qualitatively diagnosed so clients can improve their performance and reduce risk of injury/reinjury.

Strength and conditioning is a profession in which a great deal of biomechanical research has been conducted. The National Strength and Conditioning Association (NSCA) and their *Journal of Strength and Conditioning Research* publish many studies on the biomechanics of exercise. Traditionally, strength and conditioning careers were limited to coaching the physically gifted in intercollegiate and professional sports. However, more and more opportunities exist for training with a wide variety of clients in the private sector.

Strength professionals are responsible for prescribing exercises that benefit their clients. On the surface this may seem a simple task, but in reality it is quite complicated. Exercises must be selected and exercise technique monitored. Exercises must be relevant, and the intensity must be sufficient for a training response, but not too great as to cause overtraining or a high risk of injury. Biomechanics helps strength and conditioning professionals to assess these risks, benefit ratios, determine the most appropriate (sport-specific) exercises, and evaluate technique during training. As in teaching and coaching, biomechanical knowledge is important for the strength and conditioning professional so they can coordinate their efforts with sports medicine professionals.

Biomechanics also helps allied health professionals in clinical settings to determine the extent of injury and to monitor progress during rehabilitation. Many sports medicine programs have specific evaluation and diagnostic tests and visual observation systems for identification of musculoskeletal problems. The physical therapist and athletic trainer analyzing walking gait or an orthopedic surgeon evaluating function after surgery all use biomechanics to help inform their therapeutic physical medicine decisions.

These clinical applications of biomechanics in qualitative diagnosis of movement and rehabilitation exercise tend to focus more on localized muscle, skeletal, and joint issues than the examples in ▶ Chap. 9. This chapter cannot replace formal graduate training in gait analysis (Perry 1992), injury identification (Shultz et al. 2000), or medical diagnosis (Higgs and Jones 2000). It will, however, provide an introduction to the application of biomechanical principles in several sports medicine and conditioning professions. Biomechanical principles must be integrated with the clinical training, other disciplinary knowledge, knowledge of the client, and professional clinical experience.

10.1 Qualitative Diagnosis of Squat

One of the most common and important exercises in athletic conditioning is the parallel or back squat. The squat is a functional exercise used for a wide variety of sports and other fitness objectives. The squat is usually performed as a free-weight exercise, challenging balance and skill and making movement technique critical to overloading the target muscle groups and minimizing the risk of injury (Case et al. 2020). Exacting technique in free-weight exercises is best because small variations allow other muscles to contribute to the lift, diminishing overload of the muscles or movements of interest. What are the main technique points of the squat often emphasized by strength and conditioning experts? Which

D. Knudson, *Fundamentals of Biomechanics*, https://doi.org/10.1007/978-3-030-51838-7_10

10

biomechanical principles are most strongly related to those technique points?

■ Table 10.1 presents some of the typical technique points and cues for the parallel or front squat. Evaluate the strengths and weaknesses in the biomechanical principles related to the eccentric phase of the squat illustrated in ■ Fig. 10.1. Again, assume the lifter has performed a couple of repetitions this way, and you are confident you can identify stable strengths and weaknesses in application of the principles.

The lifter depicted in ■ Fig. 10.1 has very good squat technique, so there are virtually no weaknesses in application of biomechanical principles. Her stance width may be appropriate; however we cannot evaluate that in the sagittal plane. There is no indication of difficulties in terms of control of the body or the bar (Balance). The images suggest that the motion was smooth, with simultaneous coordination. The timing information in the caption indicates the squat was slow, maximizing the time the muscles were stressed (Force–Time principle). This lifter also keeps her spine straight with normal lordosis, so the spinal loads are primarily compression and are evenly applied across the disks. This more axial loading between the spinal segments is safest for the spine. Research has reported that spinal flexion reduces the extensor muscle component of force resisting anterior shear in the spine (McGill et al. 2000),

making it more difficult for the muscles to stabilize the spine. Strength and conditioning coaches would also need to be familiar with research on the effect of weight belts in squats and other heavy lifting exercises.

Our lifter completed this exercise with the appropriate full range of motion, while not apparently hyperflexing the knee. There is good trunk lean, which distributes the load on both the hip and knee extensors. The amount of trunk lean (hip flexion) in a squat is the primary factor in determining the distribution of joint moments that contribute to the exercise (McLaughlin et al. 1978; Hay et al. 1983; Escamilla 2001). The more upright posture in the front squat decreases the hip and lumbar extensor torques, while increasing the knee extensor torques required in the exercise.

A large part of the strength and conditioning professional's job is motivating and monitoring athletes. The coach needs to look for clues to the athlete's effort or a change in their ability to continue training. Some of these judgments involve application of biomechanical principles. How an athlete's balance changes over a practice or several sets of an exercise could give a strength coach clues about fatigue. Since the figure and introduction give no clues to this aspect of performance, the best intervention in this situation is to praise the good technique of the athlete and possibly provide encouragement to motivate them.

Strength and conditioning professionals also must integrate sport-specific training with other practice and competition. The next example will focus on the sport specificity of a plyometric training exercise.

10.2 Qualitative Diagnosis of Drop Jumps

Plyometrics are common exercises for improving speed and dynamic, high-speed movements in athletes. Plyometric exercises use weights and medicine balls and falls to exaggerate stretch-shortening cycle muscle ac-

■ Table 10.1 Technique points and cues for the squat	
Technique points	**Possible intervention cues**
Stance	Athletic position
Extended/neural spine	Slight arch
Slow, smooth movement	Slow and smooth
Keep thighs above horizontal	Thighs parallel to the ground

■ **Fig. 10.1** The eccentric phase of a person doing a squat. Time between images is 0.2 seconds tions. Considerable research has focused on drop jumps as a lower-body plyometric exercise for improving jumping ability (Bobbert 1990). Research has shown that drop jump exercise programs can increase bone density in children (Fuchs et al. 2001). Qualitative diagnosis of drop jumps is important in reducing the risk of injury in these exercises and monitoring technique that has been observed to vary between subjects (Bobbert et al. 1986). Qualitative diagnosis is also important because drop jumping and resistance training can affect the technique used in various jumping movements (Hunter and Marshall 2002). ◘ Table 10.2 presents important technique points and instruction cues for drop jumps. What are the strengths and weaknesses in the drop jump performance illustrated in ◘ Fig. 10.2?

The female athlete performing the drop jump illustrated in ◘ Fig. 10.2 has several good technique points and possibly one weakness. The strong points of her technique are good lower extremity positioning before touchdown, moderate countermovement, and a nearly vertical takeoff. This indicates good balance during the

exercise. It is difficult to evaluate the speed or quickness of the performance from the drawings with no temporal information in the caption. This athlete did have a short eccentric phase with a quick reversal into the concentric phase. Occasionally subjects will have a longer eccentric phase that minimizes the stretch-shortening cycle effect of drop jumps (Bobbert et al. 1986). The force–time principle applied to plyometric exercises explains why large forces and high rates of force development are created over the short time of force application in plyometrics.

The one potential weakness is not using her arms in the exercise, assuming she was not instructed to keep her hands on her hips! Most athletes should strive to utilize an arm swing with coordination similar to jumping or the specific event for which they are training. If the arms are accelerated downward as the athlete lands, this will decrease eccentric loading of the lower extremities. For jump-specific training, cue the athletes to swing their arms downward in the drop so the arms are swinging behind them during the loading phase, increasing the intensity of eccentric loading of the lower extremities. The vigorous forward and upward swing of the arms from this position increases the vertical ground-reaction force through Segmental Interaction (Feltner et al. 1999). The cue "arms down and up" could be used to remind an athlete of the technique points she should be focusing on in the following repetitions. A key conditioning principle is that the exercises selected for training should closely match the training objectives or movement that is to be improved. This matching of the exercise conditions (approximate resistance and speed) to performance

◘ Table 10.2	Technique points and cues for drop jumps
Technique points	**Possible intervention cues**
Landing position	Toe-heel landing
Rapid rebound	Quick bounce
Minimize counter-movement	Range of motion
Arm integration	Arms down and up

◘ **Fig. 10.2** An athlete doing a drop jump exercise

conditions is the application of the conditioning principle of specificity. More client and sport knowledge is necessary if the conditioning professional was to consider other intervention to balance the principles of specificity and inertia. For example, the intensity of the exercise can be increased by using a higher box or a weight vest. Another example might be that volleyball player with blocking responsibility may not utilize arm action and focus on short, quick plyometric exercises for training jumps similar to their sport skills.

10.3 **Injury Risk**

Imagine you are a strength coach at a junior college. You closely watch many of the young men in your preseason conditioning program because they have had little serious weight training in their high schools, and others may be pushing themselves too hard to meet team strength standards to qualify for competition. Suppose you see a player performing the bench press using the technique illustrated in ◘ Fig. 10.3. What are the strengths and weaknesses of performance? How would you diagnosis this performance, and what intervention would you use?

The biomechanical principles relevant to the bench press are Balance, Coordination Continuum, Force–time, and Range of Motion. When training for strength, resistance is high, the athlete must have good control of the weight (Balance), and coordination during the lift will be simultaneous. The force–time profile of strength training attempts to maintain large forces applied to the bar through as much of the range of motion as possible. The SSC nature of the movement should be minimized. This keeps the movement slow and force output near the

weight of the bar. High initial forces applied to the ball results in lower forces applied to the bar later in the range of motion (Elliott et al. 1989). The principle of Range of Motion in strength training tends toward one of two extremes. First, minimize the range of motion of joints that do not contribute to the movement and of those that allow other muscles to contribute to the movement. Second, the range of motion for joint movements or muscles that are targeted by the exercise should be maximized.

The two principles most strongly related to exercise safety in the bench press are Balance and Range of Motion. Athletes must control the weight of the bar at all times, and a lack of control will affect the range of motion used in the exercise. The athlete in ◘ Fig. 10.3 shows weaknesses in both balance and range of motion. Since the athlete is struggling to "make weight," the difference in strength between the sides of the body manifests as uneven motion of the bar and poor balance. The athlete also hyperextended his lumbar spine in straining to lift the weight.

Several aspects of this performance may have a strength coach thinking about a risk of immediate and future injury: lateral strength imbalance, poor control of bar motion, and hyperextension of the lumbar spine. Since the athlete is "maxing out," some of these weaknesses can be expected, but safety is the greatest concern. Spotters can assist lifters with poor bar control or who can complete the lift with only one side of their body. Hyperextension of the spine, however, is an immediate risk to the athlete's low back health. Hyperextension of the lumbar spine under loading is dangerous because of uneven pressures on the intervertebral disks and greater load bearing on the facet joints. The best intervention here is to terminate the lift with assistance from a spotter

◘ **Fig. 10.3** The concentric phase of a bench press from an athlete struggling to make a weight goal and return to lifting only when the athlete maintains a neutral and supported spinal posture on the bench. Here the immediate risk of injury is more important than balance, skill in the exercise, or passing a screening test.

10.4 Injury Mechanisms

Sports medicine professionals must deduce the cause of injuries from the history presented by patients or clients. Occasionally athletic trainers may be at a practice or competition where they witness an injury. Knowledge of the biomechanical causes of certain injuries can assist an athletic trainer in these situations, in that diagnosis of the particular tissues injured is facilitated. Imagine you are an athletic trainer walking behind the basket during a basketball game. You look onto the court and see one of your athletes getting injured as she makes a rebound (see ◻ Fig. 10.4). What kind of injury do you think occurred? What about the movement gave you the clues that certain tissues would be at risk of overload?

The athlete depicted in ◻ Fig. 10.4 likely sprained several knee ligaments. Landing from a jump is a high-load event for the lower extremity, where muscle activity must be built up prior to landing. It is likely the awkward landing position, insufficient pre-impact muscle activity, hip adduction, and knee twisting (internal tibial rotation) contributed to the injury. It is also likely that the anterior (ACL) and posterior (PCL) cruciate ligaments were sprained. The valgus deformation of the lower leg would also suggest potential insult to the tibial (medial) collateral ligament. Female athletes are more likely to experience a non-contact ACL injury than males (Malone et al. 1993), and the majority of ACL injuries are non-contact injuries (Griffin et al. 2000). There are many good reviews of knee ligament injury mechanisms (Bojsen-Moller and Magnusson 2000; Whiting and Zernicke 2008;).

You rush to the athlete with these injuries in mind. Unfortunately, any of these sprains are quite painful. Care must be taken to comfort the athlete, treat pain and inflammation, and prevent motion that would stress the injured ligaments. Joint tests and diagnostic imaging will eventually be used to diagnosis the exact injury. What biomechanical issue or principle do you think was most influential in this injury? Athlete diagnosis and care are the number one priority but the professional can refer to this unfortunate injury in later prehabilitation sessions with other athletes, encouraging them in their training to reduce risk of these kinds of injuries.

10.5 Exercise Specificity

The principle of specificity also applies to therapeutic exercise in rehabilitation settings. The exercises prescribed must match the biomechanical needs of the healing patient. Exercises must effectively train the muscles that have been weakened by injury and inactivity. Biomechanical research on therapeutic exercise is even more critical since athletic trainers and therapists need to know when internal loadings may exceed the mechanical strengths of normal and healing tissues.

Imagine that you are an athletic trainer treating a runner with patellofemoral pain syndrome. Patellofemoral pain syndrome (PFPS) is the current

◻ **Fig. 10.4** A basketball player injuring her knee during a rebound

10

◘ **Fig. 10.5** The leg press technique of a person trying to remediate patellofemoral pain

terminology for what was commonly called chondromalacia patella (Thomee et al. 1999). PFPS is likely inflammation of the patellar cartilage since other knee pathologies have been ruled out. It is believed that PFPS may result from misalignment of the knee, weakness in the medial components of the quadriceps, and overuse. If the vastus medialis and especially the vastus medialis obliquus (VMO) fibers are weak, it is hypothesized that the patella may track more laterally on the femur and irritate either the patellar or femoral cartilage. The exercises commonly prescribed to focus activation on the VMO are knee extensions within 30° of near complete extension, similar short-arc leg presses/squats, and isometric quadriceps setting at complete extension, and these exercises with combined hip adduction effort. While increased VMO activation for these exercises is not conclusive (Earl et al. 2001), assume you are using this therapeutic strategy when evaluating the exercise technique in ◘ Fig. 10.5. What biomechanical principles are strengths and weaknesses in this exercise.

Most biomechanical principles are well performed. Balance is not much of an issue in a leg press machine because mechanical restraints and the stronger limb can compensate for weakness in the affected limb. There is simultaneous coordination, and there appears to be slow, smooth movement (Force–Time).

The principle that is the weakest for this subject is the large knee flexion range of motion. This subject has a knee angle of about 65° at the end of the eccentric phase of the exercise. This very flexed position puts the quadriceps at a severe mechanical disadvantage, which results in very large muscle forces and the consequent large stresses on the patellofemoral and tibiofemoral joints. This exercise technique can irritate the PFPS and does not fit the therapeutic strategy, so the athletic trainer should quickly instruct this person to decrease the range of motion. Providing a cue to only slightly lower the

weight or keeping the knees extended to at least 120° would be appropriate for a patient with PFPS.

Another question for the professional to consider would be: Should this athlete even be on this leg press machine? Would it be better if they executed a different exercise? A leg press machine requires less motor control to balance the resistance than a free-weight squat exercise, so a leg press may be more appropriate than a squat. Maybe a more appropriate exercise would be a leg press machine or a cycle that allows the subject to keep the hip extended (reducing hip extensor contributions and increasing quadriceps demand) and limit the amount of knee flexion allowed. The differences in muscle involvement are somewhat similar to upright versus recumbent cycling (Gregor et al. 2002). These subtle changes in body position and direction of force application (force–motion) are important in determining the loading of muscles and joints of the body. Squats and lunges of various directions might train lower extremity muscle synergies that are both effective in reducing PFPS and being more sport-specific for the athlete. Good clinicians are knowledgeable about the biomechanical differences in various exercises and prescribe specific rehabilitation exercises in a progressive sequence to improve function.

10.6 Orthotics

Sports medicine professionals often prescribe prosthetics or orthotics to treat a variety of musculoskeletal problems. Prosthetics are artificial limbs or body parts. Orthotics are devices or braces that support, cushion, or guide the motion of the body. Shoe inserts and ankle, knee, or wrist braces are examples of orthotics. Orthotics can be bought "off the shelf" or custom-build (3D printed or fabricated) for a particular patient and part of the body.

Shoe inserts are a common orthotic treatment for excessive pronation of the subtalar joint. One origin of

◘ Fig. 10.6 Rear frontal plane view of the positions of maximum pronation in running in shoes **a** and shoes with a semi-rigid orthotic **b** on a treadmill at 5.5 m/s

excessive pronation is believed to be a low arch or flat foot. A person with a subtalar joint axis below 45° in the sagittal plane will tend to have more pronation from greater eversion and adduction of the rear foot. It has been hypothesized that the medial support of an orthotic will decrease this excessive pronation.

◘ Figure 10.6 illustrates a rear frontal plane view of the maximum pronation position in running for an athlete diagnosed with excessive rear-foot pronation. The two images show the point of maximum pronation when wearing a running shoe (a) and when wearing the same shoe with a custom semi-rigid orthotic (b).

Imagine that you are the therapist working with this runner. The runner reports that it is more comfortable to run with the orthotic, an observation that is consistent with decreased pain symptoms when using orthotics (Kilmartin and Wallace 1994). You combine this opinion with your visual and video recorded frontal plane observations of the actions of her feet in running.

Inspection of ◘ Fig. 10.6 suggests that there is essentially the same (maybe slightly less) pronation when the runner is wearing the orthotic. Biomechanical research on orthotics and rear-foot motion have not as of yet determined what amount of pronation or speed of pronation increases the risk of lower extremity injuries. The research on this intervention is also mixed, with little evidence of the immediate biomechanical effects of orthotics on rear-foot motion and the hypothesized coupling with tibial internal rotation (Heiderscheit et al. 2001). In addition, it is unclear if the small decrease in pronation (if there was one) in this case is therapeutic. The comfort and satisfaction perceived by this runner would also provide some support for continued use of this orthotic, so the therapist might encourage its continued use but ask the patient to monitor symptoms and report back.

While the logic of an orthotic medially supporting a foot that may pronate too much in absorbing the force of foot strike is sound, the clear scientific evidence for orthotics efficacy is unclear. Many soft tissue injuries have complex (multifactor) causes that make single treatment not likely to successfully influence all factors and correct problems. Rehabilitation professionals need to know the biomechanical and clinical research on therapeutic interventions and blend this knowledge with knowledge about their patient and their lifestyle. Remember the linked segment nature of the body does mean that mechanical causes of motions and injuries might originate from structures or forces away from the immediate source of pain.

10.7 Readiness

Orthopedic surgeons and athletic trainers must monitor rehabilitation progress before clearing athletes to return to their practice routine or competition. Recovery can be documented by various strengths, range of motion, and functional tests. Subjective measures of recovery include symptoms reported by the athletes (patient-reported outcome measures: PROMs) and qualitative diagnoses of movement by sports medicine professionals. Athletes will often be asked to perform various movements of increasing demands, while the professional qualitatively evaluates the athlete's control of the injured limb. A couple of common functional tests for athletes with knee injuries are multiple hops for horizontal distance or time (Fitzgerald et al. 2001; Hegedus et al. 2015).

Imagine you are an athletic trainer working with an athlete rehabilitating an ACL injury in her right knee. You ask the athlete to perform a triple hop for maximum distance. The technique of the first hop is illustrated in ◘ Fig. 10.7. As you measure the distance hopped, you go over the strengths and weaknesses in terms of the biomechanical principles of the hop in your mind. Later you will combine this assessment with the quantitative data. The distance hopped on the injured limb should not be below the common clinical standard of 80% of the unaffected limb (Fitzgerald et al. 2001). What biomechanical principles are strengths and weaknesses, and what does a diagnosis of this hopping performance tell you about her readiness to return to practice? Biomechanical technique is just one aspect of many areas that must be evaluated in making decisions on returning athletes to play (Herring et al. 2002).

Most all biomechanical principles are well performed by this athlete. This athlete is showing good hopping technique with nearly Optimal Projection for a long series of hops. She shows good coordination of arm swing, integrated with good simultaneous flexion

Fig. 10.7 An athlete doing a triple hop test

10

and extension of the lower extremity. She appears to have good balance, and her application of the Range of Motion and Force–Time principles in the right leg shows good control of eccentric and concentric muscle actions. There are no apparent signs of apprehension or lack of control of the right knee. If these qualitative observations are consistent with the distance measured for the three hops, it is likely the athletic trainer would clear this athlete to return to practice. The athletic trainer might ask the coach to closely monitor the athlete's initial practices for signs of apprehension, weakness, or poor technique as she begins more intense and sport-specific movements.

10.8 Injury Prevention

One of the most common injuries in sports is sprains of the anterior cruciate ligament (ACL). A large percentage of these ACL injures result from non-contact (landing, cutting, jumping) events in sports. The large numbers of injuries to young female athletes have resulted in considerable research on how these injuries occur and how they might be prevented. Many biomechanical factors have been hypothesized to be related to increased risk of ACL injuries in sport: peak vertical ground reaction force, knee flexion angle at landing, hamstring strength, and balance. A large prospective study of the biomechanics of landing in female adolescent athletes who then participated in high-risk sports has recently identified several variables that are associated with risk of ACL injury (Hewitt et al. 2005). The variables that

were associated with girls that became injured were greater knee abduction angle (lower leg valgus) and greater ground reaction force and knee abduction moment. It is possible that as girls enter adolescence, the increased risk of ACL injuries comes from dynamic valgus loading at the knee that results from a combination of factors. With adolescence in females, the limbs get longer and hips widen; if strength at the hip and knee, coordination, and balance do not keep up with these maturational changes, it is likely that risk of ACL injury could be increased.

While sports medicine professionals have qualitatively evaluated the strength and balance of patients in single leg stance and squats for many years, recent papers have proposed that simple two-dimensional measurements of frontal plane motion of the lower extremity in single leg squats might be a useful clinical screen tool for identifying athletes that may be at a higher risk for ACL injury (McLean et al. 2005; Wilson et al. 2006). While this test is not as dynamic as landing, it is likely a safer screening procedure that also can be qualitatively evaluated. If screening suggests an athlete may be at risk (poor control of knee in the frontal plane), research has indicated that preventative conditioning programs can decrease the risk of ACL injuries (Hewitt et al. 2006; Sadoghi et al. 2012; Lopes et al. 2018).

Figure 10.8 illustrates the position of the lower extremity at the bottom of a single leg squat for two young athletes. If you were an athletic trainer or physical therapist screening these athletes before a competitive season, which athlete would you be most concerned about for a higher risk of ACL injury? Could you draw

Fig. 10.8   Lower leg position of the bottom of a single leg squat for two young athletes

on the figure lines along the long axes of the leg and measure an angle representing the valgus orientation of the lower leg? What conditioning would you suggest for this athlete?

Would there be any special technique training you would suggest to the coach for jumping, landing, and cutting during practice?

10.9 Summary

Strength and conditioning professionals use the principles of biomechanics to qualitatively diagnose the technique of exercises, evaluate the appropriateness of exercises, and reduce the risk of injury from potentially dangerous exercise technique. Since strength training utilizes loads closer to the ultimate mechanical strength of tissues, professionals need to keep safety and exacting exercise technique in mind. Similarly, sports medicine professionals use biomechanical principles to understand injury mechanisms, select appropriate injury prevention and rehabilitation protocols, and monitor recovery. Qualitative diagnosis in sports medicine often focuses on a more focused, structural level more often than other kinesiology professions. Specialized clinical training assists biome-

chanics research with this more micro-level evaluation and diagnosis, although it is important to remember that the linked segment system of the body means causes of motion and pain may be distant to that pain/disability. Exercise and rehabilitation professionals also must integrate physiological and psychological knowledge with biomechanical principles to maximize client improvement. Other issues exercise and rehabilitation professionals must take into account beyond biomechanical principles are pain, fear, motivation, and competitive psychology.

? Discussion Questions

1. The squat and various leg press exercise stations are often used interchangeably. What biomechanical principles are more important in the squat than in the leg press, and how would you educate lifters who think that the exercises do the same thing?

2. An athlete back in the weight room after initial rehabilitation from an injury is apprehensive about resuming their conditioning program. What biomechanical principles can be modified in adapting exercises for this athlete? Suggest specific exercises and modifications.

3. What aspect of exercise specificity (muscles activated or joint motions) do you think is most important in training for a particular sport skill? Why? Does analysis of the biomechanical principles of exercises and sport movement help you with this judgment?

4. If an athlete uses unsafe technique in the weight room, should the strength coach's response be swift and negative for safety's sake or should they take a positive (teachable moment) approach in teaching safer technique? Does the athlete (age, ability, etc.) or exercise factors affect the best approach?

5. Athletes train vigorously, pushing their limits, treading a fine line between training safely and overtraining. Are there biomechanical indicators that could help the strength and conditioning professional recognize when training intensity has moved beyond overload to dangerous? Why?

6. What biomechanical principles are relevant to training overarm-throwing athletes with upper-body plyometric exercises? Be sure to integrate the muscle mechanics knowledge summarized in ▶ Chap. 4 in your answer.

7. Strength training resistances are often expressed as percentages of maximum strength (1RM). If loads on the musculoskeletal system were also expressed as percentages of mechanical strength, what training loads do you think would be safe (acceptable risk) or unsafe (unacceptable risk)?

8. What biomechanical principle do you think is more important in rehabilitating from an ankle sprain, balance or range of motion?

9. Patients recovering from knee injuries are often given braces to prevent unwanted movement and to gradually increase allowable motion. What movement characteristics would indicate that a patient is ready to exercise or function without a brace?

10. Sports medicine professionals looking for the causes of overuse injuries often evaluate joints distant from the affected area (Kibler and Livingston 2001) because of Segmental Interaction through the kinematic chain. What biomechanical principles can provide cues to potential overuse injuries in other parts of the body?

11. A major injury in athletic and sedentary populations is low back pain. What abdominal and back muscles are most specific to rehabilitation or injury prevention for an office worker? A tennis player?

12. Athletes using repetitive overarm throwing often suffer from impingement syndrome. What biomechanical principles can be applied to the function of the shoulder girdle and shoulder in analyzing the exercise and throwing performance of an injured athlete?

13. You are an athletic trainer working with an athlete recovering from a third-degree ankle sprain. You and the athlete are deciding whether to use athletic tape or an ankle brace. What does a qualitative biomechanical analysis suggest is the better of these two options? What biomechanical studies would you suggest to investigate the clinical efficacy of these options?

14. What biomechanical principles should be focused on when a therapist or trainer is working with elderly clients to prevent falls?

15. An adapted physical educator has referred a young person who might have developmental coordination disorder (DCD) to a physical therapist. Before various imaging and neurological tests are performed, what biomechanical principles should be the focus of observation, and what simple movement tests would be appropriate in the initial physical/orthopedic exam?

16. What biomechanical principles should be the focus of observation if you have a patient in rehabilitation after a hip replacement? What do you look for if the patient is walking with assistance (parallel bars, negative pressure treadmill, or suspension system)? What principles would you use in instructing this patient to use an assistive device (cane) at home?

⭐ Suggested Reading

Baechle, T. R., & Earle, R. W. (Eds.). (2000). *Essentials of strength training and conditioning* (2nd ed.). Champaign: Human Kinetics.

Dvir, Z. (Ed.). (2000). *Clinical biomechanics*. New York: Churchill Livingstone.

Fitzgerald, G. K., et al. (2001). Hop tests as predictors of dynamic knee stability. *Journal of Orthopaedic and Sports Physical Therapy, 31*, 588–597.

Garhammer, J. (1989). Weight lifting and training. In C. Vaughan (Ed.), *Biomechanics of sport* (pp. 169–211). Boca Raton: CRC Press.

Kibler, W. B., & Livingston, B. (2001). Closed chain rehabilitation of the upper and lower ex- tremity. *Journal of the American Academy of Orthopaedic Surgeons, 9*, 412–421.

Kirtley, C. (2006). *Clinical gait analysis: Theory and practice*. New York: Churchill Livingstone.

Knudson, D. (2013). *Qualitative diagnosis of human movement* (3rd ed.). Champaign: Human Kinetics.

Stone, M., Plisk, S., & Collins, D. (2002). Training principles: Evaluation of modes and methods of resistance training–A coaching perspective. *Sports Biomechanics, 1*, 79–103.

Whiting, W. C., & Zernicke, R. F. (2008). *Biomechanics of musculoskeletal injury* (2nd ed.). Champaign: Human Kinetics.

Whittle, M. (1996). *Gait analysis: An introduction* (2nd ed.). Oxford: Butterworth-Heinemann.

Wilson, G. J. (1994). Strength and power in sport. In J. Bloomfield, T. R. Ackland, & B. C. Elliott (Eds.), *Applied anatomy and biomechanics in sport* (pp. 110–208). Melbourne: Blackwell Scientific Publications.

Zatsiorsky, V. N., & Kraemer, W. J. (2006). *Science and practice of strength training* (2nd ed.). Champaign: Human Kinetics.

10

Supplementary Information

Appendix A: Glossary of Biomechanics Terms

Absolute angle An angle measured to a non-moving (inertial) frame of reference

Acceleration The rate of change of velocity (vector)

Accelerometer A device that measures acceleration

Actin The thin filaments in a myofibril that interact with myosin to create muscle tension

Accommodation A decrease in biological response to an unchanging stimulus

Action potential The electrical potential change during depolarization of nerves and activated muscle fibers

Active tension The tension created by the contractile component (actin–myosin interaction) of activated muscle

Affine scaling Image scaling technique used to measure in a plane not perpendicular to the optical axis of the camera in 2D cinematography/videography

Agonist An anatomical term referring to the concentric action of a muscle or muscle group for presumed to create a specific movement

Aliasing Distortion of a signal by an inadequate sampling rate

Analog-to-digital (A/D) conversion The process of taking a continuous signal and sampling it over time (see "sampling rate") to create a digital (discrete numbers) representation

Anatomy The study of the structure of the body

Angle–angle diagram A kinematic graph of one variable plotted against another (not time) that is useful in the study of coordination of movements

Angular acceleration The rate of change of angular velocity (vector)

Angular displacement The change in angular position (vector)

Angular momentum The quantity of angular motion, calculated as the product of the moment of inertia times the angular velocity (vector)

Angular velocity The rate of change of angular displacement (vector)

Angular impulse The angular effect of a torque acting over time: the product of the torque and the time it acts (vector)

Anisotropic Having different mechanical properties for loading in different directions

Antagonist An anatomical term referring to a muscle or muscle group that is presumed to oppose (eccentric action) a specific movement

Anthropometry The study of the physical properties of the human body

Aponeurosis Connective tissue within muscle and tendon in the form of a flat sheet

Archimedes' principle The magnitude of the buoyant force is equal to the weight of the fluid displaced

Arthrokinematics The major, freely moveable rotations allowed at joints

Balance A person's ability to control their body position relative to some base of support

Balance Principle A biomechanical application principle which states that the stability and mobility of a body position are inversely related

Ballistic Explosive, momentum-assisted movement

Bandpass filter A filter designed to pass a range (bandpass) of frequencies, removing frequencies above or below this desirable range

Bending A combination of forces on a long body that tends to bend or curve the body creating tensile loads on one side and compression loads on the other side

Bernoulli's principle The pressure a fluid can exert decreases as the velocity of the fluid increases

Bernstein's problem A theory of motor control in which skill learning involves the reduction of redundant degrees of freedom

Bilateral deficit Simultaneous activation of two limbs that causes less force generation than the sum of the two individually activated limbs

Biomechanics Study of the motion and causes of motion of living things

Boundary layer The layers of a fluid in close proximity to an object suspended in the fluid

Buoyancy The supporting or floating force of a fluid

Center of buoyancy The point at which the buoyant force acts

Center of mass/gravity The point that represents the total weight/mass distribution of a body; the mass centroid is the point where the mass of an object is balanced in all directions

Center of percussion A point on a striking object where impact with another object results in no reaction force at an associated point on the grip (see "sweet spot")

Center of pressure The location of the vertical ground reaction force vector; the center of pressure measured by a force platform represents the net forces in support, and the COP may reside in regions of low local pressure

Coactivation Simultaneous activation of agonist and antagonist muscles (co-contraction)

Coefficient of drag A measure of the relative fluid resistance between an object and a fluid

Coefficient of friction A measure of the resistance to sliding between the surfaces of two materials

Coefficient of lift A measure of the lift force that can be created between an object and a fluid

Coefficient of restitution A measure of the relative elasticity of the collision between two objects

Common mode rejection A measure of the quality of a differential amplifier in rejecting common signals (noise)

Compression A squeezing mechanical loading created by forces in opposite directions acting along a longitudinal axis

Compliance The ratio of change in length to change in applied force, or the inverse of stiffness (see "stiffness"); a material that is easily deformed has high compliance

Components The breaking up of a vector into parts, usually at right angles

Concentric muscle action The condition where activated muscles create a torque greater than the resistance torque (mio-metric)

Conservation of energy The law of conservation of energy states that energy cannot be created or destroyed; instead, energy is transformed from one form to another

Contourgram Exact tracings of the body positions of a movement from film/video images

Contractile component A part of the Hill muscle model that represents the active tension and shortening of actin and myosin

Coordination Continuum A biomechanical application principle which states that movements requiring generation of high forces tend to utilize simultaneous segmental movements, while lower-force and high-speed movements tend to use sequential movements

Couple (1) Two forces of equal size, parallel lines of actions, and opposite sense; (2) a mechanical calculation tool that is employed to represent torques without affecting linear kinetics

Creep The increase in length (strain) over time as a material is constantly loaded

Cross-talk The pick-up of EMG signals from other active muscles aside from the muscle of interest

Cut-off frequency The cutting point of a filtering technique, where frequencies above or below are removed; the lower the cut-off frequency for a low-pass filter, the greater the smoothing of the signal

Deformable body Biomechanical model that documents the forces and deformations in an object as it is loaded

Density The mass of an object divided by its volume

Degrees of freedom The number of independent movements an object may make and consequently the number of measurements necessary to document the kinematics of the object

Differential amplification EMG technique for amplifying the difference between the signals seen at two electrodes relative to a reference electrode

Digital filter A complex frequency-sensitive averaging technique used to smooth or process data

Digitize (video) The A/D conversion of an analog video signal to create the discrete picture elements (pixels) used to make a video image

Digitize (biomechanics) The process of measuring 2D locations of points on images. This tracking of points with x and y coordinates are used to calculate numerous kinematic and kinetic variables in two and three-dimensions

Direct dynamics Biomechanical simulation technique where the kinematics of a biomechanical model are iteratively calculated from muscle activation or kinetic inputs

Direct linear transformation (DLT) A short-range photogrammetric technique to create 3D coordinates (x, y, z) from the 2D coordinates (x, y) of two or more synchronized camera views of an event

Displacement Linear change in position in a particular direction (vector)

Distance Liner change in position without regard to direction (scalar)

Double differential amplification EMG technique to eliminate cross-talk

Drag The fluid force that acts parallel to the relative flow of fluid past an object

Dynamical systems Motor learning theory which argues that movement coordination emerges or self-organizes based on the dynamic properties of the body and environment rather than on a central motor program from the brain

Dynamics The branch of mechanics studying the motion of bodies under acceleration

Dynamometer A device that measures force or torque for muscular performance testing

Eccentric muscle action The condition where an activated muscle(s) creates a torque less than the resistance (plyometric) torque

Economy The amount of energy needed to do a specific amount of work

Efficiency In a system, the ratio of work done to work input

Elastic The resistance of a body to deformation (see "stiffness")

Elastic (strain) energy The potential mechanical work that can be recovered from restitution of a body that has been deformed by a force (see "hysteresis")

Electrogoniometer   A device that makes continuous measurements of joint angle(s)

Electromechanical delay   The delay between motor action potential (electric signal of muscle depolarization or EMG) and production of muscular force

Electromyography (EMG)   The amplification and recording of the electrical signal of active muscle

Energy (mechanical)   The ability to do mechanical work (potential, strain, and kinetic energy are all scalar mechanical energies)

Ergometer   Machine used to measure mechanical work

Euler angles   A way to represent the 3D motion of an object using a combination of three rotations (angles)

Excursion   The change in the length of a muscle as the joints are moved through their full range of motion

External force   A force acting on an object from its external environment

External work   Work done on a body by an external force

Fascicle   A bundle of muscle fibers (cells)

Fast Fourier transformation (FFT)   Mathematical technique to determine the frequencies present in a signal

Field (video)   Half of an interlaced video image (frame), composed of the even or odd horizontal lines of pixels

Finite difference   Calculating time derivative by discrete differences in kinematics divided by the time between data points

Finite-element model   Advanced biomechanical model to study how forces act within a deformable body

Firing rate   The number of times a motor unit is activated per second

First law of thermodynamics   Application of the law of conservation of energy to heat systems

Flexibility   Ability of joint or groups of joints to allow motion without injury. This extensibility of muscle groups about a joint or joints can be documented by linear or angular measurement of the limits of motion

Fluid   A substance, like water or gasses, that flows when acted upon by shear forces

Force   A push, pull, or tendency to distort between two bodies

Force–length relationship   Skeletal muscle mechanical property that demonstrates how muscle force varies with changes in muscle length (also called the length–tension relationship)

Force–motion principle   A biomechanical application principle which states that unbalanced forces are acting whenever one creates or modifies the movement of objects

Force platform   A complex force transducer that measures all three orthogonal forces and moments applied to a surface

Force–time principle   A biomechanical application principle which states that the time over which force is applied to an object affects the motion of that object

Force–time relationship   (See "electromechanical delay")

Force–velocity relationship   Skeletal muscle mechanical property that shows how muscle force potential depends on muscle velocity

Fourier series   A mathematical technique for summing weighted sine and cosine terms that can be used to determine frequency content or represent a time domain signal

Frame (video)   A complete video image

Free-body diagram   A technique for studying mechanics by creating a diagram that isolates the forces acting on a body

Frequency   The inverse of time or the number of cycles of an event per second

Frequency content   Time-varying signals can be modeled as sums of weighted frequencies (see "Fourier series")

Frequency response   The range of frequencies that are faithfully reproduced by an instrument

Friction   The force in parallel between two surfaces that resists sliding of surfaces past each other

Global reference frame   Measuring kinematics relative to an unmoving point on the earth

Golgi tendon organ   A muscle receptor that senses muscle tension

Goniometer   A device used to measure angular position

Gravity   The force of attraction between objects; usually referring to the vertical force of attraction between objects and the earth

Ground reaction force   The reaction (opposite) forces created by pushing against the ground (e.g., feet in running or hands in a handstand)

Harmonic   A multiple of a fundamental frequency (see "frequency content")

Helical (screw) axis motion   A way to represent the 3D motion of an object using an imaginary axis in space and rotations relative to that axis

High-pass filter   A signal-processing technique that removes the low-frequency components of a signal

Hill muscle model   A three-component model of muscle force consisting of a contractile component, a series elastic component, and a parallel elastic component

Hypertrophy   The increase in size of muscle fibers

Hysteresis The energy loss within a deformed material as it returns to its normal shape

Impulse The mechanical effect of a force acting over time (vector); $\mathbf{J} = \mathbf{F} \cdot t$

Impulse–momentum relationship Principle which states that the change in momentum of an object is equal to the net impulse applied; the original language of Newton's second law and equivalent to the instantaneous version: $\mathbf{F} = m\mathbf{a}$

Inertia The property of all matter to resist a change in its state of motion

Inertial force The mass acceleration (ma) term in Newton's second law (dynamics); the effect of inertia and acceleration on dynamic movement, but it is important to remember that its effect is not a real force acting on an object from another object

Inertia principle A biomechanical application principle which states that inertial resistance to changes in state of motion can be used to advantage in resisting motion or transferring energy

Information Observations or data with unknown accuracy/validity

In situ Latin for "in place," or structures isolated by dissection

Integrated EMG (IEMG) The area under a rectified EMG signal; correctly, the time integral reported in units of amplitude X time (mV·s); unfortunately, some studies employ outdated equipment and incorrect terminology, so that reported IEMGs are not really integrated but filtered or smoothed EMG values (mV), which is essentially a linear envelope detector

Interdisciplinary The simultaneous integrated application of several disciplines to solution of a problem

Internal force A force within an object or between the molecules of an object

Internal work Work done on body segments by internal forces (muscles, ligaments, bones)

Inverse dynamics Biomechanics research technique for estimating net forces and moments in a linked-segment model from measured kinematics and anthropometric data

In vitro Latin for "in glass," or tissues removed from the body but preserved

In vivo Latin for "in the living," or during natural movement

Isokinetic ("same, or constant, motion") The condition where activated muscles create constant joint angular velocity

Isometric ("same, or constant, length") The condition where activated muscles create a torque equal to the resistance torque, so there is no joint motion

Isotonic ("same, or constant, tension") The condition where activated muscles work against a constant gravitational resistance; muscle tension is *not* constant in these conditions

Jerk The third derivative of displacement with respect to time

Joint center An approximation of the instantaneous center of rotation of a joint

Joint reaction forces The net forces acting at joints calculated from inverse dynamics; these forces do not represent the actual bone-on-bone forces acting at joints, but a combination of bone, muscle, and ligament forces

Joule The unit of mechanical energy and work

Kinematic chain A linkage of rigid bodies; an engineering term used to simplify the degrees of freedom needed to document the mechanical behavior of a system; Steindler (1955) proposed the terminology of a kinetic chain and classifying chains as either open or closed; unfortunately, this has resulted in a great deal of confusion and an unclear manner of classifying movements/exercises: *open*, one end link is free to move; *closed*, constraints (forces) on both ends of the kinematic chain

Kinematics The branch of mechanics that describes (measures) the motion of objects relative to some frame of reference

Kinetic energy The capacity to do work due to the motion of an object

Kinetics The branch of mechanics that explains the causes of motion

Knowledge The contextual, theory-based and data-supported ideas that make the best current explanation for reality

Laminar flow Movement of fluid in smooth, parallel layers

Law of acceleration Newton's second law of motion, which states that the acceleration an object experiences is proportional to the resultant force, is in the same direction, and is inversely proportional to the object's mass (). F = ma)

Law of inertia Newton's first law of motion, which states that objects tend to resist changes in their state of motion; formally, we say an object will remain in a state of uniform motion (stillness or constant velocity) unless acted upon by an external force

Law of momentum Newton's second law written as the impulse–momentum relationship

Law of reaction Newton's third law of motion, which states that for every force there is an equal and opposite reaction force

Lever A simple machine used to magnify motion or force; a lever consists of a rigid object rotated about an axis

Lift The fluid force that acts at right angles to the relative flow of fluid

Linear envelope EMG processing technique where a rectified signal is smoothed with a low-pass filter

Linearity A measure of the accuracy of an instrument, usually expressed as a percentage of full-scale output (FSO)

Linear voltage differential transducer (LVDT) A force-measuring device

Linked-segment model A rigid body model linked together by joints

Load A force or moment applied to a material

Load cell A force-measuring device

Load-deformation curve The mechanical behavior of a material can be documented by instantaneous measurement of the deformation and load applied it

Local reference frame Measuring kinematics relative to a moving point, or nearby rigid body (joint, segment, or center of mass)

Low-pass filter A signal-processing technique that removes the high-frequency components of a signal

Magnus effect The creation of lift force on a spinning sphere

Markers High-contrast reflective materials attached to subjects to facilitate the location of segments, landmarks, or joint centers for digitizing mass the resistance of an object to linear acceleration

Maximal voluntary contraction (MVC) The maximum force/torque a person can create with a muscle group, usually under in isometric conditions

Mechanical advantage A ratio describing the effectiveness of a lever calculated by the moment arm for the force divided by the moment arm for the resistance

Mechanics The branch of physics that deals with forces and the motion they create

Mechanomyography (phonomyography, vibromyograph) The amplification and recording of the vibrations created by muscle activation

Modeling Mathematical representations of the biomechanical systems used for calculations or simulations

Moment (moment of force, torque) The rotating effect of a force

Moment arm The leverage of a force for creating a moment; the perpendicular distance from the axis of rotation to the line of action of the force

Moment of inertia The resistance to rotation (angular acceleration) of a body

Momentum The quantity of motion of an object calculated by the product of mass and velocity (vector)

Motor action potential The change in electrical charge about a muscle fiber as it is activated

Motor unit A motor neuron and the muscle fibers it innervates

Muscle action The activation of muscle to create tension that contributes to joint movement or stabilization

Muscle inhibition The inability to fully activate or achieve maximum muscle force during maximum voluntary contraction

Muscle spindle An intramuscular receptor that senses changes in muscle length

Myofibril The small cylindrical filaments that make up a muscle fiber/cell

Myosin The large filaments in a myofibril that interact with actin to create muscle tension

Myotatic reflex A short reflex arc that activates a muscle as it is stretched

Net force The resultant force or sum of all external forces acting on an object

Newton The SI unit of force; 1 Newton (N) is equal to 0.22 pounds

Normal reaction The force acting at right angles to the surfaces of objects that are in contact

Nyquist frequency A signal sampling theorem which states that the minimum digital sampling rate (Nyquist frequency) needed to accurately represent an analog signal is twice the highest frequency present in the signal

Optimal Projection principle A biomechanical application principle which states that there are ranges of optimal angles for projecting objects to achieve certain goals

Orthogonal Perpendicular (at right angles)

Orthotics Objects/braces that correct deformities or joint positioning

Overuse injury An injury created by repetitive movements below acute injury thresholds, but due to inadequate rest and/or repetitive stress, injury develops; also known as cumulative trauma disorder or repetitive motion injury

Parallel elastic component A part of the Hill muscle model that represents the passive tension from connective tissue throughout the muscle-tendon unit

Pascal The SI unit of pressure or stress (force per unit area)

Passive insufficiency The limitation of joint motion because of increases in passive tension in multiarticular muscles stretched across multiple joints

Passive tension A component of muscle tension from passive stretching of muscle, especially the connective tissue components

Pennation The angle of muscle fiber bundles relative to a tendon

Piezoelectric Crystals with electromechanical properties that can be used to measure force/acceleration

Point mass A simplified mechanical model that represents an object as a point in space with a given mass

Potential energy The capacity to do work of an object due to its vertical position in a gravitational field (gravitational potential energy) or its deformation (strain energy)

Potentiometer A device that is used to measure rotation

Power (mechanical) The rate of doing mechanical work; peak mechanical power represents the greatest mechanical effect, the ideal combination of force and velocity; power can be calculated as W/t or $\mathbf{F} \cdot \mathbf{V}$

Preamplification The amplification of small signals (EMG) close to their source before they are conducted to other devices for amplification and recording

Pressure External force divided by area over which the force acts

Projectile An object projected into space without self-propulsion capability, so the only forces acting on the object are gravity and air resistance

Proprioceptive neuromuscular facilitation (PNF) Specialized stretching procedures that utilize sequences of muscle actions to potentiate reflexes to relax muscles being stretched

Prosthetics Artificial limbs

Pythagorean theorem The two sides of a right triangle forming the right angle (a and b) and the hypotenuse (c) are related as follows: $a^2 + b^2 = c^2$

Qualitative analysis Solving a problem using a non-numerical judgment. For example, joint range of motion could be classified as limited, normal, or excessive

Quantitative analysis Solving a problem using numerical measurements and calculations

Qualitative movement diagnosis An important skill for kinesiology professionals involving the systematic observation and introspective judgment of the quality of human movement for the purpose of providing the most appropriate intervention to improve performance (Knudson 2013)

Quasistatic The state of a mechanical system where the accelerations are small enough to be assumed equal to zero for calculation/analysis purposes

Radian A dimensionless unit of rotation equal to 57.3°

Radius of gyration A convenient way to summarize an object's moment of inertia, defined as the distance from the axis of rotation at which half the object's mass must be placed in both directions to equal the object's moment of inertia

Range of motion principle A biomechanical application principle which states that the amount of linear and angular motion used will affect the speed and accuracy of human movement

Reaction change A method to calculate the center of gravity of static body postures

Reciprocal inhibition The inhibition of an opposing muscle group (antagonist) when a muscle group (agonist) is activated

Recruitment Activation of motor units of muscles by the central nervous system

Rectified EMG A processing technique that converts negative EMG voltages to positive ones

Redundancy (distribution) problem A mathematical problem with most kinetic biomechanical models, where there are more musculoskeletal unknowns than there are equations

Relative angle An angle measured between two moving objects

Residuals Difference between a smoothed and raw signal; can be used to examine the quality of the fit of the new signal to the pattern of the raw signal

Resolution (video) The number of pixels available to measure a given field of view; a video image of a 3-meter wide area with a horizontal resolution of 640 pixels has a resolution for measurement of about 5 mm

Resonance Frequency of vibration that matches the physical properties of a body so that the amplitudes of the vibration increase rather than decay over time

Resting length The middle of muscle range of motion where passive tension begins to rise

Resultant The addition of vectors to obtain their net effect (see "net force")

Right-hand rule A convention or standard for drawing the correct direction of angular velocity vectors

Rigid body Mechanical simplification (abstraction) assuming the dimensions of an object do not change during movement or loading

Root mean square (RMS) Signal processing calculation that approximates the mean absolute value of a time-varying signal

Rotator cuff The four deep, stabilizing muscles of the glenohumeral joint: the infraspinatus, supraspinatus, subscapularis, and teres minor

Sampling rate The number of discrete samples per second used to represent a signal; digital video typically has a sampling rate of 60 Hz or 60 pictures per second. Biomechanics uses higher sampling rates to document motion of fast human motions and impacts

Sarcomere The functional unit of a myofibril; a sarcomere is the region between two Z disks

Scalar Simple quantity completely defined by a single number (magnitude)

Scaling Converting image measurements to actual size

Science A systematic method for testing hypotheses with experimental evidence for the purpose of improving our understanding of reality

Second law of thermodynamics No machine can convert all the input energy into useful output energy

Segmental Interaction principle A biomechanical application principle which states that forces acting in a system of linked rigid bodies can be transferred through the links

Segmental method A research method used to calculate the center of gravity of a body using anthropometric data, joint coordinates, and static equilibrium

Series elastic component A part of the Hill muscle model that represents the passive tension of connective tissue in series with the contractile component

Shear Mechanical loading in opposite directions and at right angles to the surface of a material

Shutter speed The period of time during which a photographic or video image is captured (e.g., 1/1000 of a second); limiting this period can prevent blurring of moving objects

Simulation Use of a biomechanical model to predict motion with given input conditions in order to study the factors that affect motion (see "direct dynamics")

Size principle The orderly recruitment of motor units occurs from the smallest to the largest

Smoothing A processing technique that smooths data, removing rapid fluctuations that are not part of some biomechanical signals

Smoothing parameter An index of the amount of smoothing allowed in splines; the larger the smoothing parameter, the more smoothing (allowable deviation between the raw and fitted curve)

Snap The fourth derivative of displacement with respect to time

Speed The rate of change of distance (scalar)

Spin principle A biomechanical application principle which states that spin is put on a projectile to affect trajectory or bounce

Spline A smoothing technique that replaces the signal with several polynomials linked together; cubic (third power) and quintic (fifth power) splines are common in biomechanics

Static equilibrium When all the forces and torques acting on an object sum to zero, meaning that the object is motionless or moving at constant velocity

Statics The branch of mechanics that studies bodies at rest or in uniform motion

Stiffness The elasticity of a material, measured as the slope of the elastic (linear) region of the stress–strain curve (Young's modulus of elasticity); a material's stiffness is usually approximated using the slope of the linear region of the load-deformation curve

Strain (mechanical) The amount of deformation of a material caused by an applied force, usually expressed as a percentage change in dimensions

Strain (muscular) Muscular injury usually caused by large eccentric stretches of muscle fibers

Strain energy The capacity to do work of an object due to its deformation by an external force

Strain gauge A small array that is bonded to materials in order to sense the small changes in size (strain) as the material is loaded; usually used to measure force or acceleration

Strength (mechanical) The toughness of a material to resist loading, usually measured as the total work or peak force required to permanently deform (yield strength) or break a material (ultimate strength)

Strength (muscular) The maximum force or torque produced by a muscle group in an isometric action at a specific joint angle; research has found several domains of strength expression depending on the time, velocity, and resistance involved

Stress (mechanical) The force per unit area in a material

Stress fracture A very small fracture in cortical bone caused by repetitive loading and inadequate rest

Stress relaxation The decrease in stress in a material over time when subjected to a constant force

Stress–strain curve (see "load deformation")

Stretch-shortening cycle (SSC) A common coordination strategy where agonists for a movement are eccentrically loaded in a countermovement, immediately before the concentric action and motion in the intended direction; an SSC results in larger initial forces and greater concentric work than purely concentric actions

Synergy The combination of several muscle actions that serve to optimally achieve a motor task

Sweet spot Striking implements (bats, rackets, etc.) have zones where impact with other objects is most effective; the term *sweet spot* tends to refer to the zone with the highest coefficient of restitution, although there are zones that minimize reaction forces (center of percussion) or minimize vibration (node)

Technology The tools and methods for applying scientific knowledge to solve problems or perform tasks

Telemetry A technique to send biomechanical signals to recording devices without wires, using an FM radio transmitter and receiver

Tension A pulling apart (making longer) of mechanical loading created by forces in opposite directions acting along the longitudinal axis of a material

Tensor A complex variable that cannot be described using only magnitude and direction

Tetanus The summation or fusion of many twitches of muscle fibers into a smooth rise in tension

Thixotropy A property of a material to change passive stiffness in response to previous loading; this history-dependent behavior is apparent in the increasing stiffness of muscle with extended inactivity

Time constant Typically, an averaging/smoothing value in EMG processing; the larger the time constant the larger the time interval averaged over, meaning more smoothing

Torque (see "moment of force") The rotating effect of a force; mechanics of materials use torque to refer to torsion moments acting on an object

Torsion Opposing loads that twist an object along its longitudinal axis

Trajectory The path in space that an object follows as it moves through the air

Twitch The force response of a muscle fiber to a single stimulation

Twitch interpolation (superimposition) technique A method used to determine the maximality of a maximum voluntary action (MVC) where stimulation is provided during an MVC

Vector A complex quantity requiring description of size and direction

Viscoelastic The property of a material where force in the material is dependent on time and deformation

Weight The downward (vertical) force action on an object due to gravity

Wolff's law Bones remodel according to the stress in the tissue

Work (mechanical) Work is done when a force moves an object in the direction of the force and is calculated as the product of force and displacement

Work–energy relationship Principle in physics which states that the work done on a body is equal to the net change in energy in the body

Yield point Point on the load-deformation curve where a material continues to deform without increasing load

Young's modulus (see "stiffness")

Appendix B: Conversion Factors

Biomechanical variables are reported in traditional English units and the metric system (SI, International System). The conversion factors below appendix are useful for converting between various measurement units. It is likely you will find one unit of measurement easier to relate to, and you may need to transform some values from the literature to more convenient units of measurement.

For example, if you wanted to get a feel for how fast a person is running at 9 m/s, you could take 9 m/s times 2.23 to get 20.1 mph. If you wanted to know how fast you were running on a treadmill that reported your pace as 8.5 minutes per mile, you would first convert the pace to an average speed in miles per hour. Sixty minutes divided by 8.5 minutes would equal 7.1 mph. Next you would take 7.1 mph divided by the conversion factor (2.23) to obtain 3.2 m/s.

Variable	SI unit × factor = other unit		
Distance	m	3.28	ft
	km	0.621	Miles
	radian	57.3	Degrees
Speed	m/s	2.23	mph

Variable	SI unit × factor = other unit		
	km/hr	0.62	mph
	m/s	3.28	ft/s
	rad/s	57.3	deg/s
	rad/s	9.55	rpm
Acceleration	m/s/s	0.102	g's
Mass	kg	0.069	Slugs
Moment of inertia	$kg.m^2$	0.738	$Slugs \cdot ft^2$
Force	N	0.225	Pounds
Torque	N.m	0.738	lbs·ft
Impulse	N.s	0.225	lbs·s
Energy	Joules	0.738	ft-lbs
Work	Joules	0.738	ft-lbs
Power	Watts	1.341	Horsepower
Momentum	(kg.m)/s	0.225	(Slug·ft)/s
	$(kg\text{-}m^2)/s$	0.225	$(Slug \cdot ft^2)/s$
Stress/pressure	Pascal	0.00015	lbs/in^2

Appendix C: Suggested Answers to Odd Numbered Review Questions

This appendix provides suggested answers to, primarily, the odd-numbered review questions from Chapters 1 through 8. The purpose of review questions is to practice and rehearse key biomechanical concepts, principles, and laws. Students are encouraged to study the topics related to each question in greater depth. The discussion questions in Chapters 9 and 10 are designed for students and instructors to discuss. Discussion questions are ideal for small-group brainstorming and practice in qualitative diagnosis of human movement.

■ **Chapter 1**

1. Biomechanics is the study of how living things move using the science of mechanics. In the first half of the twentieth century, this was synonymous with kinesiology, but now kinesiology is the academic discipline of the study of human movement.

3. The advantages of qualitative biomechanical analysis are its ease of use and flexibility, but its weaknesses are related to subjectivity and lower reliability. Quantitative biomechanical analysis may have greater precision and accuracy, but its weaknesses are the high cost in terms of equipment and time.

5. A wide variety of journals publish biomechanics research. These journals include specialized biomechanics, engineering, biology, medicine, strength and conditioning, and sports-medicine journals. Peer-reviewed articles on biomechanics can be found using many persistent and skilled searchers of several bibliometric indexes/databases.

7. Biomechanics must be integrated with other kinesiology sciences because people are not robots that move without regard to environmental factors. Psychological, physiological, and perceptual issues are all examples of factors that might be more important than biomechanical factors in some situations.

■ **Chapter 2**

1. There are several anatomical terms employed to describe the location and motion of body structures. Some examples include directions (anterior/posterior, medial/lateral, superior/inferior, proximal/distal) and joint movements (flexion/extension, adduction/abduction, internal rotation/external rotation).

3. Muscle fiber types and their architectural arrangement affect muscle force and range of motion. The rise and decay of muscle tension is greatest in fast-twitch fibers and decreases the greater the oxidative or slow-twitch characteristics of the fiber. Muscle fibers arranged in parallel have greater range of motion but create less force. Pennate fiber arrangements produce greater force but have less range of motion.

5. Muscle tension has active and passive components. Passive tension does not appear to play a large role in the middle of the range of motion but does tend to limit motion when the muscle is stretched near the end of the range of motion.

7. Functional anatomy theories of muscle actions are limited as they assume more and less movable attachments, only examine one muscle at a time, and do not consider passive tension and all the other muscles and external forces that determine if the joint will rotate or in what direction. This one-joint focus also ignores forces transmitted between all joints of a linked segment biological system.

9. The axis of rotation is at right angles (normal or 90°) to the plane of motion it creates. For example, joint rotations in the frontal plane in anatomy occur about anteroposterior axes.

■ **Chapter 3**

1. Biomechanics has traditionally focused on rigid body and fluid mechanics. The majority of early biomechanical studies focused on the kinematics of movement, but there are still many studies on the causes (kinetics) of movement.

3. Scalars only require knowledge of size and units. Vector variables have size, units, and direction.

5. The nine principles of biomechanics can be subdivided into principles related to human movement and projectiles.

7. Many factors affect human movement along with the principles of biomechanics. Some factors might be performer characteristics (psychological, perceptual, or social), the physical environment, the goal of the movement, and the philosophical goals of the kinesiology professional.

9. The range of motion most often used by people depends on the mix and potential trade-off between the goals of accuracy or force/speed of motion. Biomechanics research can provide guidance for specific movements, but in general, movement for accuracy uses less range of motion, and high force or speed movements use greater range of motion (both linear and angular body movements).

■ **Chapter 4**

1. The primary loads on body tissues are compression, tension, and shear. The combined loads are bending and torsion.

3. The tensile strengths of tendon and muscle are about 14,500 and 60 lb/in^2, respectively, while the tensile strength of bone is about 18,000 lb/irt^2. These data are consistent with the higher incidence of muscle injuries compared to that for tendon or bone.

5. The force–velocity relationship has several implications for resistances and speed of movement in strength-training exercises. When training for muscular strength, large resistances should be moved slowly to train the muscle where it is strongest. Training for muscular power and endurance uses smaller resistances moved at faster speeds.

7. The force–length relationship defines the tensile force capability of muscle at different points in the range of motion. Active tension is low at short and long muscle lengths, with the greatest force potential at middle muscle lengths. Passive tension however is small in the middle of the range of motion and increase in a curvilinear fashion as muscle length increases.

9. Increasing time of force application can be limited in some activities just because of the limited time available (need to toss a ball quickly or the short time your foot is on the ground in sprinting or running jumps). Time of force application also interacts with coordination and the goal of the movement. As you increase time of force application and the speed of your motion increases, the force–velocity relationship means muscle forces that you can continue to apply will be smaller and less effective.

11. The stretch-shortening cycle is the enhancement/potentiation of concentric muscle forces when they are immediately preceded by a countermovement braked by eccentric muscle action. Mostly all movements are naturally coordinated with stretch-shortening cycles, but they are most effective in dynamic and fast movements.

13. Range of motion is often defined using either linear or angular measurements. Angles work well for specific rotations allowed by individual joints. Linear measurements are often used when the range of motion being tested involves multiple joints.

15. Therapy professionals generally use slow, passive motion with persons recovering from injury because of the viscoelastic properties of muscles and tendons. Slow increases in length make the tissue resistance also rise slowly. Moving the limb through a range of motion will help maintain this extensibility, so the patient will not lose range of motion from disuse.

■ **Chapter 5**

1. The frame of reference is the point from where motion is measured.

3. An average velocity is a velocity estimate for the middle of a time interval where displacement and time information are available ($V = d/t$). The smaller the time interval used for the calculation, the more accurate the average velocity is, and the closer it gets to true instanta-

neous velocity. An instantaneous velocity is an exact estimate of the velocity at an instant in time and is calculated using calculus.

5. With upward displacement as positive, the average vertical velocity ($V = d/t$) of the dumbbell for the concentric phase is 1.2/1.5 = 0.8 m/s, while the average vertical velocity of the eccentric phase is −1.2/2.0 = −0.6 m/s.

7. Angular kinematics is particularly suited for analysis of human movement because joint motions are primarily rotational. Markers placed on the body can be digitized to calculate the angular kinematics of the joints during human movements.

9. Since knee extension is positive (+50 deg/s), the angular acceleration of her knee ($\alpha = \omega/t$) is (0–50)/0.2 = −250 deg/s/s.

11. The coach could use a radar gun to measure maximum and warm-up throwing speeds. If the coach did not have a radar gun, they could measure off the standard distance and time of the throws with a stopwatch to calculate average velocities in each throwing condition.

13. To use the angular-to-linear velocity conversion formula ($V = \omega \cdot r$), the angular velocity must be in radian/second: 2000 deg/s divided by 57.3 deg (1 radian), which is equal to 34.9 radian/s. The velocity of the club head relative to the golfer's hands is 34.9(1.5) = 52.4 m/s.

15. The vertical acceleration of a volleyball anywhere in flight is a downward acceleration due to gravity of −9.8 m/s/s or −32.2 ft/s/s.

17. The Coordination Continuum principle states that more simultaneous coordination will be used with larger resistances and more sequential coordination with lower resistances. A healthy adult could likely lift a small (10 lbs) weight with sequential joint coordination, most likely with just passive tension in many muscles. A toddler or elderly person that would be weaker might simultaneously coordinate most of their body to squat and lift a small resistance like this.

19. The mean horizontal velocity of a spring in the first twenty meters is 6.7 m/s (V=20 m/3.0 s) and 8.6 m/s (V=60 m/7.0 s) over the first sixty meters. The mean running velocity appears to be increasing that would be consistent with a positive horizontal acceleration in the first 60 meters of a 100 meter sprint. Near this point in the race runners are a top running speed and struggle to maintain this speed and negatively accelerate.

■ **Chapter 6**

1. A 6-kg bowling ball has the same inertia in all states of motion. The ball's inertia is a fundamental property of matter and is measured by its mass, 6 kg. This will not change unless we get the ball rolling near the speed of light!

3. Increasing inertia is useful in movement when you want to maximize stability, or if there is time to get a

larger inertia moving in a desired direction. Increasing the mass of a wrestler will make it more difficult for an opponent to move the wrestler.

5. The major determining factors of dry friction are the normal reaction and the coefficient of friction. Since adding mass to a person has other effects, the best strategy is to select a shoe with a higher coefficient of friction with common flooring.

7. If we move the shearing force to the left, we create a right triangle with a 30° angle on the right and a hypotenuse of 1000 N. The longitudinal component of the joint force (F_L) is the adjacent side, so we can use the cosine relationship to calculate cos 30° = F_L/1000, so F_L = 866 N. The sine of 30° is a special value (0.5), so we can quickly see that F_s = 500 N.

9. Muscular strength is the maximum force a muscle group can create in certain conditions, usually an isometric action at a specified joint angle. Muscular power is the rate of doing muscular work. Maximum muscular power occurs at the combination of velocity and force that maximizes muscular work. This usually occurs at moderate (about a third of maximum) velocities and muscular force.

11. Given a 800-N climber has 81.6 kg (800/9.8) of inertia and upward displacement is positive, we can use Newton's second law in the vertical direction (**F = ma**) to calculate −800 + 1500 = 81.6(**a**), so **a** = 8.6 m/s/s.

13. Sequential coordination of highspeed movements is advantageous because initial proximal movement contributes to SSC muscle actions, and mechanical energy can be transferred through Segmental Interaction.

15. Given that an upward displacement is positive and a 30-kg barbell weighs −294 N (30·9.8), we can use Newton's second law in the vertical direction (**F = ma**) to calculate −294 + 4000 = 30(**a**), so **a** = 123.5 m/s/s or 12.6 g's of vertical acceleration.

■ **Chapter 7**

1. A torque or moment of force depends on the applied force and the moment arm.

3. The joints of the human body allow us to change our resistance to rotation or moment of inertia by moving the masses of the body segment toward or away from an axis of rotation. Bringing segments close to an axis of rotation decreases moment of inertia while extending segments away from an axis of rotation increases moment of inertia.

5. Newton's first angular analog says that an object will stay at rest or constant rotation unless acted upon by an external torque. Newton's second angular analog says that the angular acceleration of an object is proportional to the torque causing it, is in the same direction, and is inversely proportional to the moment of inertia. Newton's third angular analog states that for every torque acting on an object, there is an equal and opposite torque this object applies back on the other object creating the torque.

7. The center of gravity of athletes doing a lunge-and-sprint start as illustrated below is likely the positions indicated by the dot.

9. To maximize stability, a person can increase the size of the base of support, lower the center of gravity relative to the base of support, and position the center of gravity relative to anticipated forces. Maximizing stability tends to decrease the ability to move in all directions (mobility).

11. Given that the force applied by the student was 30 lb and we know the radius of the merry-go-round, it is easiest to find the rotary component (F_R) of the force to multiply by the radius (4 ft) to obtain the torque applied. We can calculate $\cos 55' = F_R/30$, so $\mathbf{F_R} = 17.2$ lb. Torque $(\mathbf{T} = \mathbf{F \cdot d_\infty})$ applied to the merry-go-round is 17.2(4) = 68.8 lb·ft. This is almost half the 120 lb·ft of torque when the force is applied at an angle that maximizes the moment arm.

13. You cannot calculate the torque because the muscle angle of pull is not known.

15. Angular momentum tends to stabilize the orientation of an object's axis of rotation.

■ **Chapter 8**

1. The major fluid forces are buoyancy, lift, and drag. Buoyancy acts upward. Drag acts parallel to and opposing the relative flow of fluid, while lift acts at right angles to the relative flow of fluid.

3. The center of gravity and center of buoyancy of the human body move in similar manner, following the mass shifts with moving segments. The center of gravity moves more than the center of buoyancy because the trunk volume dominates the volume of the rest of the body.

5. Optimal Projection angles include the effect of fluid forces as well as the release and target locations of projection activities. For example, place-kicking has an optimal angle of projection much lower than 45° because of the fluid forces of drag.

7. The centers of buoyancy of a swimmer in three flotation positions (below) are likely the positions indicated by the dot.

9. A volleyball serve with topspin dives downward because the Magnus Effect generates a downward-and-backward-directed lift force that adds to gravity.

11. Round balls tend to curve in the direction of the spin. If the front of a ball is spinning to the right (as you observe it as it is coming toward you), the lift force will act to the right and make the ball curve to the right.

13. Swimmers and cyclists shave so as to decrease surface drag, which resists their motion, while a rougher surface of a spinning baseball will create a greater lift force. The greater Magnus Effect and lift force acting on the baseball is more important than the minor effect the roughness will have on drag.

Appendix D: Right-Angle Trigonometry Review

Trigonometry is a branch of mathematics that is particularly useful in dealing with right-angle triangles and making transformations between linear coordinates and angles. This is important in the study of biomechanics because vectors are usually resolved into right-angle components. This appendix provides a brief review of four trigonometric relationships for two-dimensional analysis in the first quadrant. There are many more trigonometric relationships that are fully defined for all 360° of a circle. The four relationships will be defined relative to right triangle illustrated below.

The sides of a triangle are traditionally labeled in two ways, with letters and names describing their position relative to one of the acute angles of interest (θ). The longest side of the triangle is the *hypotenuse* or *c*. The side next to the angle of interest is usually labeled *a* or the *adjacent* side. The last side is the *opposite* side or *b*.

The first relationship is the *Pythagorean theorem*, which describes the relationship between the lengths of the sides in all right triangles. If you have knowledge of any two of the three sides of a triangle you can apply the formula $c^2 = a^2 + b^2$ to solve for the magnitude of the other side.

The sine, cosine, and tangent are the most commonly used trigonometric relationships, because they define the relationships between the acute angles and the dimensions of right triangles. The abbreviation and formula for each relationship is:

$$\sin \theta = b/c \quad \cos \theta = a/c \quad \tan \theta = b/a$$

Suppose the right triangle depicted below corresponds to the following data on the release conditions of a soccer kick: c = 40 m/s and θ_0 = 35°. A biomechanist wanting to determine the vertical velocity (b) in order to determine the time of flight could write:

$$\sin 35° = \mathbf{V}_v / 40,$$

and solving could yield

$$\mathbf{V}_v = 22.9 \text{ m / s}$$

Now use the cosine, tangent, or Pythagorean theorem to see if you can confirm if the horizontal velocity of the ball is 32.8 m/s.

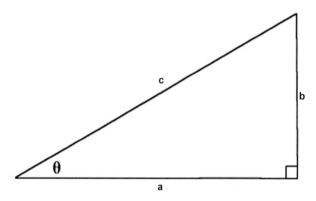

Appendix E: Qualitative Evaluation of Biomechanical Principles

Principle	Body part	Evaluative rating (inadequate-normal-excessive)
Balance		
Coordination		
Force–Motion		
Force–Time		
Inertia		
Range of Motion		
Segmental Interaction		
Optimal Projection		
Spin		

Lab Activities

This section of the book provides applied, active learning laboratory experiences. These labs are designed to illustrate key points from the chapters of the text. The labs are also designed to be flexible enough to be used and expanded as full labs for universities with four-credit courses or as short active learning exercise/demonstrations for three-credit courses. The emphasis is on using actual human movements with minimal, high-cost research equipment. Instructors can certainly substitute quantitative research equipment if it is available. While quantitative measurements and calculations are part of some labs, most of them focus on students' conceptual understanding of biomechanics and their ability to qualitatively analyze human movement. Most labs are structured for work in small groups of three to five students.

Citations of background information are provided for students to prepare for the labs or further explore their scientific basis. Space does not allow for all relevant research citations to be included on each lab. If instructors assign background reading prior to labs, they might consider assigning specific sections of the resources suggested. I am indebted to many of my peers who have shared their teaching ideas at biomechanics meetings, especially those who have attended and contributed to the last few national conferences on teaching biomechanics and continue to publish research on leaning biomechanical concepts using the Biomechanics Concept Inventory (Knudson et al. 2003).

Lab Activity 1: Finding Biomechanical Sources

Biomechanics is the study of the causes of biological movement. Biomechanics is a core subdiscipline of kinesiology, the academic study of human movement/physical activity. All kinesiology professions use biomechanical knowledge to inform their practice. Both scholarly and professional journals publish biomechanical research. There are many people interested in biomechanics, so biomechanical literature is spread out across many traditional scholarly and professional areas. This lab will help you appreciate the breadth of biomechanics in your chosen career and provide you with experience in finding biomechanical sources.

Background Reading

Chapter 1 herein: "Introduction to Biomechanics of Human Movement"
Ciccone, C. D. (2002). Evidence in practice. *Physical Therapy*, *82*, 84–88.
Minozzi, S. et al. (2000). Searching for rehabilitation articles on Medline and Embase: An example with cross-over design. *Archives of Physical Medicine and Rehabilitation*, *81*, 720–722.

Tasks

1. Identify one kinesiology career/professional area and human movement of interest.
2. Review 1 year of a journal related to this area of interest for biomechanical articles.
3. Identify a potential biomechanical topic of interest from your professional interests.
4. Search at least two computer database (Google Scholar, Medline, SportDiscus, Web of Science, etc.) for peer-reviewed biomechanical articles on your topic.
5. Answer the questions.

Finding Biomechanical Sources

1. What is your professional/career area of interest and give a human movement topic you have a biomechanical interest in?

2. Report the name of the journal, number of articles published in a particular year, and the approximate percentage of articles related to biomechanics of your movement interest.

3. Summarize the results of two searches of at least two literature databases (Google Scholar, Medline, SportDiscus). Be sure to specify the exact search you used and the number and quality of citations you obtained.

4. Based on all your searches, list the two citations you believe to be most relevant to your professional interests.

5. Comment on the diversity of sources you observed in your search.

6. Rate the quality of the sources you found based on the hierarchy of evidence presented in Chapter 1.

Lab Activity 2: Functional Anatomy

Anatomy is the study of the structure of the human body. The joint motions created by muscles in humans have been studied by anatomists in several ways: cadaver dissection and manipulation, observation, and palpation. Historically, anatomy and anatomical kinesiology have used the mechanical method of muscle action analysis to hypothesize the agonists for specific movements. This requires a detailed knowledge of the planes of movement, joint axes, attachments, courses of the muscles, and the classification of joints. Anatomy provides only part (qualitative terminology) of the prerequisite knowledge necessary to determine how muscles create movement. A century of EMG research has clearly shown the inadequacy of functional anatomy to explain how muscles act to create human movement (Hellebrandt 1963). Chapter 2 summarized how several areas of biomechanics research must be integrated (EMG, kinetics, simulation) to understand all the actions of muscles in human movement. This lab will review the mechanical method of muscle action analysis in functional anatomy and show why biomechanics is needed to determine the actions of muscles using a few demonstrations/activities.

Background Reading

Chapter 2 herein: "Anatomical Description and Its Limitations"
Hellebrandt, F. A. (1963). Living anatomy. *Quest*, *1*, 43–58.
Herbert, R. et al. (1993). Making inferences about muscles forces from clinical observations. *Australian Journal of Physiotherapy*, *39*, 195–202.
Maas, H. et al. (2004). Muscle force is determined by muscle relative position: Isolated effects. *Journal of Biomechanics*, *37*, 99–110.

Tasks

1. For the anatomical plane and joint(s) specified, use functional anatomy to hypothesize a muscle involved and the muscle action responsible for the following demos and record them on the lab report.

Demo 1 —	Sagittal plane	Elbow joint	Arm curl
Demo 2 —	Sagittal plane	Lumbar vertebrae	Trunk flexion
Demo 3 —	Sagittal plane	Metacarpophalangeal	Passive wrist flexion
Demo 4 —	Frontal plane	Hip joint	Left hip adduction

2. Perform the demos:
   Demo 1: Lie supine with a small dumbbell in your right hand and slowly perform arm curls. Have your lab partner palpate your upper arm, being sure to note differences in muscle activation in the first 80 and last 80° of the range of motion. Analyze only the lifting phase.
   Demo 2: Lie supine with your hips flexed to 90° and your quadriceps relaxed. Cross your arms over your chest and tighten your abdominal muscles. Make a note of which end of your body is elevated. See if you can make either or both sides of your body rise.
   Demo 3: In the anatomical position, pronate your right forearm and flex your elbow completely. Totally relax your right hand and wrist. In this position (hand roughly horizontal), use your left hand to extend your relaxed right wrist and let gravity passively flex the wrist. Note the motion of the fingers during wrist extension and flexion.
   Demo 4: From the anatomical position, stand on your left foot (flexing the right knee) and abduct your shoulders so that your arms are horizontal. Smoothly lower and raise your right hip (left hip adduction and then abduction) as many times as you can in 1 minute. Note the muscles that feel fatigued.
3. Answer the questions.

Functional Anatomy?

For the anatomical plane and joint(s) specified, use functional anatomy to hypothesize a muscle and the muscle action responsible for the following activities

	Plane	Joint	Movement	Muscle	Action
Demo 1 —	Sagittal	Elbow joint	Arm curl	_____	_____
Demo 2 —	Sagittal	Lumbar vertebrae	Trunk flexion	_____	_____
Demo 3 —	Sagittal	Metacarpophalangeal	Wrist flexion	_____	_____
Demo 4 —	Frontal	Hip joint	Hip adduction	_____	_____

1. Functional anatomy does not consider the action of other forces (other muscles or external forces) in hypothesizing muscle actions. Describe the muscle actions throughout the range of motion in the horizontal plane arm curl, and note why an external force changes the muscle activation strategy.

2. Classifying muscle attachments as an "origin" or "insertion" is not always clear. What muscle(s) are active in the abdominal exercise, and what attachments are being pulled?

3. What muscle(s) created metacarpophalangeal extension when the wrist was passively flexed in Demo 3? What muscle(s) created metacarpophalangeal flexion when the wrist was passively extended? How does the muscle create this motion without activation?

4. Was there discomfort in the left hip adductors in Demo 4? What muscle and action was responsible for controlling left hip adduction?

5. Give a movement example (be specific) where functional anatomy may be incorrect because of:
   - External forces
   - Muscle synergy
   - Passive tension
   - Attachment stability changes

Lab Activity 3: Qualitative and Quantitative Analysis of Range of Motion

This text summarizes many biomechanical variables and concepts into nine principles of biomechanics. The analysis of human movements using these biomechanical principles can be qualitative (subjective) or quantitative (based on numerical measurements). All kinesiology professions have used both qualitative and quantitative analyses of human movement, but qualitative analysis is used most often. This lab will explore the range of motion principle of biomechanics, using a variety of static flexibility tests common in physical education and physical therapy. This lab will show you there are a variety of ways to quantify range of motion and that there are strengths and weaknesses of both qualitative and quantitative analyses of human movement.

Physical therapists used to perform a standing toe touch to screen for persons with limited hamstring flexibility. Patients either passed the test by being able to touch their toes with their fingers while keeping their legs straight or they failed to touch their toes, indicating poor hamstring flexibility. Flexible hamstrings allow a person to tilt their pelvis forward more, making it easier to touch their toes. Recently, more accurate field tests of static flexibility have been developed. The tests that will be used are the sit-and-reach test (SRT), active knee extension (AKE), and the modified Schober test (MST). The results of these flexibility tests can be analyzed qualitatively (judging if the subject has adequate flexibility) or quantitatively. Quantitative analysis can either be norm-referenced (comparing scores to all other people) or criterion-referenced. Criterion-referenced testing compares test scores to some standard of what should be. Criteria or standards are usually based on evidence on what correlates with health (health-related fitness) or with physical abilities to perform jobs safely (occupational screening). For example, physical therapists studying the sit-and-reach test suggested that subjective observation of the forward tilt of the rear of the pelvis is an effective assessment of hamstring flexibility as the SRT score (Cornbleet and Woolsey 1996).

Background Reading

Chapter 3 herein: "Fundamentals of Biomechanics and Qualitative Movement Diagnosis"
Cornbleet, S., & Woolsey, N. (1996). Assessment of hamstring muscle length in school-aged children using the sit-and-reach test and the inclinometer measure of hip joint angle. *Physical Therapy*, *76*, 850–855.
Gajdosik, R., & Lusin, G. (1983). Hamstring muscle tightness: Reliability of an active–knee-extension test. *Physical Therapy*, *63*, 1085–1088.
Gleim, G. W., & McHugh, M. P. (1997). Flexibility and its effects on sports injury and performance. *Sports Medicine*, *24*, 289–299.
Knudson, D. et al. (2000, June). Current issues in flexibility fitness. *The President's Council on Physical Fitness and Sports Research Digest*, 1–8.

Tasks

1. Select three volunteers for flexibility testing
2. Learn how to use a sit-and-reach box, inclinometer, goniometer, and tape measure for SRT, AKE, and MST.
3. Collect the following quantitative assessments of lumbar and hamstring range of motion for one side of the body: SRT, AKE, and MST. While these measurements are being taken, have people in your lab group do a qualitative/categorical assessment (hypoflexible/ankylosis, normal, hyperflexible) of the subject being tested.
4. Answer the questions.

Qualitative and Quantitative Analysis of Range of Motion

Ratings of Hamstring Flexibility

	Qualitative	SRT	AKE
Subject 1	_____	_____	_____
Subject 2	_____	_____	_____
Subject 3	_____	_____	_____

Ratings of Lumbar Flexibility

	Qualitative	Schober
Subject 1	_____	_____
Subject 2	_____	_____
Subject 3	_____	_____

1. Given that the healthy standard for adult (>17 years) males and females in the SRT are 17.5 and 20 cm, respectively, and a passing AKE is $\theta_K = 160°$, how well did your qualitative and quantitative ratings of hamstring flexibility agree?

2. Given that the passing score for the MST is 7 cm, how well did your qualitative and quantitative ratings of lumbar flexibility agree?

3. List the characteristics of the range of motion you evaluated in your qualitative ratings of hamstring flexibility.

4. Range of motion is a kinematic (descriptive) variable and does not provide kinetic (muscle-tendon resistance) information about the passive tension in stretching. Static flexibility measurements like these have been criticized for their subjectivity related to a person's tolerance for stretch discomfort (Gleim and McHugh 1997). Are there kinetic aspects of stretching performance that can be qualitatively judged by your observations of these flexibility tests?

5. Compare and contrast the strengths and weaknesses of a qualitative versus quantitative assessment of static flexibility.

Lab Activity 4: Muscle Actions and the Stretch-Shortening Cycle (SSC)

The forces (tension) muscles exert to create movement vary dramatically across their length, velocity (shortening and lengthening), and timing of activation. The classic in vitro muscle mechanical characteristics interact with other factors (activation, leverage, connective tissue stiffness, etc.) to determine the amount of torque a muscle group can create. The torque a muscle group creates naturally affects muscular strength, endurance, and other performance variables. The purpose of this lab is to demonstrate the performance consequence of muscle actions and the stretch-shortening cycle (SSC). The endurance of the elbow flexors will be examined in concentric and eccentric actions to review the force–velocity relationship. Two kinds of vertical jumps will be examined to determine the functional consequences of the SSC.

Background Reading

Chapter 4 herein: "Mechanics of the Musculoskeletal System"
Komi, P. V. (Ed.) (1992). *Strength and power in sport*. New York: Blackwell Science.
Kubo, K. et al. (1999). Influence of elastic properties of tendon structures on jump performance in humans. *Journal of Applied Physiology*, *87*, 2090–2096.
Lieber, R. L., & Bodine-Fowler, S. (1993). Skeletal muscle mechanics: Implications for rehabilitation. *Physical Therapy*, *73*, 844–856.

Tasks

1. Select five volunteers for elbow flexor endurance testing. For each subject select a dumbbell with submaximal resistance (between 50% and 80% 1RM). Record the number or concentric-only repetitions (partners lower the dumbbell) for the person's stronger limb and the number or eccentric-only (partners lift the dumbbell) for their weaker limb. Attempt to keep a similar cadence for each test.
2. Perform and measure the maximum height for the countermovement jump (CMJ) and an equivalent static jump (SJ) for everyone in the lab. The SJ begins using isometric muscle actions to hold a squat position that matches the lowest point of the CMJ for that person. Observe jumps carefully since it is difficult to match starting positions, and it is difficult (unnatural) for subjects to begin the concentric phase of the SJ with virtually no countermovement.
3. Perform the calculations and answer the questions.

Muscle Actions and the Stretch-Shortening Cycle (SSC)

Maximal repetitions with a submaximal resistance

	Concentric–stronger side	Eccentric–weaker side
Subject 1	_____	_____
Subject 2	_____	_____
Subject 3	_____	_____
Subject 4	_____	_____
Subject 5	_____	_____

Prestretch augmentation in SSC
CMJ _____
SJ _____
PA (%) = ((CMJ − SJ)/SJ) · 100 (Kubo et al. 1999)
My PA _____
Class mean PA _____

Questions

1. Did the stronger side of the body have the most endurance? Explain the results of this comparison of concentric and eccentric muscles actions based on the force–velocity relationship of muscle.

2. Hypothesize the likely lower extremity muscle actions in the SJ and the CMJ.

3. How much improvement in vertical jump could be attributed to using a SSC?

4. What aspects of coaching jumps and other explosive movements must be emphasized to maximize performance? Explain why your technique points may improve performance based on muscle mechanics or principles of biomechanics.

Lab Activity 5A: Horizonal Velocity in Sprinting

Linear kinematics in biomechanics is used to create precise descriptions/measurement of human motion. It is important for teachers and coaches to be familiar with many kinematic variables (like speeds, pace, or times) that are representative of various levels of performance. Most importantly, professionals need to understand that velocity is a vector quantity that varies over time, as well as have an intuitive understanding of where peak velocities and accelerations in a particular direction occur in movement. This lab will focus on your own sprinting data in a 40-meter dash and a world class 100-meter sprint performance to examine the relationship between displacement, velocity, and acceleration. These activities provide the simplest examples of linear kinematics since the body is modeled as a point mass and motion of the body is measured in one direction that does not change (so speed and velocity are the same).

Background Reading

Chapter 5 herein: "Linear and Angular Kinematics"
Haneda, Y., et al. (2003). Changes in running velocity and kinetics of the lower limb joints in the 100 m sprint running. *Japanese Journal of Biomechanics in Sports and Exercise*, *7*, 193–205.
Mero, A. et al. (1992). Biomechanics of sprint running: A review. *Sports Medicine*, *13*, 376–392.
Murase, Y., et al. (1976). Analysis of the changes in progressive speed during the 100-meter dash. In P. V. Komi (Ed.), *Biomechanics V-B* (pp. 200–207). Baltimore: University Park Press.

Tasks

1. Estimate how fast you can run in mph _____
2. Following a warm-up, perform a maximal 40 meter sprint. Obtain times with four stopwatches for times at the 10, 20, 30, and 40 meter marks.
3. Perform the calculations and answer the questions.

Kinesiology Major Normative Data
Time (s)

	Females				Males			
	10	20	30	40	10	20	30	40
Mean	2.3	3.9	5.4	7.0	2.0	3.3	4.6	5.9
SD	0.2	0.4	0.5	0.7	0.2	0.2	0.3	0.5

Choose a champion sprinter to compare yourself to in the first 40 meters. The times below are 10-meter interval times (in seconds) and the total race time for three elite sprinters.

1. Mr. Maurice Greene: 1999WR 1.86, 1.03, 0.92, 0.88, 0.86, 0.84, 0.85, 0.85, 0.85, 0.86, Race = 0.967
2. Mr. Usain Bolt: 2009WR 1.88, 1.0, 0.9, 0.84, 0.83, 0.82, 0.81, 0.82, 0.82, 0.84, Race = 9.58
3. Ms. Tori Bowie 2017 2.07, 1.15, 1.02, 0.94, 0.93, 0.93, 0.94, 0.95, 0.95, 0.97, Race = 10.85

Lab Activity 5A

Horizontal Velocity in Sprinting

Record *your* times in the spaces below.

   10 m 20 m   30 m   40 m

$t_1 =$ _____

$t_2 =$ _____

$t_3 =$ _____

$t_4 =$ _____

Questions

1. Calculate the average horizontal velocity in each of the 10 m intervals of your 40 m sprint ($V = \Delta d/\Delta t$). Report your answers in m/s and mph (m/s · 2.237 = mph).

2. Calculate the average horizontal velocities for the intervals for one of the elite athletes' 100 m sprint. Note that the times in the table represent the change in time (time to run the interval: Δt), not the cumulative time, as in your 40 m sprint data. Average velocities are usually assigned to the midpoints of the interval used for the calculation.

Velocity (m/s) at the

5 _____ meter points.

15 _____

25 _____

35 _____

45 _____

55 _____

65 _____

75 _____

85 _____

95 _____

3. Plot an elite athlete's (Green, Bolt, or Bowie) and your horizontal velocities on the following velocity-displacement graph.

4. Give a qualitative description of the general slopes of the Greene/Bolt/Bowie horizontal velocity graph in question 3 (the general pattern would be same if this were a true velocity–time graph) that determine the major phases of horizontal acceleration in maximal sprinting. Where is acceleration the largest and why?

Lab Activity 5B: Accuracy of Throwing Speed Measurements

Linear kinematics in biomechanics are used to create precise descriptions of human motion. It is important for teachers and coaches to be familiar with many kinematic variables (like speeds, pace, or times) and the accuracy and consistency of these measurements. The accuracy of a speed calculated from the formula $s = l/t$ strongly depends on the time interval used and errors in measurement. The speed calculated is also an average over the time interval used for the calculation. The reliability of a measurement of speed decreases with greater variation from measurement errors and subject performance. This lab will allow you to explore accuracy and consistency issues in the measurement of ball speed in softball throwing.

Background Reading

Chapter 5 herein: "Linear and Angular Kinematics"
Atwater, A. E. (1979). Biomechanics of overarm throwing movements and of throwing injuries. *Exercise and Sport Sciences Reviews, 7*, 75–80.
Brody, H. (1991, March/April). How to more effectively use radar guns. *TennisPro*, 4–5.

Tasks

1. Estimate how fast you can throw a softball: _____
2. Following a warm-up, perform maximal and 75% effort throws to a partner or chain link fence 20 m away. Measure and record the speed of the throws two ways: with a radar gun and by flight times averaged from four stopwatches. Be sure to note the variation in times measured by stopwatch operators and record all time and radar data for all throws for everyone in the lab. Average speed of the throw will assume the distance of ball flight was 20 m, not the actual flat parabolic trajectory/path. Remember safety, and only throw with maximum effort after warm-up and when all class partners are ready.
3. Perform the calculations to calculate the average speed of your throws and answer the questions.

Kinesiology major normative data for maximum effort throws

	Females speed (mph)	Males speed (mph)
Mean	43.3	64.0
SD	8.6	8.7

Lab Activity 5B: Accuracy of Throwing Speed Measurements

Record *your* times in the spaces below.

Maximal throw	75% Effort throw
Speed = _____	Speed = _____
$t =$ _____	$t =$ _____
Speed = _____	Speed = _____
$t =$ _____	$t =$ _____

1. Calculate the average speed of your maximal and 75% effort throw ($s = \Delta l / \Delta t$) from the stopwatch data. Report your answers in m/s and mph (m/s $* 2.237$ = mph). What factors would account for differences you observed between the radar and stopwatch measurements of ball speed?

2. Comment on the typical differences in stopwatch times for the four timers for maximal throws and 75% effort throws. About how accurate are stopwatches for estimating softball throwing speed?

3. Comment on how consistent were the radar measurements of your maximal and 75% effort throws. Given that reliability, how much of a difference would you consider meaningful?

4. Coaches sometimes ask athletes to perform warm-ups, drills, or practice at submaximal speeds. How effective were you and the persons in your lab at throwing at 75% of maximal speed?

Lab Activity 6A: Top Gun Kinetics: Force-Motion Principle

Newton's laws of motion explain how forces create motion in objects. The application principle related to Newton's second law is the force–motion principle. The purpose of this lab is to improve your understanding of Newton's laws of motion. As a candidate for the prestigious "Top Gun" kinetic scooter pilot in biomechanics class, you must not only perform the missions but use kinetics to explain your scooter's flight. Biomechanics Top Gun is like a Naval Top* Gun in that skill and knowledge are required to earn the honor. It is important that you follow the instructions for each mission explicitly. Care should be taken by pilots and their ground crew to perform the task correctly and safely. Note that your "multimillion-dollar" scooters provide *low* (not quite zero) friction conditions, so you need to move/push briskly so you can ignore the initial effects of rolling friction. Note the horizontal motions mean the effects of gravity on the observed motions can be ignored, just like the negligible forces of air resistance (you are not flying near the speed of sound). Kinetics explains all motion: from scooters, braces, rackets, jump shots, to muscle actions. Think about the forces, what directions they act, and the motion observed in each mission. This lab is roughly based on a lab developed by Larry Abraham (Abraham 1991) at the University of Texas. *This phrase references an older (1986) movie about the hyper-competitive world of naval aviators.

Background Reading

Chapter 6 herein: "Linear Kinetics"
Abraham, L. D. (1991). *Lab manual for KIN 326: Biomechanical analysis of movement*. Austin.

Tasks

1. Using your multimillion-dollar scooters, ropes, and spring/bathroom scales, perform the following training missions:
   - Sit on the scooter and maximally push off from a wall (afterburner check). Experiment with various body positions and techniques.
   - Sit on your scooter and push off from a partner on another scooter.
   - Loop a rope over a bathroom scale held by a partner on a scooter. Sit on your scooter and pull your partner, who passively holds the scale, and note the largest force exerted.
   - Repeat the last mission, but have your partner also vigorously pull on the scale.
2. Answer the questions.

Lab Activity 6A

Top Gun Kinetics: Force–Motion Principle

1. How far were you able to glide by pushing off from the wall? What is the relationship between the direction of your push and the direction of motion?

2. How far were you able to glide by pushing off from another scooter pilot? Explain any differences from task 1 using Newton's Laws of Motion.

3. How much force was applied to pull a passive partner? Which scooter pilot moved the most and why?

4. How much force was applied when both partners vigorously pulled on the rope? Explain any differences in the observed motion from task 3 using Newton's laws.

5. Assume the mass of your scooter cannot be modified, but you are charged with recommending technique that maximizes scooter speed and agility. Use the force–motion principle to suggest why a certain body position and propulsion technique is best.

Lab Activity 6B: Impulse-Momentum: Force-Time Principle

The timing of force application to objects affects the stress and motion created. Newton's second law applied to forces acting over time is the impulse–momentum relationship. The change in momentum of an object is equal to the impulse of the resultant force. This activity will allow you to experience some interesting real-life examples of the impulse–momentum relationship. The purpose of this lab is to improve your understanding of changing the motion of an object (specifically, it's momentum) by applying force over a period of time. In some ways body tissues are similar to water balloons in that too much force can create stresses and strains that lead to injury. It is important for teachers/coaches to understand how movement technique affects the impulse and peak force that can be applied to an object. This lab is modified from a lab proposed by McGinnis and Abendroth-Smith (1991).

Background Reading

Chapter 6 herein: "Linear Kinetics"

McGinnis, P., & Abendroth-Smith, J. (1991). Impulse, momentum, and water balloons. In J. Wilkerson, E. Kreighbaum, & C. Tant, (Eds.), *Teaching kinesiology and biomechanics in sports* (pp. 135–138). Ames: Iowa State University.

Knudson, D. (2001c). Accuracy of predicted peak forces during the power drop exercise. In J. R. Blackwell (Ed.) *Proceedings of oral sessions: XIX international symposium on biomechanics in sports* (pp. 135–138). San Francisco: University of San Francisco.

Tasks

1. Estimate how far you can throw a softball-sized water balloon. _____
2. Estimate the maximum distance you could catch a similar water balloon. _____
3. Fill several water balloons to approximately softball size ($\approx$7–10 cm in diameter).
4. Measure the maximal distance you can throw the water balloon. _____
5. Measure the maximal distance you and a partner can throw and catch a water balloon. _____
6. Answer the questions.

Lab Activity 6B

Impulse–Momentum: Force–Time Principle

Distance of Throw _____

    Distance of Toss and Catch _____

1. What technique factors were important in the best water balloon throws?

2. What technique factors were most important in successfully catching a water balloon?

3. Theoretically, if you could throw a water balloon 25 m, could you catch it? Why?

4. How are the mechanical behaviors of water balloons similar to muscles and tendons?

5. Below is a graph of the vertical force (Newtons) measured when a medicine ball was dropped from the same height and bounced (●) or was caught and thrown back up in a power drop exercise (◆). Use the force–time principle to explain the differences in the forces applied to the medicine ball. Data from Knudson (2001c).

Lab Activity: 7A: Angular Kinetics of Exercise

The positions of body segments relative to gravity determine the gravitational torques that must be balanced by the muscles of the body. The purpose of this lab is to improve your understanding of torque, summation of torques, lifting, and center of gravity. These biomechanical parameters are extremely powerful in explaining the causes of human movement because of the angular motions of joints. Several classic lifting and exercise body positions are analyzed because the slow motion (very small or zero acceleration) in these movements comprise a quasi-static condition. In static conditions, Newton's second law can be simplified to static equilibrium: $\Sigma F = 0$ and $\Sigma T = 0$. Remember that a torque ($\mathbf{T}$) or moment of force is the product of the force and the perpendicular distance between the line of action of the force and the axis of rotation ($\mathbf{T} = \mathbf{F} \cdot \mathbf{d}\perp$).

Background Reading

Chapter 7: "Angular Kinetics"

Chaffin, B. D. et al. (1999). *Occupational biomechanics* (3rd ed.). New York: Wiley.

Hay, J. G. et al. (1983). Load, speed and equipment effects in strength-training exercises. In H. Matsui & K. Kobayashi (Eds.), *Biomechanics III-B* (pp. 939–950). Champaign: Human Kinetics.

van Dieen, J. H. et al. (1999). Stoop or squat: A review of biomechanical studies on lifting technique. *Clinical Biomechanics*, *14*, 685–696.

Tasks

1. If an athlete doubled his/her trunk lean in a squat exercise, how much more resistance would their back feel? Estimate the extra load on the lower back if a person performed a squat with a 40° trunk lean compared to a 20° trunk lean. _____%
2. Obtain height, weight, and trunk length (greater trochanter to shoulder joint) data for a person in the lab.
3. The amount of trunk lean primarily determines the stress placed on the back and hip extensors (Hay et al. 1993). Perform two short endurance tests to see how trunk lean affects muscle fatigue. Use a standard bodyweight squat technique. Hold the squats with hands on hips in an isometric position for 30 seconds and subjectively determine which muscle groups were stressed the most. Test 1 is a squat with a nearly vertical trunk and a knee angle of approximately 120°. Test 2 is a squat with a trunk lean of about 45° and a knee angle of approximately 120°. Wait at least 5 minutes between tests.
4. Perform calculations on the following simple free-body diagrams of exercise and body positions to examine how gravitational torques vary across body configurations and answer the questions.

Lab Activity 7A

Angular Kinetics of Exercise

1. Where were the sites of most fatigue in the two squat tests? What muscle group feels more fatigue in a nearly vertical trunk orientation? Why?

2. Kinematic measurements from film/video and anthropometric data are often combined to make angular kinetic calculations. A static analysis can be done when the inertial forces and torques (dynamic loading from high-speed movement) are small. Assume the figure below is an image of you captured from video while performing bodyweight squats. Calculate a gravitational torque of *your* upper body about the hip (M/L) axis. Assume your head, arms, and trunk (HAT) have mass equal to 0.679 of body mass. The center of gravity of your HAT acts at 62.6% up from the hip to the shoulder.

3. Calculate the gravitational torque about the hip if the bottom of your squat exercise has a trunk lean of 40°. (Show free-body diagram and work.)

4. If the weight of the head, arms, and trunk do not change during the squat exercise, what mechanical change increases the gravitational torque as the person leans forward?

5. How different is the load on the back/hip extensors when you double your trunk lean? Is the size of this difference what you expected? Why is it different?

Lab Activity 7B: Calculating Center of Gravity Using Angular Kinetics

The purpose of this lab is to improve your understanding of torque, summation of torques, and center of gravity. Torque is a useful kinetic variable explaining the causes of human movement because of the angular motions of joints. Locating the center of gravity of an object and tracking its motion is useful in understanding how the force of gravity affects movement and balance. The reaction board method will be used with the angular analog and static form of Newton's second law ($\Sigma T = 0$). Remember that torque (T) or moment of force is the product of the force and the perpendicular distance between the line of action of the force and the axis of rotation ($T = F \cdot d_\infty$).

Background Reading

Chapter 7: "Angular Kinetics"
Gard, S. A. et al. (2004). Comparison of kinematic and kinetics methods for computing the vertical motion of the body center of mass during walking. *Human Movement Science*, *22*, 597–610.

Tasks

1. Estimate the height of your center of gravity as a percentage of your height: _____.

2. Record your height and weight. Measure the length of the reaction board from one supporting edge to the other.

3. Measure the reaction force lying on the reaction board in your normal standing position, and in another sport/activity relevant position of interest to you. Think about where you should you put your feet to make the calculation easier to express relative to your body.

4. Perform calculations to calculate the location of your center of gravity and answer the questions.

Lab Activity 7B

Angular Kinetics of Exercise

Record your data in the space below:

    Height _____
    Weight _____
    Reaction standing _____
    Reaction other _____

1. Draw a free-body diagram of you on the reaction board and calculate the location of your center of gravity.

2. Calculate the height of your center of gravity as a percentage of your height and discuss any differences from normative data for your gender.

3. Calculate the location of your center of gravity in the other body position (show free-body diagram and work).

4. Explain the difference in the center of gravity location between the two body postures you studied and how it might affect stability and mobility.

Lab Activity 8: Magnus Effect in Baseball Pitching

Fluid forces have dramatic effects in many human movements. Fluid dynamics is of vital interest to coaches of swimming, cycling, running, and sports where wind or ball velocities are great. The fluid forces of lift and drag increase with the square of velocity. The purpose of this lab is to improve your understanding of how fluid forces (specifically lift) can be used to affect a thrown balls trajectory. The example is in baseball pitching, although the Spin Principle applies to other ball sports. Pitching technique and the Magnus Effect are explored in the "rise" of a fastball, the "break" of a slider, and the "drop" of a curveball. Skilled performance in many sports involves appropriate application of rotation to a ball to create fluid forces for an advantageous trajectory.

Background Reading

Chapter 8 herein: "Fluid Mechanics"
Allman, W. F. (1984). Pitching rainbows: The untold physics of the curve ball. In E. W. Schrier & W. F. Allman (Eds.), *Newton at the bat: The science in sports* (pp. 3–14). New York: Charles Scribner & Sons.
Knudson, D. (1997). The Magnus effect in baseball pitching. In J. Wilkerson, K. Ludwig, & M. Butcher (Eds.), *Proceedings of the 4th national symposium on teaching biomechanics* (pp. 121–125). Denton: Texas Woman's University Press.

Tasks

1. Set up a mock baseball pitching situation indoors with a pitching rubber and home plate about 7 m apart. Warm up the shoulder and arm muscles and gradually increase throwing intensity with whiffle balls. Exchange the whiffle ball for a Styrofoam ball.
2. Hitters often perceive that a well-thrown fastball "rises" (seems to jump over their bat). The fastball is usually thrown with the index and middle fingers spread and laid across the seams of a ball, with the thumb providing opposition from the front of the ball. At release, the normal wrist flexion and radioulnar pronation of the throwing motion create downward and forward finger pressure on the ball. These finger forces create backspin on the ball. Try to increase the rate of backspin to determine if the ball will rise or just drop less than a similar pitch. Be careful to control the initial direction of the pitches by using visual references in the background. Estimate the rise or drop of the pitch relative to the initial trajectory at release.
3. A pitch that is easy to learn after the basic fastball is a slider. A slider creates a lateral "break" that can be toward or away from a batter, depending on the handedness of the pitcher and batter. The grip for a slider (right-handed pitcher) has the index and middle finger together and shifted to the right side of the ball (rear view). The thumb provides opposition from the left side of the ball. Normal wrist flexion and pronation at release now create a final push to the right side of the ball, imparting a sidespin rotation. A typical right-handed pitcher (facing a right-handed hitter) would usually direct this pitch initially toward the center to the outside corner of home plate, so the ball would break out of reach.
4. A pitch that can make a batter look foolish is the curveball. The common perception of hitters watching a well-thrown curveball is that the ball seems to "drop off the table." The ball looks like it is rolling along a horizontal table toward you and suddenly drops off the edge. The grip for a curveball is similar to a fastball grip, but with a different orientation of the seams. At release the index and middle fingers are on top of the ball, making a final push forward and downward. Common teaching cues are to pull down at release like pulling down a shade or snapping your fingers. Research has shown that radioulnar pronation is delayed in the curveball, so that at release the forearm is still in a slightly supinated position. Curveballs are thrown with forearm pronation just like other pitches; it is just delayed to near the moment of release.
5. If time is available, students can do some "show and tell" with other pitch variations. These include variations in release (sidearm, windmill softball pitch, grips, screwball, knuckleball, etc.).
6. Answer the questions.

Magnus Effect in Baseball Pitching

1. Were you able to make a Styrofoam fastball rise? Draw a free-body diagram of your fastball, showing all relevant forces, and explain how it relates to the vertical motion of the ball you observed.

2. Could you make a Styrofoam slider break sideways? If so, how much?

3. Draw a rear view of the ball from the pitcher's (your) perspective, and draw on the ball the axis of ball rotation and Magnus force for your slider.

4. Draw a rear view of the ball from the pitcher's (your) perspective and draw on the ball the axis of ball rotation and Magnus force for your curveball. In what direction(s) did your curveball break?

5. Did your curveball have more lateral or downward break? Why?

6. To get a ball to curve or break to the right with the Spin Principle, describe how force is applied to the ball? Would this be the same for curves to the right in other impact and release sports?

Lab Activity 9: Qualitative Diagnosis of Lead-Up Activities

An effective teaching strategy for many sports skills is to provide a sequence of lead-up activities that are similar to and build up to the skill of interest. How biomechanically similar the lead-up activities are to the sport skill of interest is important to physical educators. A qualitative answer to the similarity question will be explored in a sport skill selected by the instructor. The present lab will allow you to practice qualitative diagnosis of human movements using the biomechanical principles.

Background Reading

Chapter 9 herein: "Applying Biomechanics in Teaching and Coaching"
Knudson, D. V. (2013). *Qualitative diagnosis of human movement* (3rd ed.). Champaign: Human Kinetics.

Tasks

1. For the sport skill identified by the instructor, identify two lead-up skills, activities, or drills.
2. Select a volunteer to perform these movements.
3. Videotape several repetitions of the movements from several angles.
4. Observe and evaluate the performance of the biomechanical principles in each movement using videotape replay.
5. Answer the questions.

Qualitative Diagnosis of Lead-Up Activities

1. What biomechanical principles are most relevant to the sport skill of interest?

2. What was the first lead-up movement? What biomechanical principles are related to performance of this lead-up movement?

3. What was the second lead-up movement? What biomechanical principles are related to performance of this lead-up movement?

4. For the volunteer in your lab, what lead-up movement was most sport-specific? What biomechanical principles were most similar to the sport skill?

Lab Activity 10: Comparison of Skilled and Novice Performance

Coaching strives to maximize the performance of an athlete or team in competition. A key ingredient of athletic success is motor skill. Most aspects of skill are related to the biomechanical principles of human movement. A good way to practice the qualitative diagnosis of sport skills is to compare the application of biomechanical principles of a novice and those of a skilled performer. The purpose of this lab is to compare the application of biomechanical principles in a skilled performer and a novice performer in a common sport skill.

Background Reading

Chapter 9 herein: "Applying Biomechanics in Teaching and Coaching"
Hay, J. G. (1993). *The biomechanics of sports techniques* (4th ed.). Englewood Cliffs: Prentice-Hall.
Knudson, D. V. (2013). *Qualitative diagnosis of human movement* (3rd ed.). Champaign: Human Kinetics.

Tasks

1. Select a sport skill where a novice and a skilled performer can be found from students in the lab.
2. Select two volunteers (one novice and one skilled) to perform the skill.
3. Videotape several repetitions of the skill from several angles
4. Observe and evaluate performance of the biomechanical principles in each movement using videotape replay.
5. Answer the questions.

Comparison of Skilled and Novice Performance

1. What are the biomechanical principles most relevant to the sport skill of interest?

2. What biomechanical principles are strengths and weaknesses for the novice performer?

3. What biomechanical principles are strengths and weaknesses for the skilled performer?

4. What intervention would you recommend for the novice performer and why?

5. What intervention would you recommend for the skilled performer and why?

Lab Activity 11: Comparison of Training Modes

Strength and conditioning coaches prescribe exercises to improve performance based on the principle of specificity. This is often called the "SAID" principle: specific adaptation to imposed demands. There are a variety of free-weight, elastic, and mechanical resistances that coaches can prescribe to train the neuromuscular system. Qualitative diagnosis of exercise technique based on biomechanical principles can help a strength coach make two important evaluations: is the exercise technique safe and is it sport-specific? This lab will focus on the latter. The purpose of this lab is to compare the specificity of exercise technique in training for a sport skill.

Background Reading

Chapter 10 herein: "Applying Biomechanics in Exercise and Rehabilitation"
Knudson, D. V. (2013). *Qualitative diagnosis of human movement* (3rd ed.).Champaign: Human Kinetics.

Tasks

1. Select a sport skill of interest.
2. Select three exercises that will train the main agonists for the propulsive phase of the skill.
3. Be sure to select an elastic resistance, inertial resistance (free weight), and an exercise machine. Strive to make the resistances about equal in these exercises.
4. Select a volunteer to perform the exercises.
5. Video several repetitions of the exercises perpendicular to the primary plane of movement.
6. Observe and evaluate the performance of the biomechanical principles in each exercise using video replay.
7. Answer the questions.

Comparison of Training Modes

1. What are the biomechanical principles most relevant to the sport skill of interest?

2. What was the first exercise? What biomechanical principles of this exercise are similar to the sport skill?

3. What was the second exercise? What biomechanical principles of this exercise are similar to the sport skill?

4. What was the third exercise? What biomechanical principles of this exercise are similar to the sport skill?

5. Which exercise was most sport-specific? Why? (Be sure to explain based on the importance of certain biomechanical principles in terms of performance in the sport.)

Lab Activity 12: Qualitative Diagnosis of Walking Gait

Rehabilitation professionals qualitatively diagnose movement to find clues to injury and to monitor recovery from injury. Walking is a well-learned movement that athletic trainers, physical therapists, and physicians all qualitatively diagnose, most often focusing on lower extremity function. There is a variety of qualitative and quantitative systems of gait analysis. This lab will focus on the qualitative diagnosis of two walking gaits based on biomechanical principles. Professionals qualitatively diagnosing gait must remember that quantitative biomechanical analyses are needed in order to correctly estimate the loads in musculoskeletal structures, so assumptions about muscle actions in gait from body positioning alone are unwise (Herbert et al. 1993). This more focused (specific joints and muscles) qualitative diagnosis in clinical settings than teaching/coaching settings often uses specific gait analysis models/systems (Whittle 2001).

Background Reading

Chapter 10 herein: "Applying Biomechanics in Exercise and Rehabilitation"
Herbert, R. et al. (1993). Making inferences about muscles forces from clinical observations. *Australian Journal of Physiotherapy*, *39*, 195–202.
Knudson, D. V. (2013). *Qualitative diagnosis of human movement* (3rd ed.).Champaign: Human Kinetics.
Kirtley, C. (2006). *Clinical gait analysis: Theory and practice*. New York: Churchill Livingstone.

Tasks

1. Select a volunteer to perform the walking trials.
2. Have the volunteer walk in three conditions: their natural gait, as fast as they comfortably can, and simulating an injury. Injury can be easily simulated by restricting joint motion with athletic tape or a brace. Antalgic (painful) gait can be simulated by placing a small stone/object in a shoe.
3. Video several cycles of each waking gait in the frontal and sagittal planes
4. Observe and evaluate performance related to the biomechanical principles in each gait using videotape replay.
5. Answer the questions.

Qualitative Analysis of Walking Gait

1. What biomechanical principles are most evident in natural walking gait?

2. What biomechanical principles increased or decreased in importance relative to normal gait, during fast gait?

3. What injury did you simulate? What biomechanical principles increased or decreased in importance relative to normal gait, during injured gait?

4. What musculoskeletal structures are affected in your simulated injury? Hypothesize the likely changes in muscular actions and joint angular kinematics because of this injury, and note where you might find biomechanical literature to confirm your diagnosis.

Bibliography

Aagaard, P. (2003). Training induced changes in neural function. *Exercise and Sport Sciences Reviews, 31*, 61–67.

Aagaard, P., et al. (1998). A new concept for isokinetics hamstring: Quadriceps muscle strength ratio. *American Journal of Sports Medicine, 26*, 231–237.

Aagaard, P., et al. (2000). Antagonist muscle coactivation during isokinetic knee extension. *Scandinavian Journal of Medicine & Science in Sports, 10*, 58–67.

Abe, T., et al. (2000). Fasicle length of leg muscles is greater in sprinters than distance runners. *Medicine & Science in Sports & Exercise, 32*, 1125–1129.

Abernethy, P., et al. (1995). Strength and power assessment: Issues, controversies and challenges. *Sports Medicine, 19*, 401–417.

Abraham, L. D. (1991). *Lab manual for KIN 326: Biomechanical analysis of movement*. Austin: University of Texas at Austin.

Abraham, L. D., et al. (2018). *Guidelines for undergraduate biomechanics* (4th ed. [Guidance document] ed.). Reston: SHAPE America. https://www.shapeamerica.org/uploads/pdfs/2018/guidelines/Guidelines-for-UG-Biomechanics.pdf.

Adair, R. (1990). *The physics of baseball*. New York: Harper & Row.

Adamson, G. T., & Whitney, R. J. (1971). Critical appraisal of jumping as a measure of human power. In J. Vredenbregt & J. Wartenweiler (Eds.), *Biomechanics II* (pp. 208–211). Baltimore: University Park Press.

Adrian, M. J., & Cooper, J. M. (1995). *Biomechanics of human movement* (2nd ed.). Madison: Brown & Benchmark.

Ait-Haddou, R., et al. (2000). Theoretical considerations on cocontraction of sets of agonist and antagonist muscles. *Journal of Biomechanics, 33*, 1105–1111.

Akeson, W. H., et al. (1987). Effects of immobilization on joints. *Clinical Orthopaedics, 219*, 28–37.

Alaways, L. W., et al. (2001). Identification of release conditions and aerodynamic forces in pitched-baseball trajectories. *Journal of Applied Biomechanics, 17*, 63–76.

Alentorn-Geli, E. et al. (2009). Prevention of non-contact anterior cruciate ligament injuries in soccer players: Part 1: Mechanism of injury and underlying risk factors. *Knee Surgery, Sport Traumatology and Arthroscopy, 17*, 705–729.

Aleshinsky, S. Y. (1986a). An energy "sources" and "fractions" approach to the mechanical energy expenditure problem, I: Basic concepts, description of the model, analysis of one–link system movement. *Journal of Biomechanics, 19*, 287–293.

Aleshinsky, S. Y. (1986b). An energy "sources" and "fractions" approach to the mechanical energy expenditure problem, IV: Criticism of the concepts of "energy" transfers within and between segments. *Journal of Biomechanics, 19*, 307–309.

Alexander, R. M. (1992). *The human machine*. New York: Columbia University Press.

Alexander, R. M. (2002). Tendon elasticity and muscle function. Comparative Biochemistry and Physiology Part A: *Molecular & Integrative Physiology, 133*, 1001–1011.

Allard, P., et al. (Eds.). (1995). *Three-dimensional analysis of human motion*. Champaign: Human Kinetics.

Allman, W. F. (1984). Pitching rainbows: The untold physics of the curve ball. In E. W. Schrier & W. F. Allman (Eds.), *Newton at the bat: The science in sports* (pp. 3–14). New York: Charles Scribner & Sons.

Anderson, F. C., & Pandy, M. G. (1993). Storage and utilization of elastic strain energy during jumping. *Journal of Biomechanics, 26*, 1413–1427.

Apriantono, T., et al. (2006). The effect of muscle fatigue on instep kicking kinetics and kinematics in association football. *Journal of Sports Sciences, 24*, 951–960.

Aragon-Vargas, L. F., & Gross, M. M. (1997a). Kinesiological factors in vertical jumping performance: Differences among individuals. *Journal of Applied Biomechanics, 13*, 24–44.

Aragon-Vargas, L. F., & Gross, M. M. (1997b). Kinesiological factors in vertical jumping performance: Differences within individuals. *Journal of Applied Biomechanics, 13*, 45–65.

Arampatzis, A., et al. (2004). Differences between measured and resultant joint moments during voluntary and artificially elicited isometric knee extension contractions. *Clinical Biomechanics, 19*, 277–283.

Arellano, R. (1999). Vortices and propulsion. In R. Sanders & J. Linsten (Eds.), *SWIMMING: Applied proceedings of the XVII international symposium on biomechanics in sports* (Vol. 1, pp. 53–66). Perth: Edith Cowan University.

Ariel, G. B. (1983). Resistive training. *Clinics in Sports Medicine, 2*, 55–69.

Arndt, A. N., et al. (1998). Individual muscle contributions to the *in vivo* achilles tendon force. *Clinical Biomechanics, 13*, 532–541.

Arokoski, J. P. A., et al. (2000). Normal and pathological adaptations of articular cartilage to joint loading. *Scandinavian Journal of Medicine and Science in Sports, 10*, 186–198.

Aruin, A. S. (2001). Simple lower extremity two-joint synergy. *Perceptual and Motor Skills, 92*, 563–568.

Ashton-Miller, J. A., & Schultz, A. B. (1988). Biomechanics of the human spine and trunk. *Exercise and Sport Sciences Reviews, 16*, 169–204.

Atha, J. (1981). Strengthening muscle. *Exercise and Sport Sciences Reviews, 9*, 1–73.

Atwater, A. E. (1979). Biomechanics of overarm throwing movements and of throwing injuries. *Exercise and Sport Sciences Reviews, 7*, 43–85.

Atwater, A. E. (1980). Kinesiology/biomechanics: Perspectives and trends. *Research Quarterly for Exercise and Sport, 51*, 193–218.

Atwater, A. E. (1982). Basic biomechanics applications for the coach. In J. Terauds (Ed.), *Biomechanics in sports: Proceedings of the international symposium of biomechanics in sports* (pp. 21–30). Del Mar: Academic Publishers.

Babault, N., et al. (2001). Activation of human quadriceps femoris during isometric, concentric, and eccentric contractions. *Journal of Applied Physiology, 91*, 2628–2634.

Baechle, T. R., & Earle, R. W. (Eds.). (2008). *Essentials of strength training and conditioning* (3rd ed.). Champaign: Human Kinetics.

Bahamonde, R. (2000). Changes in angular momentum during the tennis serve. *Journal of Sports Sciences, 18*, 579–592.

Bahamonde, R., & Knudson, D. (2003). Kinetics of the upper extremity in the open and square stance tennis forehand. *Journal of Science and Medicine in Sport, 6*, 90–103.

Bahill, A. T., & Freitas, M. M. (1995). Two methods of recommending bat weights. *Annals of Biomedical Engineering, 23*, 436–444.

Baltzopoulos, V. (2021). Inverse dynamics, joint reaction forces and loading in the musculoskeletal system: Guidelines for correct mechanical terms and recommendations for accurate reporting of results. Sports Biomechanics, https://doi.org/10.1080/14763141.2020.1841826.

Barclay, C. J. (1997). Initial mechanical efficiency in cyclic contractions of mouse skeletal muscle. *Journal of Applied Biomechanics, 13*, 418–422.

Barfield, W. R. (1998). The biomechanics of kicking in soccer. *Clinics in Sports Medicine, 17*, 711–728.

Barlow, D. A. (1971). Relation between power and selected variables in the vertical jump. In J. M. Cooper (Ed.), *Selected topics on biomechanics* (pp. 233–241). Chicago: Athletic Institute.

Barr, A. E., & Barbe, M. F. (2002). Pathophysiological tissue changes associated with repetitive movement: A review of the evidence. *Physical Therapy, 82*, 173–187.

Bartlett, R. (2012). *Sports biomechanics: Reducing injury and improving performance* (2nd ed.). New York: Routledge.

Bartlett, R. M., & Best, R. J. (1988). The biomechanics of javelin throwing: A review. *Journal of Sports Sciences, 6*, 1–38.

Bartlett, R., et al. (1996). Three-dimensional evaluation of the kinematics release parameters for javelin throwers of different skill levels. *Journal of Applied Biomechanics, 12*, 58–71.

Bartlett, R. M., et al. (2007). Is movement variability important for sports biomechanists? *Sports Biomechanics, 6*, 224–243.

Basmajian, J. V., & De Luca, C. J. (1985). *Muscles alive: Their functions revealed by electromyography* (5th ed.). Baltimore: Williams & Wilkins.

Basmajian, J. V., & Wolf, S. L. (Eds.). (1990). *Therapeutic exercise* (5th ed.). Baltimore: Williams & Wilkins.

Bates, B. T., et al. (1983). An assessment of subject variability, subject-shoe interaction, and the evaluation of running shoes using ground reaction forces. *Journal of Biomechanics, 16*, 181–191.

Bates, N. A., et al. (2017). Novel mechanical impact simulator designed to generate clinically relevant anterior cruciate ligament ruptures. *Clinical Biomechanics, 44*, 36–44.

Bauer, J. J., et al. (2001). Quantifying force magnitude and loading rate from drop landings that induce osteogenesis. *Journal of Applied Biomechanics, 17*, 142–152.

Bawa, P. (2002). Neural control of motor output: Can training change it? *Exercise and Sport Sciences Reviews, 30*, 59–63.

Beattie, P. F., & Meyers, S. P. (1998). Magnetic resonance imaging in low back pain: General principles and clinical issues. *Physical Therapy, 78*, 738–753.

Beer, F., et al. (2018). *Vector mechanics for engineers: Statics and dynamics* (12th ed.). New York: McGraw-Hill.

Benjanuvatra, N., et al. (2002). Comparison of buoyancy, passive and net active drag forces between Fastskin™ and standard swimsuits. *Journal of Science and Medicine in Sport, 5*, 115–123.

Berger, M. A. M., et al. (1995). Hydrodynamic drag and lift force on human hand/arm models. *Journal of Biomechanics, 28*, 125–133.

Bernstein, N. A. (1967). *The coordination and regulation of movements*. Oxford: Pergamon Press.

Berthoin, S., et al. (2001). Predicting sprint kinematic parameters from anaerobic field tests in physical education students. *Journal of Strength and Conditioning Research, 15*, 75–80.

Biewener, A. A. (1998). Muscle function *in vivo*: A comparison of muscles used for elastic energy savings versus muscles used to generate mechanical power. *American Zoologist, 38*, 703–717.

Biewener, A. A., & Blickhan, R. (1988). Kangaroo rat locomotion: Design for elastic energy storage or acceleration. *Journal of Experimental Biology, 140*, 243–255.

Biewener, A. A., & Roberts, T. J. (2000). Muscle and tendon contributions to force, work, and elastic energy savings: A comparative perspective. *Exercise and Sport Sciences Reviews, 28*, 99–107.

Bird, M., & Hudson, J. (1998). Measurement of elastic-like behavior in the power squat. *Journal of Science and Medicine in Sport, 1*, 89–99.

Bjerklie, D. (1993). High-tech olympians. *Technology Review, 96*(1), 22–30.

Blackard, D. O., et al. (1999). Use of EMG analysis in challenging kinetic chain terminology. *Medicine and Science in Sports & Exercise, 31*, 443–448.

Bloomfield, J., et al. (1994). *Applied anatomy and biomechanics in sport*. Melbourne: Blackwell Scientific Publications.

Bobbert, M. F. (1990). Drop jumping as a training method for jumping ability. *Sports Medicine, 9*, 7–22.

Bobbert, M. F., & van Ingen Schenau, G. J. (1988). Coordination in vertical jumping. *Journal of Biomechanics, 21*, 249–262.

Bobbert, M. F., & van Soest, A. J. (1994). Effects of muscle strengthening on vertical jump height: A simulation study. *Medicine and Science in Sports and Exercise, 26*, 1012–1020.

Bobbert, M. F., & van Zandwijk, J. P. (1999). Dynamics of force and muscle stimulation in human vertical jumping. *Medicine & Science in Sports & Exercise, 31*, 303–310.

Bobbert, M. F., et al. (1986). Biomechanical analysis of drop and countermovement jumps. *European Journal of Applied Physiology, 54*, 566–573.

Bobbert, M. F., et al. (1996). Why is countermovement jump height greater than squat jump height? *Medicine & Science in Sports & Exercise, 28*, 1402–1412.

Bober, T., et al. (1987). Factors influencing the angular velocity of a human limb segment. *Journal of Biomechanics, 20*, 511–521.

Boden, B. P., et al. (2000). Etiology and prevention of noncontact ACL injury. *Physican and Sportsmedicine, 28*(4), 53–60.

Bojsen-Moller, F., & Magnusson, S. P. (2000). Basic science of knee joint injury mechanisms [special issue]. *Scandinavian Journal of Medicine & Science in Sports, 10*(2), 57.

Bouisset, S. (1973). EMG and muscle force in normal motor activities. In J. E. Desmedt (Ed.), *New developments in electromyography and clinical neurophysiology* (pp. 502–510). Basel: Karger.

Brancazio, P. (1984). *Sport science: Physical laws and optimum performance*. New York: Simon & Schuster.

Bressler, B., & Frankel, J. P. (1950). The forces and moments in the leg during level walking. *Transactions of the American Society of Mechanical Engineers, 72*, 27–36.

Brody, H. (1987). *Tennis science for tennis players*. Philadelphia: University of Pennsylvania Press.

Broer, M. R., & Zernicke, R. F. (1979). *Efficiency of human movement* (4th ed.). Philadelphia: W.B. Saunders.

Brondino, L., et al. (2002). Elbow flexor inhibition as a function of muscle length. *Journal of Applied Biomechanics, 18*, 46–56.

Brown, L. E. (Ed.). (2000). *Isokinetics in human performance*. Champaign: Human Kinetics.

Brown, I. E., & Loeb, G. L. (1998). Post-activation potentiation—A clue for simplifying models of muscle dynamics. *American Zoologist, 38*, 743–754.

Brown, J. M. M., et al. (2007). Muscles within muscles: Coordination of 19 muscle segments within three shoulder muscles during isometric motor tasks. *Journal of Electromyography and Kinesiology, 17*, 57–73.

Buszard, T., et al. (2020). Children's coordination of the "sweet spot" when striking a forehand is shaped by the equipment used. *Scientific Reports, 10*, 21003. https://doi.org/10.1038/s41598-020-59475-5.

Buckalew, D. P., et al. (1985). Biomechanical profile of elite women marathoners. *International Journal of Sport Biomechanics, 1*, 330–347.

Bunn, J. W. (1972). *Scientific principles of coaching* (2nd ed.). Englewood Cliffs: Prentice-Hall.

Burke, R. E. (1986). The control of muscle force: Motor unit recruitment and firing patterns. In N. L. Jones et al. (Eds.), *Human muscle power* (pp. 97–106). Champaign: Human Kinetics.

Burkholder, T. J., et al. (1994). Relationship between muscle fiber types and sizes and muscle architectural properties in the mouse hindlimb. *Journal of Morphology, 221*, 177–190.

Callaghan, M. J., & Oldham, J. A. (1996). The role of quadriceps exercise in the treatment of patellofemoral pain syndrome. *Sports Medicine, 21*, 384–391.

Calvano, N. J., & Berger, R. E. (1979). Effects of selected test variables on the evaluation of football helmet performance. *Medicine and Science in Sports, 11*, 293–301.

Cappozzo, A., et al. (1992). *Biolocomotion: A century of research using moving pictures*. Rome: Promograph.

Carlstedt, C. A., & Nordin, M. (1989). Biomechanics of tendons and ligaments. In M. Nordin & V. H. Frankel (Eds.), *Basic biomechanics of the musculoskeletal system* (2nd ed., pp. 59–74). Philadelphia: Lea & Febiger.

Caruthers, E. J., et al. (2016). Muscle forces and their contributions to vertical and horizontal acceleration of the center of mass during sit-to-stand transfer in young, healthy adults. *Journal of Applied Biomechanics, 32*, 487–503.

Case, M. J. et al. (2020). Barbell squat relative strength as an identifier for lower extremity injury in collegiate athletes. *Journal of Strength and Conditioning Research, 34*, 1249–1253.

Cavagna, G. A., et al. (1965). Effect of negative work on the amount of positive work performed by an isolated muscle. *Journal of Applied Physiology, 20*, 157–158.

Cavanagh, P. R. (1988). On "muscle action" vs "muscle contraction". *Journal of Biomechanics, 21*, 69.

Cavanagh, P. R. (1990). Biomechanics: A bridge builder among the sport sciences. *Medicine & Science in Sports & Exercise, 22*, 546–557.

Cavanagh, P. R., & Kram, R. (1985). The efficiency of human movement–A statement of the problem. *Medicine & Science in Sports & Exercise, 17*, 304–308.

Cavanagh, P. R., & LaFortune, M. A. (1980). Ground reaction forces in distance running. *Journal of Biomechanics, 15*, 397–406.

Cavanagh, P. R., et al. (1985). An approach to biomechanical profiling of elite distance runners. *International Journal of Sport Biomechanics, 1*, 36–62.

Chaffin, B. D., et al. (2006). *Occupational biomechanics* (4th ed.). New York: Wiley.

Challis, J. H. (1998). An investigation of the influence of bi-lateral deficit on human jumping. *Human Movement Science, 17*, 307–325.

Chalmers, G. (2002). Do golgi tendon organs really inhibit muscle activity at high force levels to save muscles from injury, and adapt with strength training? *Sports Biomechanics, 1*, 239–249.

Chaloupka, E. C., et al. (2000). Metabolic and cardiorespiratory responses to continuous box lifting and lowering in nonimpaired subjects. *Journal of Orthopaedic and Sports Physical Therapy, 30*, 249–262.

Chapman, A. E. (1985). The mechanical properties of human muscle. *Exercise and Sport Sciences Reviews, 13*, 443–501.

Chapman, A. E., & Caldwell, G. E. (1983). Kinetic limitations of maximal sprinting speed. *Journal of Biomechanics, 16*, 79–83.

Chapman, A. E., & Sanderson, D. J. (1990). Muscular coordination in sporting skills. In J. M. Winters & S. L. Y. Woo (Eds.), *Multiple muscle systems: Biomechanics and movement organization* (pp. 608–620). New York: Springer-Verlag.

Charlton, W. P. H., et al. (2002). Differences in femoral notch anatomy between men and women: A magnetic resonance imaging study. *American Journal of Sports Medicine, 30*, 329–333.

Chatard, J., & Wilson, B. (2003). Drafting distance in swimming. *Medicine & Science in Sports & Exercise, 35*, 1176–1181.

Cholewicki, J., & McGill, S. M. (1992). Lumbar posterior ligament involvement during extremely heavy lifts estimated from fluoroscopic measurements. *Journal of Biomechanics, 25*, 17–28.

Chow, J. W., et al. (1999). Determining the force–length–velocity relations of the quadriceps muscles, I: Anatomical and geometric parameters. *Journal of Applied Biomechanics, 15*, 182–190.

Chowdhary, A. G., & Challis, J. H. (2001). The biomechanics of an overarm throwing task: A simulation model examination of optimal timing of muscle activations. *Journal of the Theoretical Biology, 211*, 39–53.

Ciccone, C. D. (2002). Evidence in practice. *Physical Therapy, 82*, 84–88.

Ciullo, J. V., & Zarins, B. (1983). Biomechanics of the musculotendinous unit: Relation to athletic performance and injury. *Clinics in Sports Medicine, 2*, 71–86.

Clarke, C. H. (1966). *Muscular strength and endurance in man.* Englewood Cliff: Prentice-Hall.

Cleather, D. J., et al. (2015). The role of biarticular hamstrings and gastrocnemius muscles in closed chain lower limb extension. *Journal of Theoretical Biology, 365*, 217–225.

Corbin, C. B., & Eckert, H. M. (Eds.). (1990). *Resolution: The evolving undergraduate major.* AAPE Academy Papers No. 23 (p. 104). Champaign: Human Kinetics.

Corcos, D. M., et al. (1993). Principles for learning single-joint movements, I: Enhanced performance by practice. *Experimental Brain Research, 94*, 499–513.

Cormie, P., et al. (2010). Changes in the eccentric phase contribute to improve stretch-shorten cycle performance after training. *Medicine & Science in Sports & Exercise, 42*, 1731–1744.

Cornbleet, S., & Woolsey, N. (1996). Assessment of hamstring muscle length in school-aged children using the sit-and-reach test and the inclinometer measure of hip joint angle. *Physical Therapy, 76*, 850–855.

Counsilman, J. E. (1971). The application of Bernoulli's Principle to human propulsion in water. In L. Lewillie & J. Clarys (Eds.), *First international symposium on biomechanics of swimming* (pp. 59–71). Brussels: Université Libre de Bruxelles.

Coutts, A. (2017). Challenges in developing evidence-based practice in high-performance sport. *International Journal of Sports Physiology and Performance, 12*, 717–718.

Cox, V. M., et al. (2000). Growth induced by incremental static stretch in adult rabbit latissimus dorsi muscle. *Experimental Physiology, 85*(3), 193–202.

Cronin, J., et al. (2001a). Developing explosive power: A comparison of technique and training. *Journal of Science and Medicine in Sport, 4*, 59–70.

Cronin, J. B., et al. (2001b). Magnitude and decay of stretch-induced enhancement of power output. *European Journal of Applied Physiology, 84*, 575–581.

Cronin, J. B., et al. (2010). Magnitude and decay of stretch-induced enhancement of power output. *European Journal of Applied Physiology, 84*, 575–581.

Cross, R. (2000). The coefficient of restitution for collisions of happy balls, unhappy balls, and tennis balls. *American Journal of Physics, 68*, 1025–1031.

Cross, R. (2002). Measurements of the horizontal coefficient of restitution for a super ball and a tennis ball. *American Journal of Physics, 70*, 482–489.

Crotin, R. L., et al. (2014). Altered stride length in response to increased exerting among baseball pitchers. *Medicine & Science in Sports & Exercise, 46*, 565–571.

Daish, C. B. (1972). *The physics of ball games.* London: The English Universities Press.

Darling, W. G., & Cooke, J. D. (1987). Linked muscle activation model for movement generation and control. *Journal of Motor Behavior, 19*, 333–354.

Davids, K., et al. (2000). Understanding and measuring coordination and control in kicking skills in soccer: Implications for talent identification and skill acquisition. *Journal of Sports Sciences, 18*, 703–714.

Davids, K., et al. (2003). Movement systems as dynamical systems: The functional role of variability and its implications for sports medicine. *Sports Medicine, 33*, 245–260.

Davis, K. G., & Marras, W. S. (2000). Assessment of the relationship between box weight and trunk kinematics: Does a reduction in box weight necessarily correspond to a decrease in spinal loading? *Human Factors, 42*, 195–208.

De Deyne, P. G. (2001). Application of passive stretch and its implications for muscle fibers. *Physical Therapy, 81*, 819–827.

De Koning, F. L., et al. (1985). The force–velocity relationship of arm flexion in untrained males and females and arm-trained athletes. *European Journal of Applied Physiology, 54*, 89–94.

De Koning, J. J., et al. (2000). From biomechanical theory to application in top sports: The Klapskate story. *Journal of Biomechanics, 33*, 1225–1229.

De Luca, C. J. (1997). The use of surface electromyography in biomechanics. *Journal of Applied Biomechanics, 13*, 135–163.

Delp, S. L., et al. (1990). An interactive graphic-based model of the lower extremity to study orthopaedic surgical procedures. *Journal of Biomedical Engineering, 37*, 757–767.

Denny-Brown, D., & Pennybacker, J. B. (1938). Fibrillation and fasciculation in voluntary muscle. *Brain, 61*, 311–344.

Desmedt, J. E., & Godaux, D. (1977). Ballistic contractions in man: Characteristic recruitment pattern of single motor units of the tibialis anterior muscle. *Journal of Physiology (London), 264*, 673–694.

DeVita, P., & Skelly, W. A. (1992). Effect of landing stiffness on joint kinetics and energetics in the lower extremity. *Medicine & Science in Sports & Exercise, 24*, 108–115.

Di Fabio, R. P. (1999). Making jargon from kinetic and kinematic chains. *Journal of Orthopaedic and Sports Physical Therapy, 29*, 142–143.

Dickerman, R. D., et al. (2000). The upper range of lumbar spine bone mineral density? An examination of the current world record holder in the squat lift. *International Journal of Sports Medicine, 21*, 469–470.

Dickinson, M. H., et al. (2000). How animals move: A integrated view. *Nature, 288*, 100–106.

Dingenen, B. et al. (2019). Two-dimensional video analysis can discriminate differences in running kinematics between recreational runners with and without running-related knee injury. *Physical Therapy in Sport, 38*, 184–191.

Dillman, C. J., et al. (1994). Biomechanical differences of open and closed chain exercises with respect to the shoulder. *Journal of Sport Rehabilitation, 3*, 228–238.

Dixon, S. J., et al. (1999). Artificial playing surfaces research: A review of medical, engineering and biomechanical aspects. *International Journal of Sports Medicine, 20*, 209–218.

Doorenbosch, C. A. M., et al. (1997). On the effectiveness of force application in guided leg movements. *Journal of Motor Behavior, 29*, 27–34.

Dorge, H. C., et al. (2002). Biomechanical differences in soccer kicking with the preferred and the non–preferred leg. *Journal of Sports Sciences, 20*, 293–299.

Dotson, C. O. (1980). Logic of questionable density. *Research Quarterly for Exercise and Sport, 51*, 23–36.

Dowell, L. J. (1978). Throwing for distance: Air resistance, angle of projection, and ball size and weight. *Motor Skills: Theory into Practice, 3*, 11–14.

Dowling, J. J., & Vamos, L. (1993). Identification of kinetic and temporal factors related to vertical jump performance. *Journal of Applied Biomechanics, 9*, 95–110.

Duck, T. A. (1986). Analysis of symmetric vertical jumping using a five segment computer simulation. In M. Adrian & H. Deutsch (Eds.), *Biomechanics: The 1984 olympic scientific congress proceedings* (pp. 173–178). Eugene: Microform Publications.

Dufek, J. S., et al. (1995). Interactive effects between group and single–subject response patterns. *Human Movement Science, 14*, 301–323.

Dvir, Z. (Ed.). (2000). *Clinical biomechanics*. New York: Churchill Livingstone.

Dyson, G. (1986). *Mechanics of athletics* (8th ed.). New York: Holmes & Meier.

Earl, J. E., et al. (2001). Activation of the VMO and VL during dynamic mini-squat exercises with and without isometric hip adduction. *Journal of Electromyography and Kinesiology, 11*, 381–386.

Ebenbichler, G. R., et al. (2001). Sensory-motor control of the lower back: Implications for rehabilitation. *Medicine & Science in Sports & Exercise, 33*, 1889–1898.

Ecker, T. (1996). *Basic track and field biomechanics* (2nd ed.). Los Altos: Tafnews Press.

Edgerton, V. R., et al. (1986). Morphological basis of skeletal muscle power output. In N. L. Jones et al. (Eds.), *Human muscle power* (pp. 43–64). Champaign: Human Kinetics.

Edman, K. A., et al. (1997). The biphasic force velocity relationship in force muscle fibers and its evaluation in terms of cross-bridge function. *Journal of Physiology, 503*, 141–156.

Eksstrand, J., et al. (1982). Lower extremity goniometric measurements: A study to determine their reliability. *Archives of Physical Medicine and Rehabilitation, 63*, 171–175.

Elby, A. (2001). Helping physics students learn how to learn. *American Journal of Physics, 69*, S54–S64.

Elftman, H. (1939). Forces and energy changes in the leg during walking. *American Journal of Physiology, 125*, 339–356.

Elliott, B. (1981). Tennis racquet selection: A factor in early skill development. *Australian Journal of Sport Sciences, 1*, 23–25.

Elliott, B. (1983). Spin and the power serve in tennis. *Journal of Human Movement Studies, 9*, 97–104.

Elliott, B. (1999). Biomechanics: An integral part of sport science and sport medicine. *Journal of Science and Medicine and Sport, 2*, 299–310.

Elliott, B., & Bartlett, R. (2006). Sports biomechanics: Does it have a role in coaching? *International Journal of Sport Science and Coaching, 1*, 177–183.

Elliott, B. C., & Mester, J. (Eds.). (1998). *Training in sport: Applying sport science*. New York: John Wiley & Sons.

Elliott, B. C., et al. (1989). A biomechanical analysis of the sticking region in the bench press. *Medicine & Science in Sports & Exercise, 21*, 450–462.

Elliott, B. C., et al. (1999). Internal rotation of the upper-arm segment during a stretch-shortening cycle movement. *Journal of Applied Biomechanics, 15*, 381–395.

Eng, J. J., et al. (1997). Intralimb dynamics simplify reactive control strategies during locomotion. *Journal of Biomechanics, 30*, 581–588.

Englehorn, R. A. (1983). Agonist and antagonist muscle EMG activity changes with skill acquisition. *Research Quarterly for Exercise and Sport, 54*, 315–323.

Enoka, R. M. (1996). Eccentric contractions require unique activation strategies by the nervous system. *Journal of Applied Physiology, 81*, 2339–2346.

Enoka, R. (2015). *The neuromechanical basis of human movement* (5th ed.). Champaign: Human Kinetics.

Eriksson, E. (1996). The scientific basis of rehabilitation. *American Journal of Sports Medicine, 24*, S25–S26.

Eriksson, E., & Helgesson, G. (2018). Time to stop talking about 'predatory journals'. *Learned Publishing, 31*, 181–183.

Escamilla, R. F. (2001). Knee biomechanics of the dynamic squat exercise. *Medicine & Science in Sports & Exercise, 33*, 127–141.

Escamilla, R. F., et al. (2000). Effects of throwing overweight and under-weight baseballs on throwing velocity and accuracy. *Sports Medicine, 29*, 259–272.

Escamilla, R. F., et al. (2007). Pitching biomechanics as a pitcher approaches muscular fatigue during a simulated baseball game. *American Journal of Sports Medicine, 35*, 23–33.

Ettema, G. J. C. (2001). Muscle efficiency: The controversial role of elasticity and mechanical energy conversion in stretch–shortening cycles. *European Journal of Applied Physiology, 85*, 457–465.

Falla, D., & Gallina, A. (2020). New insights into pain-related changes in muscle activation revealed by high-density surface electromyography. *Journal of Electromyography and Kinesiology, 52*, 102422. https://doi.org/10.1016/j.jelekin.2020.102422.

Farrally, M. R., & Cochran, A. J. (Eds.). (1999). *Science and golf, III*. Champaign: Human Kinetics.

Faulkner, J. A. (2003). Terminology for contractions of muscles during shortening, while isometric, and during lengthening. *Journal of Applied Physiology, 95*, 455–459.

Faulkner, G., et al. (2006). Exercise science and the development of evidence-based practice: A "better practices" framework. *European Journal of Sport Science, 6*, 117–126.

Feldman, A. G., et al. (1998). Multi-muscle control in human movements. *Journal of Electromyography and Kinesiology, 8*, 383–390.

Feltner, M. E. (1989). Three-dimensional interactions in a two-segment kinetic chain, II: Application to the throwing arm in baseball pitching. *International Journal of Sport Biomechanics, 5*, 420–450.

Feltner, M., & Dapena, J. (1986). Dynamics of the shoulder and elbow joints of the throwing arm during a baseball pitch. *International Journal of Sport Biomechanics, 2*, 235–259.

Feltner, M. E., et al. (1999). Upper extremity augmentation of lower extremity kinetics during counter-movement vertical jumps. *Journal of Sports Sciences, 17*, 449–466.

Feltner, M. E., et al. (2004). Segmental and kinetic contributions in vertical jumps performed with and without and arm swing. *Research Quarterly for Exercise and Sport, 75*, 216–230.

Finni, T. (2006). Structural and functional feature of the human muscle-tendon unit. *Scandinavian Journal of Medicine & Science in Sports, 16*, 147–158.

Finni, T., et al. (2000). In vivo human triceps surae and quadriceps femoris muscle function in a squat jump and counter movement jump. *European Journal of Applied Physiology, 83*, 416–426.

Finni, T., et al. (2001). Concentric force enhancement during human movement. *Acta Physiological Scandinavica, 173*, 369–377.

Fitts, R. H., & Widrick, J. L. (1996). Muscle mechanics: Adaptations with exercise-training. *Exercise and Sport Sciences Reviews, 24*, 427–473.

Fitzgerald, G. K., et al. (2001). Hop tests as predictors of dynamic knee stability. *Journal of Orthopaedic and Sports Physical Therapy, 31*, 588–597.

Flaxman, T. E., et al. (2017). Predicting the functional roles of knee joint muscles from internal joint moments. *Medicine & Science in Sports & Exercise, 49*, 527–537.

Fleisig, G. S., et al. (1995). Kinetics of baseball pitching with implications about injury mechanisms. *American Journal of Sports Medicine, 23*, 233–239.

Fleisig, G. S., et al. (1999). Kinematic and kinetic comparison of baseball pitching among various levels of development. *Journal of Biomechanics, 32*, 1371–1375.

Fleisig, G. S., et al. (2018). Do baseball pitchers improve mechanics after biomechanical evaluations. *Sports Biomechanics, 17*, 314–321.

Fong, D., et al. (2009). Biomechanics of supination ankle sprain: A case report of an accidental injury event in the laboratory. *American Journal of Sports Medicine, 37*, 822–827.

Forrester, S. E. (2015). Setting the number of trials in experimental biomechanics studies. *International Biomechanics, 2*, 62–72.

Fowles, J. R., et al. (2000a). Reduced strength after passive stretch of the human plantar flexors. *Journal of Applied Physiology, 89*, 1179–1188.

Fowles, J. R., et al. (2000b). The effects of acute passive stretch on muscle protein synthesis in humans. *Canadian Journal of Applied Physiology, 25*, 165–180.

Fox, M. D., & Delp, S. L. (2010). Contributions of muscles and passive dynamics to swing initiation over a range of walking speeds. *Journal of Biomechanics, 43*, 1450–1455.

Fradkin, A. J., et al. (2006). Does warming up prevent injury in sport: The evidence from randomized controlled trials? *Journal of Science and Medicine in Sport, 9*, 214–220.

Frank, K., et al. (2000). The effect of strength shoes on muscle activity during quiet standing. *Journal of Applied Biomechanics, 16*, 204–209.

Frederick, E. C. (1986). Biomechanical consequences of sport shoe design. *Exercise and Sport Sciences Reviews, 14*, 375–400.

Freeman, S., et al. (2014). Active learning increases student performance in science, engineering, and mathematics. *Proceedings of the National Academy of Sciences, 111*, 8319–8320. https://doi.org/10.1073/pnas.1319030111.

Fuchs, R. K., Bauer, J. J., & Snow, C. M. (2001). Jumping improves hip and lumbar spine bone mass in prepubescent children: A randomized control trial. *Journal of Bone and Mineral Research, 16*, 148–156.

Fujii, N., & Hubbard, M. (2002). Validation of a three-dimensional baseball pitching model. *Journal of Applied Biomechanics, 18*, 135–154.

Fukashiro, S., & Komi, P. V. (1987). Joint moment and mechanical power flow of the lower limb during vertical jump. *International Journal of Sports Medicine, 8*(Supp), 15–21.

Fukashiro, S., et al. (2006). Biomechanical behavior of muscle-tendon complex during dynamic human movements. *Journal of Applied Biomechanics, 22*, 131–147.

Fukunaga, T., et al. (1997). Determination of fasicle length and pennation in a contracting human muscle in vivo. *Journal of Applied Physiology, 82*, 354–358.

Fukunaga, T., et al. (2002). Muscle and tendon interaction during human movements. *Exercise and Sport Sciences Reviews, 30*, 106–110.

Fullagar, H., et al. (2019). Practitioner perceptions of evidence-based practice in elite sport in the United States of America. *Journal of Strength and Conditioning Research, 33*(11), 2897–2904. https://doi.org/10.1519/JSC.000000000000334.

Funato, K., et al. (1996). Specific movement power related to athletic performance in weight lifting. *Journal of Applied Biomechanics, 12*, 44–57.

Funato, K., et al. (2000). Measurement of specific movement power application: Evaluation of weight lifters. *Ergonomics, 43*, 40–54.

Fung, Y. C. (1981). *Biomechanics: Mechanical properties of living tissues*. New York: Springer-Verlag.

Gabriel, D. A., & Boucher, J. P. (2000). Practicing a maximal performance task: A cooperative strategy for muscle activity. *Research Quarterly for Exercise and Sport, 71*, 217–228.

Gajdosik, R., & Lusin, G. (1983). Hamstring muscle tightness: Reliability of an active–knee extension test. *Physical Therapy, 63*, 1085–1088.

Galloway, J. C., & Koshland, G. F. (2002). General coordination of shoulder, elbow and wrist dynamics during multijoint arm movements. *Experimental Brain Research, 142*, 163–180.

Gambetta, V. (1995, April). Following a functional path. *Training & Conditioning, 5*(2), 25–30.

Gambetta, V. (1997, Winter). Leg strength for sport performance. *Sports Coach, 25*, 8–10.

Gandevia, S. C. (1999). Mind, muscles and motoneurons. *Journal of Science and Medicine in Sport, 2*, 167–180.

Gareis, H., et al. (1992). The isometric length-force models of nine different skeletal muscles. *Journal of Biomechanics, 25*, 903–916.

Garhammer, J. (1989). Weight lifting and training. In C. Vaughan (Ed.), *Biomechanics of sport* (pp. 169–211). Boca Raton: CRC Press.

Garrett, W. E. (1996). Muscle strain injuries. *American Journal of Sports Medicine, 24*(6), S2–S8.

Gatton, M. L., & Pearcy, M. J. (1999). Kinematics and movement sequencing during flexion of the lumbar spine. *Clinical Biomechanics, 14*, 376–383.

Gielen, S. (1999). What does EMG tell us about muscle function? *Motor Control, 3*, 9–11.

Gleim, G. W., & McHugh, M. P. (1997). Flexibility and its effects on sports injury and performance. *Sports Medicine, 24*, 289–299.

Gottlieb, G. L. (1996). Muscle compliance: Implications for the control of movement. *Exercise and Sport Sciences Reviews, 24*, 1–34.

Gottlieb, G. L., et al. (1989). Strategies for the control of single mechanical degree of freedom voluntary movements. *Behavior and Brain Sciences, 12*, 189–210.

Grace, T. G. (1985). Muscle imbalance and extremity injury: A perplexing relationship. *Sports Medicine, 2*, 77–82.

Greathouse, D. G., et al. (2004). Terminologia anatomica: Revised anatomical terminology. *Journal of Orthopaedic and Sports Physical Therapy, 34*, 363–367.

Greenwald, R. M., et al. (2001). Differences in batted ball speed with wood and aluminum baseball bats: A batting cage study. *Journal of Applied Biomechanics, 17*, 241–252.

Gregersen, C. S., et al. (2006). How changing the inversion/eversion foot angle affects the nondriving intersegmental knee moments and the relative activation of the vastii muscles in cycling. *Journal of Biomechanical Engineering, 128*, 391–398.

Gregor, R. J. (1997). Mechanics and energetics of the stretch-shortening cycle [special issue]. *Journal of Applied Biomechanics, 13*, 4.

Gregor, R. J., & Abelew, T. A. (1994). Tendon force measurements and movement control: A review. *Medicine & Science in Sports & Exercise, 26*, 1359–1372.

Gregor, S. M., et al. (2002). Lower extremity general muscle moment patterns in healthy individuals during recumbent cycling. *Clinical Biomechanics, 17*, 123–129.

Grieco, A., et al. (1998). Epidemiology of musculoskeletal disorders due to biomechanical overload. *Ergonomics, 41*, 1253–1260.

Griffin, L. Y., et al. (2000). Noncontact anterior cruciate ligament injuries: Risk factors and prevention strategies. *Journal of the American Academy of Orthopaedic Surgeons, 8*, 141–150.

Griffiths, R. I. (1989). The mechanics of the medial gastrocnemius muscle in the freely hopping wallaby. *Journal of Experimental Biology, 147*, 439–456.

Griffiths, R. I. (1991). Shortening of muscle fibers during stretch of the active cat medial gastrocnemius muscle: The role of tendon compliance. *Journal of Physiology (London), 436*, 219–236.

Groves, R. & Camaione, D. N. (1975). *Concepts in kinesiology*. Philadelphia: Sanders.

Gruen, A. (1997). Fundamentals of videography–A review. *Human Movement Science, 16*, 155–187.

Gulch, R. W. (1994). Force–velocity relations in human skeletal muscle. *International Journal of Sports Medicine, 15*, S2–S10.

Hadorn, D. C., et al. (1996). Rating the quality of evidence for clinical practice guidelines. *Journal of Clinical Epidemiology, 49*, 749–754.

Hagen, K. B., et al. (1993). Physiological and subjective responses to maximal repetitive lifting employing stoop and squat technique. *European Journal of Applied Physiology, 67*, 291–297.

Hake, R. R. (1998). Interactive-engagement versus traditional methods: A six thousand student survey of mechanics test data for introductory physics. *American Journal of Physics, 66*, 64–74. https://doi.org/10.1119/1.18809.

Hakkinen, K., et al. (1985). Effect of explosive type strength training on isometric force– and relaxation–time, electromyographic and muscle fibre characteristics of leg extensor muscles. *Acta Physiologiica Scandinavica, 125*, 587–600.

Hamill, J., et al. (2014). *Biomechanical basis of human movement* (4th ed.). Philadelphia: Wolters Kluwer.

Harman, E. A., et al. (1990). The effects of arm and countermovement on vertical jumping. *Medicine & Science in Sports & Exercise, 22*, 825–833.

Harris, J. C. (1993). Using kinesiology: A comparison of applied veins in the subdisciplines. *Quest, 45*, 389–412.

Hatze, H. (1974). The meaning of the term: "Bio-mechanics". *Journal of Biomechanics, 7*, 189–190.

Hatze, H. (1993). The relationship between the coefficient of restitution and energy losses in tennis rackets. *Journal of Applied Biomechanics, 9*, 124–142.

Hatze, H. (2000). The inverse dynamics problem of neuromuscular control. *Biological Cybernetics, 82*, 133–141.

Hawkins, D., & Metheny, J. (2001). Overuse in-juries in youth sports: Biomechanical considerations. *Medicine & Science in Sports & Exercise, 33*, 1701–1707.

Hay, J. G. (1987). *A bibliography of biomechanics literature* (5th ed.). Iowa City: University of Iowa Press.

Hay, J. G. (1993). *The biomechanics of sports techniques* (4th ed.). Englewood Cliffs: Prentice-Hall.

Hay, J. G., & Reid, J. G. (1982). *Anatomy, mechanics, and human motion* (2nd ed.). Englewood Cliffs: Prentice-Hall.

Hay, J. G., et al. (1981). Technique and performance: Identifying the limiting factors. In A. Morecki (Ed.), *Biomechanics VII-B* (pp. 511–520). Baltimore: University Park Press.

Hay, J. G., et al. (1983). Load, speed and equipment effects in strength-training exercises. In H. Matsui & K. Kobayashi (Eds.), *Biomechanics III-B* (pp. 939–950). Champaign: Human Kinetics.

Hay, J. G., et al. (1986). The techniques of elite male long jumpers. *Journal of Biomechanics, 19*, 855–866.

Hayes, W. C. (1986). Bone mechanics: From tissue mechanical properties to an assessment of structural behavior. In G. W. Schmid-Schonbein et al. (Eds.), *Frontiers in biomechanics* (pp. 196–209). New York: Springer-Verlag.

Heckman, C. J., & Sandercock, T. G. (1996). From motor unit to whole muscle properties during locomotor movements. *Exercise and Sport Sciences Reviews, 24*, 109–133.

Heerkens, Y. F., et al. (1986). Passive resistance of the human knee: The effect of immobilization. *Journal of Biomedical Engineering, 8*, 95–104.

Hegedus, E. J., et al. (2015). Clinician-friendly lower extremity physical performance measures in athletes: A systematic review of measurement properties and correlation with injury, part 1. The tests for knee function including hop tests. *British Journal of Sports Medicine, 49*, 642–648.

Heiderscheit, B., et al. (2001). A biomechanical perspective: Do foot orthoses work? *British Journal of Sports Medicine, 35*, 4–5.

Hellebrandt, F. A. (1963). Living anatomy. *Quest, 1*, 43–58.

Henneman, E., et al. (1965). Excitability and inhibitability of moto-neurons of different sizes. *Journal of Neurophysiology, 28*, 599–620.

Henrich, J., et al. (2010). Most people are not WEIRD. *Nature, 466*, 29.

Herbert, R., et al. (1993). Making inferences about muscles forces from clinical observations. *The Australian Journal of Physiotherapy, 39*, 195–202.

Herring, S. W., et al. (1984). Regulation of sarcomere number in skeletal muscle: A comparison of hypotheses. *Muscle and Nerve, 7*, 161–173.

Herring, S. A., et al. (2002). The team physician and return-to-play issues: A consensus statement. *Medicine & Science in Sports & Exercise, 34*, 1212–1214.

Herzog, W. (1996a). Muscle function in movement and sports. *American Journal of Sports Medicine, 24*, S14–S19.

Herzog, W. (1996b). Force-sharing among synergistic muscles: Theoretical considerations and experimental approaches. *Exercise and Sport Sciences Reviews, 24*, 173–202.

Herzog, W. (2000). Muscle properties and coordination during voluntary movement. *Journal of Sports Sciences, 18*, 141–152.

Herzog, W. (2019). Passive force enhancement in striated muscle. *Journal of Applied Physiology, 126*, 1782–1789.

Herzog, W., & Leonard, T. R. (2000). The history dependence of force production in mammalian skeletal muscle following stretch-shortening and shortening-stretch cycles. *Journal of Biomechanics, 33*, 531–542.

Herzog, W., et al. (2000). Specificity and plasticity of mammalian skeletal muscles. *Journal of Applied Biomechanics, 16*, 98–109.

Hewitt, T. E., et al. (2005). Biomechanical measures of neuromuscular control and valgus loading of the knee predict anterior cruciate ligament injury risk in female athletes. *American Journal of Sports Medicine, 33*, 492–501.

Hewitt, T. E., et al. (2006). Anterior cruciate ligament injuries in female athletes, part 2: A meta-analysis of neuromuscular interventions aimed at injury prevention. *American Journal of Sports Medicine, 34*, 490–498.

Higgs, J., & Jones, M. (2000). *Clinical reasoning in the health professions*. London: Butterworth-Heinemann.

Hiley, M. J., et al. (2015). Optimal technique for maximal forward rotating vaults in men's gymnastics. *Human Movement Science, 42*, 117–131.

Hill, A. V. (1926, April). The scientific study of athletics. *Scientific American, 21*(4), 224–225.

Hill, A. V. (1927, August). Are athletes machines? *Scientific American, 137*(2), 124–126.

Hill, A. V. (1970). *The first and last experiments in muscle mechanics.* Cambridge: Cambridge University Press.

Hinson, M. N., et al. (1979). Isokinetics: A clarification. *Research Quarterly, 50*, 30–35.

Hintermeister, R. A., et al. (1998). Quantification of elastic resistance knee rehabilitation exercises. *Journal of Orthopaedic and Sports Physical Therapy, 28*, 40–50.

Hirashima, M., et al. (2002). Sequential muscle activity and its functional role in the upper extremity and trunk during overarm throwing. *Journal of Sports Sciences, 20*, 301–310.

Hirashima, M., et al. (2008). Kinetic chain of overarm throwing in terms of joint rotations revealed by induced acceleration analysis. *Journal of Biomechanics, 41*, 2874–2883.

Hochmuth, G., & Marhold, G. (1978). The further development of biomechanical principles. In E. Asmussen & K. Jorgensen (Eds.), *Biomechanics VI-B* (pp. 93–106). Baltimore: University Park Press.

Hof, A. L. (2001). The force resulting from the action of mono- and biarticular muscle in a limb. *Journal of Biomechanics, 34*, 1085–1089.

Hof, A. L., & Otten, E. (2005). Assessment of two-dimensional induced accelerations from measured kinematic and kinetic data. *Gait & Posture, 22*, 182–188.

Hoffman, S. J., & Knudson, D. V. (Eds.). (2018). *Introduction to kinesiology* (5th ed.). Champaign: Human Kinetics.

Holder-Powell, H. M., & Rutherford, O. M. (1999). Reduction in range of movement can increase maximum voluntary eccentric forces for the human knee extensor muscles. *European Journal of Applied Physiology, 80*, 502–504.

Holt, J., et al. (1996). Flexibility redefined. In T. Bauer (Ed.), *Biomechanics in sports XIII* (pp. 170–174). Thunder Bay: Lakehead University.

Hong, D., & Roberts, E. M. (1993). Angular movement characteristics of the upper trunk and hips in skilled baseball pitching. In T. R. Hamill et al. (Eds.), *Biomechanics in sports XI* (pp. 339–343). Amherst: International Society of Biomechanics in Sports.

Hong, D., et al. (2000). A three-dimensional, six-segment chain analysis of forceful overarm throwing. *Journal of Electromyography and Kinesiology, 11*, 95–112.

Hore, J., et al. (1996). Finger flexion does not contribute to ball speed in overarm throws. *Journal of Sports Sciences, 14*, 335–342.

Horrigan, J. M., et al. (1999). Magnetic resonance imaging evaluation of muscle usage associated with three exercises for rotator cuff rehabilitation. *Medicine & Science in Sports & Exercise, 31*, 1361–1366.

Houston, C. S., & Swischuk, L. E. (1980). Varus and valgus–No wonder they are confused. *New England Journal of Medicine, 302*(8), 471–472.

Hubbard, M. (1989). The throwing events in track and field. In C. Vaughan (Ed.), *Biomechanics of sport* (pp. 213–238). Boca Raton: CRC Press.

Hubbard, M. (1993). Computer simulation in sport and industry. *Journal of Biomechanics, 26*(S1), 53–61.

Hubbard, M., & Alaways, L. W. (1987). Optimum release conditions for the new rules javelin. *International Journal of Sport Biomechanics, 3*, 207–221.

Hubbard, M., et al. (2001). Dependence of release variables in the shot put. *Journal of Biomechanics, 34*, 449–456.

Hubley, C. L., & Wells, R. P. (1983). A work–energy approach to determine individual joint contributions to vertical jump performance. *European Journal of Applied Physiology, 50*, 247–254.

Hudson, J. L. (1986). Coordination of segments in the vertical jump. *Medicine & Science in Sports & Exercise, 18*, 242–251.

Hudson, J. L. (1989). *Guidelines and standards: Application of concepts.* Paper presented to the AAHPERD National Convention, Boston.

Hudson, J. L. (1990). The value visual variables in biomechanical analysis. In E. Kreighbaum & A. McNeill (Eds.), *Proceedings of the VIth international symposium of the international society of biomechanics in sports* (pp. 499–509). Bozeman: Montana State University.

Hudson, J. L. (1995). Core concepts in kinesiology. *JOPERD, 66*(5), 54–55, 59–60.

Hudson, J. L. (1997). The biomechanics body of knowledge. In J. Wilkerson, K. Ludwig, & M. Bucher (Eds.), *Proceedings of the IVth national symposium on teaching biomechanics* (pp. 21–42). Denton: Texas Woman's University Press.

Hudson, J. L. (2000). Sports biomechanics in the year 2000: Some observations on trajectory. In Y. Hong & D. P. Johns (Eds.), *Proceedings of the XVIth international symposium of the international society of biomechanics in sports* (Vol. 2, pp. 525–528). Hong Kong: The Chinese University of Hong Kong.

Hughs, M. (2019). *Essentials of performance analysis in sport* (3rd ed.). New York: Routledge.

Huijing, P. A. (1999). Muscular force transmission: A unified, dual or multiple system? A review and some explorative experimental results. *Archives of Physiology and Biochemistry, 107*, 292–311.

Huijing, P. A., & Baan, G. C. (2001). Extramuscular myofacial force transmission within the rat anterior tibial compartment: Proximo-distal differences in muscle force. *Acta Physiological Scandinavica, 173*, 297–311.

Hukins, D. W., et al. (1990). Comparison of structure, mechanical properties, and functions of lumbar spinal ligaments. *Spine, 15*, 787–795.

Hunter, J. P., & Marshall, R. N. (2002). Effects of power and flexibility training on vertical jump technique. *Medicine & Science in Sports & Exercise, 34*, 478–486.

Huston, R. L., & Grau, C. A. (2003). Basketball shooting strategies–The free throw, direct shot and layup. *Sports Engineering, 6*, 49–64.

Hutton, R. S. (1993). Neuromuscular basis of stretching exercise. In P. Komi (Ed.), *Strength and power in sports* (pp. 29–38). London: Blackwell Scientific Publications.

Huxham, F. E., et al. (2001). Theoretical considerations in balance assessment. *The Australian Journal of Physiotherapy, 47*, 89–100.

Ichinose, Y., et al. (2000). In vivo estimation of contraction velocity of human vastus lateralis muscle during "isokinetic" action. *Journal of Applied Physiology, 88*, 851–856.

Inaba, Y., et al. (2019). Computation of optimal release parameters of jump shots in basketball. In J. Cabri et al. (Eds.), *Sport Science Research and Technology Support, 975*, 164–175. New York: Springer.

International Society of Biomechanics Subcommittee on Standards and Terminology. (1987, Winter). *ISB Newsletter*, 9.

Iossifidou, A. N., & Baltzopoulos, V. (2000). Inertial effects on moment development during isokinetic concentric knee extension testing. *Journal of Orthopaedic and Sports Physical Therapy, 30*, 317–327.

Ito, M., et al. (1998). Nonisometric behavior of fascicles during isometric contractions of a human muscle. *Journal of Applied Physiology, 85*, 1230–1235.

Ives, J. C., & Knudson, D. (2007). Improving professional practice in exercise science: The need for greater disciplinary balance. *Sports Medicine, 37*, 103–115.

Izquierdo, M., et al. (1999). Maximal strength and power characteristics in isometric and dynamic actions of the upper and lower extremities in middle-aged and older men. *Acta Physiological Scandinavica, 167*, 57–68.

Jackson, A. S., & Frankiewicz, R. J. (1975). Factorial expressions of muscular strength. *Research Quarterly, 46*, 206–217.

Jakobi, J. M., & Cafarelli, E. (1998). Neuromuscular drive and force production are not altered during bilateral contractions. *Journal of Applied Physiology, 84*, 200–206.

James, C. R., et al. (2000). Effects of injury proneness and task difficulty on joint kinetic variability. *Medicine & Science in Sports & Exercise, 32*, 1833–1844.

Jenkins, M. (2002). Advanced materials and sporting performance. *Interdisciplinary Science Reviews, 27*, 61–65.

Jensen, J. L., et al. (1994). For young jumpers, differences are in movement's control, not its coordination. *Research Quarterly for Exercise and Sport, 65*, 258–268.

Johnson, R. (1990). *Effects of performance principle training upon skill analysis competency* [Ph.D. dissertation]. Ohio State University.

Johnson, D. L., & Bahamonde, R. (1996). Power output estimate in university athletes. *Journal of Strength and Conditioning Research, 10*, 161–166.

Jorgensen, T. P. (1994). *The physics of golf.* New York: American Institute of Physics.

Judge, L. W., et al. (2020). Pre- and post-activity stretching practices of collegiate soccer coaches in the United States. *International Journal of Exercise Science, 13*(6), 260–272.

Kamibayashi, L. K., & Richmond, F. J. R. (1998). Morphometry of human neck muscles. *Spine, 23*, 1314–1323.

Kaneko, M., et al. (1983). Training effect of different loads on the force–velocity and mechanical power output in human muscle. *Scandinavian Journal of Sports Sciences, 5*, 50–55.

Kasprisin, J. E., & Grabiner, M. D. (2000). Joint angle-dependence of elbow flexor activation levels during isometric and isokinetics maximum voluntary contractions. *Clinical Biomechanics, 15*, 743–749.

Kawakami, Y., & Fukunaga, T. (2006). New in- sights into in vivo human skeletal muscle function. *Exercise and Sport Sciences Reviews, 34*, 16–21.

Kawakami, Y. et al. (2002). In vivo muscle-fibre behaviour during counter-movement exercise in human reveals significant role for tendon elasticity. *Journal of Physiology, 540*, 635–646.

Kawakami, Y., et al. (1998). Architectural and functional features of human triceps surae muscles during contraction. *Journal of Applied Physiology, 85*, 398–404.

Kawakami, Y., et al. (2000). Architecture of contracting human muscles and its functional significance. *Journal of Applied Biomechanics, 16*, 88–98.

Kay, A. D., et al. (2015). Effects of contract-relax, static stretching, and isometric contractions on muscle-tendon mechanics. *Medicine & Science in Sports & Exercise, 47*, 2181–2190.

Keating, J. L., & Matyas, T. A. (1998). Unpredictable error in dynamometry measurements: A quantitative analysis of the literature. *Isokinetics and Exercise Science, 7*, 107–121.

Kellis, E., & Baltzopoulos, V. (1997). The effects of antagonistic moment on the resultant knee joint moment during isokinetics testing of the knee extensors. *European Journal of Applied Physiology, 76*, 253–259.

Kellis, E., & Baltzopoulos, V. (1998). Muscle activation differences between eccentric and concentric isokinetic exercise. *Medicine & Science in Sports & Exercise, 30*, 1616–1623.

Kelly, B. T., et al. (1996). The manual muscle examination for rotator cuff strength: An electromyographic investigation. *American Journal of Sports Medicine, 24*, 581–588.

Kendall, F. P., et al. (2005). *Muscles: Testing and function* (5th ed.). Baltimore: Lippincott Williams & Wilkins.

Khan, K. M., et al. (2000). Overuse tendinosis, not tendinitis, I: A new paradigm for a difficult clinical problem. *Physican and Sportsmedicine, 28*(5), 38–48.

Kibler, W. B., & Livingston, B. (2001). Closed-chain rehabilitation of the upper and lower extremity. *Journal of the American Academy of Orthopaedic Surgeons, 9*, 412–421.

Kilmartin, T. E., & Wallace, A. (1994). The scientific basis for the use of biomechanical foot orthoses in the treatment of lower limb sports injuries–A review of the literature. *British Journal of Sports Medicine, 28*, 180–184.

Kim, E., & Pak, S. (2002). Students do not over-come conceptual difficulties after solving 1000 traditional problems. *American Journal of Physics, 70*, 759–765.

Kinesiology Academy. (1992, Spring). Guidelines and standards for undergraduate kinesiology/biomechanics. *Kinesiology Academy Newsletter, 51*(2), 3–6.

Kingma, I., et al. (1995). Optimizing the determination of the body center of mass. *Journal of Biomechanics, 28*, 1137–1142.

Kirkendall, B. T., & Garrett, W. E. (1997). Function and biomechanics of tendons. *Scandinavian Journal of Medicine & Sciences in Sports, 7*, 62–66.

Kirtley, C. (2006). *Clinical gait analysis: Theory and practice.* New York: Churchill Livingstone.

Klee, A., et al. (2000). Correlation between muscular function and posture–Lowering the degree of pelvic inclination with exercise. In Y. Hong & D. P. Johns (Eds.), *Proceedings of XVIIIth international symposium on biomechanics in sports* (Vol. 1, pp. 162–165). Hong Kong: The Chinese University of Hong Kong.

Kleissen, R. F. M., et al. (1998). Electromyography in the biomechanical analysis of human movement and its clinical application. *Gait and Posture, 8*, 143–158.

Klute, G. K., et al. (2001). Mechanical properties of prosthetic limbs: Adapting to the patient. *Journal of Rehabilitation Research and Development, 38*, 299–307.

Knudson, D. (1991). The tennis topspin forehand drive: Technique changes and critical elements. *Strategies, 5*(1), 19–22.

Knudson, D. (1993). Biomechanics of the basket-ball jump shot: Six key teaching points. *JOPERD, 64*(2), 67–73.

Knudson, D. (1996). A review of exercises and fitness tests for the abdominal muscles. *Sports Medicine Update, 11*(1), 4–5, 25–30.

Knudson, D. (1997). The Magnus Effect in base-ball pitching. In J. Wilkerson, K. Ludwig, & M. Butcher (Eds.), *Proceedings of the 4th national symposium on teaching biomechanics* (pp. 121–125). Denton: Texas Woman's University Press.

Knudson, D. (1998). Stretching: Science to practice. *JOPERD, 69*(3), 38–42.

Knudson, D. (1999a). Issues in abdominal fitness: Testing and technique. *JOPERD, 70*(3), 49–55, 64.

Knudson, D. (1999b). Stretching during warm-up: Do we have enough evidence? *JOPERD, 70*(7), 24–27, 51.

Knudson, D. (1999c). Validity and reliability of visual ratings of the vertical jump. *Perceptual and Motor Skills, 89*, 642–648.

Knudson, D. (2000). What can professionals qualitatively analyze? *JOPERD, 71*(2), 19–23.

Knudson, D. (2001a, July). Improving stroke technique using biomechanical principles. *Coaching and Sport Science Review, 24*, 11–13.

Knudson, D. (2001b). A lab to introduce biomechanics using muscle actions. In J. R. Black-well & D. V. Knudson (Eds.), *Proceedings: Fifth national symposium on teaching biomechanics in sports* (pp. 36–41). San Francisco: University of San Francisco.

Knudson, D. (2001c). Accuracy of predicted peak forces during the power drop exercise. In J. R. Blackwell (Ed.), *Proceedings of oral sessions: XIX international symposium on biomechanics in sports* (pp. 135–138). San Francisco: University of San Francisco.

Knudson, D. (2005). Evidence-based practice in kinesiology: The theory to practice gap revisited. *The Physical Educator, 62*, 212–221.

Knudson, D. (2007a). Warm-up and flexibility. In J. Chandler & L. Brown (Eds.), *Introduction to strength and conditioning.* Baltimore: Lippincott Williams & Wilkins.

Knudson, D. (2007b). Qualitative biomechanical principles for application in coaching. *Sports Biomechanics, 6*, 108–117.

Knudson, D. (2009). Correcting the use of the term "power" in the strength and conditioning literature. *Journal of Strength and Conditioning Research, 23*, 1902–1908.

Knudson, D. (2013). *Qualitative diagnosis of human movement: Improving performance in sport and exercise* (3rd ed.). Champaign: Human Kinetics.

Knudson, D. (2019). Kinesiology's tower of babel: Advancing the field with consistent nomenclature. *Quest, 71*, 42–50.

Knudson, D., & Bahamonde, R. (2001). Effect of endpoint conditions on position and velocity near impact in tennis. *Journal of Sports Sciences, 19*, 839–844.

Knudson, D., & Blackwell, J. (1997). Upper extremity angular kinematics of the one-handed backhand drive in tennis players with and without tennis elbow. *International Journal of Sports Medicine, 18*, 79–82.

Knudson, D., & Kluka, D. (1997). The impact of vision and vision training on sport performance. *JOPERD, 68*(4), 17–24.

Knudson, D., & Morrison, C. (1996). An integrated qualitative analysis of overarm throwing. *JOPERD, 67*(6), 31–36.

Knudson, D., & Morrison, C. (2002). *Qualitative analysis of human movement* (2nd ed.). Champaign: Human Kinetics.

Knudson, D., & Noffal, G. (2005). Time course of stretch-induced isometric strength deficits. *European Journal of Applied Physiology, 94*, 348–351.

Knudson, D., & Wallace, B. (2019). Student perceptions of low-tech active learning and mastery of introductory biomechanics concepts. *Sports Biomechanics, 20*, 1–11. https://doi.org/10.1080/14763141.2019.1570322.

Knudson, D., Magnusson, P., & McHugh, M. (2000). Current issues in flexibility fitness. *President's Council on Physical Fitness and Sports Research Digest, 3*(10), 1–8.

Knuttgen, H. G., & Kraemer, W. J. (1987). Terminology and measurement in exercise performance. *Journal of Applied Sport Science Research, 1*, 1–10.

Koh, T. J. (1995). Do adaptations in serial sarcomere number occur with strength training? *Human Movement Science, 14*, 61–77.

Koike, S., & Harada, Y. (2014). Dynamic contribution analysis of tennis-serve-motion in consideration of torque generating mode. *Procedia Engineering, 72*, 97–102.

Koike, S., et al. (2019). Direct and indirect effects of joint torque inputs during an induced speed analysis of a swinging motion. *Journal of Biomechanics, 86*, 8–16.

Komi, P. V. (1984). Physiological and biomechanical correlates of muscle function: Effects of muscle structure and stretch-shortening cycle on force and speed. *Exercise and Sport Sciences Reviews, 12*, 81–121.

Komi, P. V. (1986). The stretch-shortening cycle and human power output. In N. L. Jones et al. (Eds.), *Human muscle power* (pp. 27–39). Champaign: Human Kinetics.

Komi, P. V. (Ed.). (1992). *Strength and power in sport*. London: Blackwell Science.

Komi, P. V., & Gollhofer, A. (1997). Stretch reflexes can have an important role in force enhancement during the SSC exercise. *Journal of Applied Biomechanics, 13*, 451–460.

Komi, P. V., et al. (1996). Optic fibre as a transducer of tendomuscular forces. *European Journal of Applied Physiology, 72*, 278–280.

Kong, P. W. (2009). Gastrocnemius injury during running: A case report. *International Journal of Sports Medicine, 30*, 46–52.

Kongsgaard, M., et al. (2006). Decline eccentric squats increases patellar tendon loading compared to standard eccentric squats. *Clinical Biomechanics, 21*, 748–754.

Kornecki, S., et al. (2001). Muscular co-operation during joint stabilization, as reflected by EMG. *European Journal of Applied Physiology, 84*, 453–461.

Kouzaki, M., et al. (2002). Alternate muscle activity observed between knee extensor synergists during low-level sustained contractions. *Journal of Applied Physiology, 93*, 675–684.

Kovacs, I., et al. (1999). Foot placement modifies kinematics and kinetics during drop jumping. *Medicine & Science in Sports & Exercise, 31*, 708–716.

Kraan, G. A., et al. (2001). Starting from standing: Why step backwards? *Journal of Biomechanics, 34*, 211–215.

Kraemer, W. J., et al. (1996). Strength and power training: Physiological mechanisms of adaptation. *Exercise and Sport Sciences Reviews, 24*, 363–397.

Kredo, T., et al. (2016). Guide to clinical practice guidelines: The current state of play. *International Journal for Quality in Health Care, 28*, 1220128.

Kreighbaum, E., & Barthels, K. M. (1996). *Biomechanics: A qualitative approach to studying human movement*. Boston: Allyn & Bacon.

Kreighbaum, E. F., & Smith, M. A. (Eds.). (1996). *Sports and fitness equipment design*. Champaign: Human Kinetics.

Krugh, J., & LeVeau, B. (1999). Michael Jordan's vertical jump [abstract]. *Journal of Orthopaedic and Sports Physical Therapy, 20*, A10.

Kubo, K., et al. (1999). Influence of elastic properties of tendon structures on jump performance in humans. *Journal of Applied Physiology, 87*, 2090–2096.

Kubo, K., et al. (2000a). Elasticity of tendon structures of the lower limbs in sprinters. *Acta Physiologica Scandinavica, 168*, 327–335.

Kubo, K., et al. (2000b). *In vivo* dynamics of human medial gastrocnemius muscle–tendon complex during stretch-shortening cycle exercise. *Acta Physiologica Scandinavica, 170*, 127–135.

Kubo, K., et al. (2001a). Influence of static stretching on viscoelastic properties of human tendon *in vivo*. *Journal of Applied Physiology, 90*, 520–527.

Kubo, K., et al. (2001b). Effects of isometric training on elasticity of human tendon structures *in vivo*. *Journal of Applied Physiology, 91*, 26–32.

Kubo, K., et al. (2002). Effect of stretching training on the viscoelastic properties of human tendon structures *in vivo*. *Journal of Applied Physiology, 92*, 595–601.

Kuhman, D. J., & Hurt, C. P. (2019). Lower extremity joints and muscle group in the human locomotor system alter mechanical functions to meet task demand. *Journal of Experimental Biology, 222*(Pt 20), jeb206383. https://doi.org/10.1242/jeb.206383.

Kulig, K., et al. (1984). Human strength curves. *Exercise and Sport Sciences Reviews, 12*, 417–466.

Kulmala, J.-P., et al. (2016). Walking and running require greater effort from the ankle than the knee extensor muscles. *Medicine & Science in Sports & Exercise, 48*, 2181–2189.

Kumagai, K., et al. (2000). Sprint performance is related to muscle fascicle length in male 100-m sprinters. *Journal of Applied Physiology, 88*, 811–816.

Kumar, S. (1999). *Biomechanics in ergonomics*. London: Taylor & Francis.

Kurokawa, S., et al. (2001). Behavior of fasicles and tendinous structures of human gastrocnemius during vertical jumping. *Journal of Applied Physiology, 90*, 1349–1358.

Kwon, Y. H., & Han, K. (2016). Bridging the gap: Key principles in biomechanically good golf swing. In *Proceedings of the 34th International Conference on Biomechanics in Sports* (pp. 1214–1217). Tsukuba: ISBS. https://ojs.ub.uni-konstanz.de/cpa/article/view/6827.

Lafortune, M. A., & Hennig, E. M. (1991). Contribution of angular motion and gravity to tibial acceleration. *Medicine & Science in Sports & Exercise, 23*, 360–363.

Lake, M. J., & Cavanagh, P. R. (1996). Six weeks of training does not change running mechanics or improve running economy. *Medicine & Science in Sports & Exercise, 28*, 860–869.

Lamontagne, A., et al. (2000). Contribution of passive stiffness to ankle plantar flexor moment during gait after stroke. *Archives of Physical Medicine and Rehabilitation, 81,* 351–358.

Laplaud, D., et al. (2006). Reproducibility of eight lower limb muscles activity level in the course of an incremental pedaling exercise. *Journal of Electromyography and Kinesiology, 16,* 158–166.

Larkins, C., & Snabb, T. E. (1999). Positive versus negative foot inclination for maximum height two–leg vertical jumps. *Clinical Biomechanics, 14,* 321–328.

Latash, M. L., & Zatsiorsky, V. M. (Eds.). (2001). *Classics in movement science.* Champaign: Human Kinetics.

Lawson, R. A., & McDermott, L. C. (1987). Student understanding of work–energy and impulse–momentum theorems. *American Journal of Physics, 55,* 811–817.

Lees, A. (1999). Biomechanical assessment of individual sports for improved performance. *Sports Medicine, 28,* 299–305.

Lees, A., & Barton, G. (1996). The interpretation of relative momentum data to assess the contribution of the free limbs to the generation of vertical velocity in sports activities. *Journal of Sports Sciences, 14,* 503–511.

Lees, A., & Fahmi, E. (1994). Optimal drop heights for plyometric training. *Ergonomics, 37,* 141–148.

Lees, A., et al. (2004). Understanding how an arm swing enhances performance in the vertical jump. *Journal of Biomechanics, 37,* 1929–1940.

Legwold, G. (1984). Can biomechanics produce olympic medals? *Physician and Sportsmedicine, 12*(1), 187–189.

Lehman, G. J., & McGill, S. M. (2001). Quantification of the differences in electromyographic activity magnitude between the upper and lower portions of the rectus abdominis muscle during selected trunk exercises. *Physical Therapy, 81,* 1096–1101.

LeVeau, B. (1992). *Williams and Lissner's: Biomechanics of human motion* (3rd ed.). Philadelphia: W. B. Sanders.

Li, L. (2012). How can sport biomechanics contribute to the advance of world record and best athletic performance? *Measurement in Physical Education and Exercise Science, 16,* 194–202.

Lieber, R. L., & Bodine-Fowler, S. C. (1993). Skeletal muscle mechanics: Implications for rehabilitation. *Physical Therapy, 73,* 844–856.

Lieber, R. L., & Friden, J. (1993). Muscle damage is not a function of muscle force but active muscle strain. *Journal of Applied Physiology, 74,* 520–526.

Lieber, R. L., & Friden, J. (1999). Mechanisms of muscle injury after eccentric contraction. *Journal of Science and Medicine in Sport, 2,* 253–265.

Lieber, R. L., & Friden, J. (2000). Functional and clinical significance of skeletal muscle architecture. *Muscle and Nerve, 23,* 1647–1666.

Lindbeck, L., & Kjellberg, K. (2000). Gender differences in lifting technique. *Ergonomics, 44,* 202–214.

Linthorne, N. P., et al. (2005). Optimum take-off angle in the long jump. *Journal of Sports Sciences, 23,* 703–712.

Liu, W., et al. (2013). Multi-joint coordination of functional arm reaching: Induced position analysis. *Journal of Applied Biomechanics, 29,* 235–240.

Lopes, T. J. A., et al. (2018). The effects of injury prevention programs on the biomechanics of landing tasks: A systematic review with meta-analysis. *American Journal of Sports Medicine, 46,* 1492–1499.

Luthanen, P., & Komi, P. V. (1978a). Segmental contributions to forces in vertical jump. *European Journal of Applied Physiology, 38,* 181–188.

Luthanen, P., & Komi, P. V. (1978b). Mechanical factors influencing running speed. In E. Asmussen & K. Jorgensen (Eds.), *Biomechanics VI–B* (pp. 23–29). Baltimore: University Park Press.

Luttgens, K., & Wells, K. F. (1982). *Kinesiology: Scientific basis of human motion* (7th ed.). Philadelphia: Saunders.

Lutz, G. J., & Lieber, R. L. (1999). Skeletal muscle myosin II– Structure and function. *Exercise and Sport Sciences Reviews, 27,* 63–77.

Lutz, G. E., et al. (1993). Comparison of tibiofemoral joint forces during open-kinetic-chain and closed-kinetic-chain exercises. *Journal of Bone and Joint Surgery, 75A,* 732–739.

Maas, H., & Finni, T. (2018). Mechanical coupling between muscle-tendon units reduces peak stresses. *Exercise and Sport Sciences Reviews, 46,* 26–33.

Maas, H., et al. (2004). Muscle force is determined by muscle relative position: Isolated effects. *Journal of Biomechanics, 37,* 99–110.

Maganaris, C. N. (2001). Force–length characteristics of *in vivo* human skeletal muscle. *Acta Physiologica Scandinavica, 172,* 279–285.

Maganaris, C. N., & Paul, J. P. (2000). Load-elongation characteristics of the *in vivo* human tendon and aponeurosis. *Journal of Experimental Biology, 203,* 751–756.

Magnusson, P., & Renstrom, P. (2006). The European College of Sports Sciences position statement: The role of stretching exercises in sports. *European Journal of Sport Science, 6,* 87–91.

Magnusson, S. P., et al. (2000). Passive tensile stress and energy of the human hamstring muscles *in vivo. Scandinavian Journal of Medicine & Science in Sports, 10,* 351–359.

Maisel, I. (1998, August). MadDash. *Sports Illustrated,* 39–43.

Malinzak, R. A., et al. (2001). A comparison of knee joint motion patterns between men and women in selected athletic tasks. *Clinical Biomechanics, 16,* 438–445.

Malone, T. R., et al. (1993). Relationship of gender to anterior cruciate ligament injuries in intercollegiate basketball players. *Journal of the Southern Orthopaedic Association, 2,* 36–39.

Mann, R. V. (1981). A kinetic analysis of sprinting. *Medicine and Science in Sports and Exercise, 13,* 325–328.

Margaria, R., Aghemo, P., & Rovelli, E. (1966). Measurement of muscular power (anaerobic) in man. *Journal of Applied Physiology, 21,* 1662–1664.

Marshall, R. N., & Elliott, B. C. (2000). Long-axis rotation: The missing link in proximal-to-distal segmental sequencing. *Journal of Sports Sciences, 18,* 247–254.

Martin-Delgado, M. A. (2020). The new SI and the fundamental constants of nature. *European Journal of Physics,* 41(6), 063003. doi:10.1088/1361-6404/abab5e.

Matanin, M. J. (1993). *Effects of performance principle training on correct analysis and diagnosis of motor skills* [Ph.D. dissertation]. Ohio State University. Dissertation Abstracts International, 54, 1724A.

Matsuo, T., et al. (2018). Middle finger and ball movements around ball release during baseball fastball pitching. *Sports Biomechanics, 17,* 180–191.

McBride, J. M., et al. (1999). A comparison of strength and power characteristics between power lifters, olympic lifters, and sprinters. *Journal of Strength and Conditioning Research, 13,* 58–66.

McGhee, D. E., & Steele, J. R. (2020). Breast biomechanics: What do we really know. *Physiology, 35,* 144–156.

McGill, S. M. (1998). Low back exercises: Evidence for improving exercise regimens. *Physical Therapy, 78,* 754–765.

McGill, S. M. (2001). Low back stability: From formal description to issues for performance and rehabilitation. *Exercise and Sport Sciences Reviews, 29,* 26–31.

McGill, S. M. (2016). *Low back disorders: Evidence-based prevention and rehabilitation* (3rd ed.). Champaign: Human Kinetics.

McGill, S. M., & Norman, R. W. (1985). Dynamically and statically determined low back moments during lifting. *Journal of Biomechanics, 18,* 877–886.

McGill, S. M., et al. (2000). Changes in lumbar lordosis modify the role of the extensor muscles. *Clinical Biomechanics, 15,* 777–780.

McGinnis, P., & Abendroth-Smith, J. (1991). Impulse, momentum, and water balloons. In J. Wilkerson et al. (Eds.), *Teaching kinesiology and biomechanics in sports* (pp. 135–138). Ames: Iowa State University.

McGregor, R. R., & Devereux, S. E. (1982). *EEVeTeC.* Boston: Houghton Mifflin Co.

McLaughlin, T. M., et al. (1978). Kinetics of the parallel squat. *Research Quarterly, 49,* 175–189.

McLean, S. P., & Hinrichs, R. N. (2000a). Influence of arm position and lung volume on the center of buoyancy of competitive swimmers. *Research Quarterly for Exercise and Sport, 71,* 182–189.

McLean, S. P., & Hinrichs, R. N. (2000b). Buoyancy, gender, and swimming performance. *Journal of Applied Biomechanics, 16,* 248–263.

McLean, S. P., & Reeder, M. S. (2000). Upper extremity kinematics of dominant and non–dominant side batting. *Journal of Human Movement Studies, 38,* 201–212.

McLean, S. G., et al. (2005). Evaluation of a two dimensional analysis method as a screening and evaluation tool for anterior cruciate ligament injury. *British Journal of Sports Medicine, 39,* 355–362.

McNair, P. J., et al. (2000). Decreasing landing forces: Effect of instruction. *British Journal of Sports Medicine, 34,* 293–296.

McPoil, T. G., et al. (1995). A comparison of two in-shoe plantar pressure measurement systems. *The Lower Extremity, 2,* 95–103.

Mehta, R. D. (1985). Aerodynamics of sports balls. *Annual Review of Fluid Mechanics, 17,* 151–189.

Mehta, R. D., & Pallis, J. M. (2001a). The aerodynamics of a tennis ball. *Sports Engineering, 4,* 177–189.

Mehta, R. D., & Pallis, J. M. (2001b). Sports ball aerodynamics: Effects of velocity, spin and surface roughness. In F. H. Froes & S. J. Haake (Eds.), *Materials and science in sports* (pp. 185–197). Warrendale: The Minerals, Metals and Materials Society [TMS].

Meinel, K., & Schnabel, G. (1998). *Bewegungslehre–Sportmotorik.* Berlin: Sportverlag Berlin.

Mena, D., et al. (1981). Analysis and synthesis of human swing leg motion during gait and its clinical applications. *Journal of Biomechanics, 14,* 823–832.

Mero, A., et al. (1981). Relationships between the maximal running velocity, muscle fiber characteristics, force production, and force relaxation of sprinters. *Scandinavian Journal Sports Sciences, 3,* 16–22.

Mero, A., et al. (1992). Biomechanics of sprint running: A review. *Sports Medicine, 13,* 376–392.

Messier, S. P., & Cirillo, K. J. (1989). Effects of verbal and visual feedback system on running technique, perceived exertion and running economy in female novice runners. *Journal of Sports Sciences, 7,* 113–126.

Miller, D. I. (1980). Body segment contributions to sport skill performance: Two contrasting approaches. *Research Quarterly for Exercise and Sport, 51,* 219–233.

Miller, D. I. (2000). *Biomechanics of competitive diving.* Fort Lauderdale: U.S. Diving.

Mills, K. (2015). Motion analysis in the clinic: There's an app for that. *Journal of Physiotherapy, 61,* 49–50.

Minetti, A. E. (2004). Passive tools for enhancing muscle-driven motion and locomotion. *Journal of Experimental Biology, 207,* 1265–1272.

Minozzi, S., et al. (2000). Searching for rehabilitation articles on Medline and Embase: An example with cross-over design. *Archives of Physical Medicine and Rehabilitation, 81,* 720–722.

Mirka, G., et al. (1997). Selective activation of the external oblique musculature during axial torque production. *Clinical Biomechanics, 12,* 172–180.

Mizera, F., & Horvath, G. (2002). Influence of environmental factors on shot put and hammer throw range. *Journal of Biomechanics, 35,* 785–796.

Montgomery, J. & Knudson, D. (2002). A method to determine stride length for baseball pitching. *CCIS: Applied Research in Coaching and Athletics Annual, 17,* 75–84.

Moore, S. P., & Marteniuk, R. G. (1986). Kinematic and electromyographic changes that occur as a function of learning a time-constrained aiming task. *Journal of Motor Behavior, 18,* 397–426.

Morgan, D. L., et al. (2000). Tension changes in the cat soleus muscle following slow stretch or shortening of the contracting muscle. *Journal of Physiology (London), 522,* 503–513.

Moritani, T. (1993). Neuromuscular adaptations during the acquisition of muscle strength, power, and motor tasks. *Journal of Biomechanics, 26*(S1), 95–107.

Moritani, T., & Yoshitake, Y. (1998). The use of electromyography in applied physiology. *Journal of Electromyography and Kinesiology, 8,* 363–381.

Morris, G. (1977). Dynamic visual acuity: Implications for the physical educator and coach. *Motor Skills: Theory into Practice, 2,* 15–20.

Morrish, G. (1999). Surface electromyography: Methods of analysis, reliability, and main applications. *Critical Reviews in Physical and Rehabilitation Medicine, 11,* 171–205.

Morriss, C., et al. (2001). The function of blocking in elite javelin throws: A re–evaluation. *Journal of Human Movement Studies, 41,* 175–190.

Mountjoy, M., et al. (2014). The IOC consensus statement: Beyond the female athlete triad–Relative energy deficiency in sports (RED-S). *British Journal of Sports Medicine, 48,* 491–497.

Mow, V. C., & Hayes, W. C. (Eds.). (1991). *Basic orthopaedic biomechanics.* New York: Raven Press.

Mujika, I., & Taipale, R. S. (2019). Sport science on women, women in sport science. *International Journal of Sports Physiology and Performance, 14,* 1013–1014.

Mullally, E. M. et al. (2021). Situations and mechanisms of non-contact knee injury in adult netball: A systematic review. *Physical Therapy in Sport, 47,* 193–200.

Muraoka, T., et al. (2002). Length change of human gastrocnemius aponeurosis and tendon during passive joint motion. *Cells, Tissues, and Organs, 171,* 260–268.

Muraoka, T., et al. (2004). Influence of tendon slack on electromechanical delay in the human medial gastrocnemius in vivo. *Journal of Applied Physiology, 96,* 540–544.

Muraoka, T., et al. (2005). Estimation of passive ankle joint moment during standing and walking. *Journal of Applied Biomechanics, 21,* 72–84.

Mureika, J. R. (2000). The legality of wind and altitude assisted performances in the sprints. *New Studies in Athletics, 15*(3/4), 53–58.

Murray, M. P., et al. (1967). Center of gravity, center of pressure, and supportive forces during human activities. *Journal of Applied Physiology, 23,* 831–838.

Murray, W. M., et al. (2000). The isometric functional capacity of muscles that cross the elbow. *Journal of Biomechanics, 33,* 943–952.

Myers, D. C., et al. (1993). The dimensions of human physical performance: Factor analysis of strength, stamina, flexibility, and body composition measures. *Human Performance, 6,* 309–344.

Nagano, A., & Gerritsen, K. G. M. (2001). Effects of neuromuscular strength training on vertical jumping performance–A computer simulation study. *Journal of Applied Biomechanics, 17,* 113–128.

Nagano, A., et al. (1998). Comparison of new approaches to estimate mechanical output of individual joints in vertical jumps. *Journal of Biomechanics, 31*, 951–955.

Nagano, A., et al. (2005). Contribution of non-extensor muscle of the let to maximal-effort countermovement jumping. *Biomedical Engineering Online, 4*, 52), 1–52),10.

Naito, K., & Maruyama, T. (2008). Contributions of the muscular torques and motion-dependent torques to generate rapid elbow extension during overhand baseball pitching. *Sports Engineering, 11*, 47–56.

Nakano, N., et al. (2020). Basketball players minimize the effect of motor noise by using near-minimum release speed in free-throw shooting. *Human Movement Science, 70*, 102583.

Nakazawa, K., et al. (1993). Differences in activation patterns in elbow flexor muscles during isometric, concentric, and eccentric contractions. *European Journal of Applied Physiology, 66*, 214–220.

Neal, R. J., & Wilson, B. D. (1985). 3D kinematics and kinetics of the golf swing. *International Journal of Sport Biomechanics, 1*, 221–232.

Neptune, R. R., & Kautz, S. A. (2001). Muscle activation and deactivation dynamics: The governing properties in fast cyclical human movement performance? *Exercise and Sport Sciences Reviews, 29*, 76–81.

Neptune, R. R., & van den Bogert, A. J. (1998). Standard mechanical energy analyses do not correlate with muscle work in cycling. *Journal of Biomechanics, 31*, 239–245.

Newell, K. M., et al. (1989). Search strategies and the acquisition of coordination. In S. A. Wallace (Ed.), *Perspectives on the coordination of movement* (pp. 85–122). New York: Elsevier.

Newton, R. U., et al. (1996). Kinematics, kinetics, and muscle activation during explosive upper body movements. *Journal of Applied Biomechanics, 12*, 31–43.

Newton, R. U., et al. (1997). Influence of load and stretch shortening cycle on the kinematics, kinetics and muscle activation that occurs during explosive upper-body movements. *European Journal of Applied Physiology, 75*, 333–342.

Nichols, T. R. (1994). A biomechanical perspective on spinal mechanisms of coordinated muscular action: An architecture principle. *Acta Anatomica, 151*, 1–13.

Nielsen, A. B., & Beauchamp, L. (1992). The effect of training in conceptual kinesiology on feedback provision patterns. *Journal of Teaching in Physical Education, 11*, 126–138.

Nigg, B. M. (Ed.). (1986). *Biomechanics of running shoes*. Champaign: Human Kinetics.

Nigg, B. M., et al. (1989). The tennis shoe–Biomechanical design criteria. In B. Segesser & W. Pforringer (Eds.), *The shoe in sport* (pp. 39–46). Chicago: Year Book Medical Publishers.

Nordin, M., & Frankel, V. (2001). *Basic biomechanics of the musculoskeletal system* (3rd ed.). Baltimore: Williams & Wilkins.

Nordin, M., & Frankel, V. (2021). *Basic biomechanics of the musculoskeletal system* (5th ed). Baltimore: Wolters Kluwer.

Norkin, C. C., & White, D. J. (2017). *Measurement of joint motion–A guide to goniometry* (7th ed.). Philadelphia: F. A. Davis.

Norman, R. (1975). Biomechanics for the community coach. *JOPERD, 46*(3), 49–52.

Norman, R. W. (1983). Biomechanical evaluation of sports protective equipment. *Exercise and Sport Sciences Reviews, 11*, 232–274.

Novacheck, T. F. (1998). The biomechanics of running. *Gait and Posture, 7*, 77–95.

Nunome, H., et al. (2002). Three-dimensional kinetic analysis of side-foot and instep soccer kicks. *Medicine & Science in Sports & Exercise, 34*, 2028–2036.

Nunome, H., Ikegami, Y., et al. (2006a). Segmental dynamics of soccer instep kicking with the preferred and non-preferred leg. *Journal of Sports Sciences, 24*, 529–541.

Nunome, H., Lake, M., et al. (2006b). Impact phase kinematics of instep kicking in soccer. *Journal of Sports Sciences, 24*, 11–22.

Nussbaum, M. A., & Chaffin, D. B. (1997). Pattern classification reveals intersubject group differences in lumbar muscle recruitment during static loading. *Clinical Biomechanics, 12*, 97–106.

Nuzzo, J. L. (2019). The case for retiring flexibility as a major component of physical fitness. *Sports Medicine, 50*(5), 853–870. https://doi.org/10.1007/s40279-019-01248-w.

Olds, T. (2001). Modelling of human locomotion: Applications to cycling. *Sports Medicine, 31*, 497–509.

Ostarello, J. Z. (1986). A visit to the shrines of the vertical jump and the 40 yard dash. In J. Terauds, B. A. Gowitzke, & L. E. Holt (Eds.), *Proceedings of the 4th International Symposium on Biomechanics in Sports* (pp. 292–299). Halifax: Dalhousie University.

Owens, M. S., & Lee, H. Y. (1969). A determination of velocities and angles of projection for the tennis serve. *Research Quarterly, 40*, 750–754.

Ozkaya, N. et al. (2017). *Fundamentals of biomechanics: equilibrium, motion, and deformation*. 4th ed. New York: Springer.

Pandy, M. G. (1999). Moment arm of a muscle force. *Exercise and Sport Sciences Reviews, 27*, 79–118.

Pandy, M. G., et al. (1990). An optimal control model for maximum-height human jumping. *Journal of Biomechanics, 23*, 1185–1198.

Panjabi, M. M., & White, A. A. (2001). *Biomechanics in the musculoskeletal system*. New York: Churchill Livingstone.

Pappas, G. P., et al. (2002). Nonuniform shortening in the biceps brachii during elbow flexion. *Journal of Applied Physiology, 92*, 2381–2389.

Patel, T. J., & Lieber, R. L. (1997). Force transmission in skeletal muscle: From actomyosin to external tendons. *Exercise and Sport Sciences Reviews, 25*, 321–364.

Patla, A. E. (1987). Some neuromuscular strategies characterising adaptation process during prolonged activity in humans. *Canadian Journal of Sport Sciences, 12*(Supp), 33S–44S.

Paton, M. E., & Brown, J. M. N. (1994). An electromyographic analysis of functional differentiation in human pectoralis major muscle. *Journal of Electromyography and Kinesiology, 4*, 161–169.

Paton, M. E., & Brown, J. M. N. (1995). Functional differentiation within latissimus dorsi. *Electromyography and Clinical Neurophysiology, 35*, 301–309.

Peach, J. P., et al. (1998). Three-dimensional kinematics and trunk muscle myoelectric activity in the young lumbar spine: A database. *Archives of Physical Medicine and Rehabilitation, 79*, 663–669.

Pedotti, A., & Crenna, P. (1999). Individual strategies of muscle recruitment in complex natural movements. In J. M. Winters & S. L. Woo (Eds.), *Multiple muscle systems* (pp. 542–549). Berlin: Springer-Verlag.

Perrin, D. H. (1993). *Isokinetic exercise and assessment*. Champaign: Human Kinetics.

Perry, J. (1992). *Gait analysis: Normal and pathological function*. Thorofare: Slack Inc.

Phillips, S. J., et al. (1983). Quantification of intersegmental reactions during rapid swing motion. *Journal of Biomechanics, 16*, 411–417.

Phillips, B. A., et al. (2000). Muscle force measured using "break" testing with a hand-held myometer in normal subjects aged 20 to 69 years. *Archives of Physical Medicine and Rehabilitation, 81*, 653–661.

Piazza, S. J., & Cavanagh, P. R. (2000). Measurement of the screw-home mechanism of the knee is sensitive to errors in axis alignment. *Journal of Biomechanics, 33*, 1029–1034.

Pinniger, G. J., et al. (2000). Tension regulation during lengthening and shortening actions of the human soleus muscle. *European Journal of Applied Physiology, 81*, 375–383.

Plagenhoef, S. (1971). *Patterns of human motion: A cinematographic analysis*. Englewood Cliffs: Prentice-Hall.

Plagenhoef, S., et al. (1983). Anatomical data for analyzing human motion. *Research Quarterly for Exercise and Sport, 54*, 169–178.

Poyhonen, T., et al. (2001). Electromyographic and kinematic analysis of therapeutic knee exercises under water. *Clinical Biomechanics, 16*, 496–504.

Preuss, R., & Fung, J. (2005). Can acute low back pain result from segmental spinal buckling during sub-maximal activities? A review of the current literature. *Manual Therapy, 10*, 14–20.

Prilutsky, B. I. (2000). Coordination of two– and one-joint muscles: Functional consequences and implications for motor control. *Motor Control, 4*, 1–44.

Prilutsky, B. I., & Zatsiorsky, V. M. (1994). Tendon action of two-joint muscles: Transfer of mechanical energy between joints during jumping, landing, and running. *Journal of Biomechanics, 27*, 25–34.

Pugh, L. (1971). The influence of wind resistance in running and walking and the mechanical efficiency of work against horizontal or vertical forces. *Journal of Physiology, 213*, 255–276.

Putnam, C. (1983). Interaction between segments during a kicking motion. In H. Matsui & K. Kobayashi (Eds.), *Biomechanics VIII-B* (pp. 688–694). Champaign: Human Kinetics.

Putnam, C. (1991). A segment interaction analysis of proximal-to-distal sequential segment motion patterns. *Medicine & Science in Sports & Exercise, 23*, 130–144.

Putnam, C. (1993). Sequential motions of the body segments in striking and throwing skills: Descriptions and explanations. *Journal of Biomechanics, 26*(S1), 125–135.

Rabita, G., et al. (2000). Differential effect of knee extension isometric training on the different muscles of the quadriceps femoris in humans. *European Journal of Applied Physiology, 83*, 531–538.

Rall, J. A. (2017). Nobel Laureate A. V. Hill and the refugee scholars, 1933–1945. *Advances in Physiology Education, 41*, 248–259.

Rassier, D. E., et al. (1999). Length dependence of active force production in skeletal muscle. *Journal of Applied Physiology, 86*, 1445–1457.

Ravn, S., et al. (1999). Choice of jumping strategy in two standard jumps, squat and countermovement jump–Effect of training background or inherited preference? *Scandinavian Journal of Medicine & Science in Sports, 9*, 201–208.

Richmond, F. J. R. (1998). Elements of style in neuromuscular architecture. *American Zoologist, 38*, 729–742.

Ridderikhoff, A., et al. (1999). Jumping for distance: Control of the external force in squat jumps. *Medicine & Science in Sports & Exercise, 31*, 1196–1204.

Rigby, B. J., Hirai, N., Spikes, J. D., & Eyring, H. (1959). The mechanical properties of rat tail tendon. *Journal of General Physiology, 43*, 265–283.

Roberton, M. A., & Halverson, L. E. (1984). *Developing children–Their changing movement*. Philadelphia: Lea & Febiger.

Roberts, E. M. (1991). Tracking velocity in motion. In C. L. Tant et al. (Eds.), *Biomechanics in sports IX* (pp. 3–25). Ames: Iowa State University.

Roberts, T. J., et al. (1997). Muscular force in running turkeys: The economy of minimizing work. *Science, 275*, 1113–1115.

Robinovitch, S. N., et al. (2000). Prevention of falls and fall–related fractures through biomechanics. *Exercise and Sport Sciences Reviews, 28*, 74–79.

Rodano, R., & Squadrone, R. (2002). Stability of selected lower limb joint kinetic parameters during vertical jump. *Journal of Applied Biomechanics, 18*, 83–89.

Rogers, M. M., & Cavanagh, P. R. (1984). Glossary of biomechanical terms, concepts, and units. *Physical Therapy, 64*, 82–98.

Roncesvalles, M. N. C., et al. (2001). Development of lower extremity kinetics for balance control in infants and young children. *Journal of Motor Behavior, 33*, 180–192.

Ross, A. L., & Hudson, J. L. (1997). Efficacy of a mini-trampoline program for improving the vertical jump. In J. D. Wilkerson et al. (Eds.), *Biomechanics in sports XV* (pp. 63–69). Denton: Texas Woman's University.

Rowlands, L. K., et al. (1995). Kinesiology of the empty can test. *American Journal of Physical Medicine and Rehabilitation, 74*, 302–304.

Rushall, B. S., et al. (1994). A re-evaluation of forces in swimming. *Journal of Swimming Research, 10*, 6–30.

Sackett, D. L., et al. (1996). Evidence based medicine: What it is and what it isn't. *British Medical Journal, 312*, 71–72.

Sadoghi, P., et al. (2012). Effectiveness of anterior cruciate ligament injury prevention training programs. *Journal of Bone and Joint Surgery, 94*, 769–776.

Sale, D. (1987). Influence of exercise and training on motor unit activation. *Exercise and Sport Sciences Reviews, 16*, 95–151.

Sale, D. (1992). Neural adaptation to strength training. In P. Komi (Ed.), *Strength and power in sport* (pp. 249–265). London: Blackwell Scientific Publications.

Sale, D. (2002). Postactivation potentiation: Role in human performance. *Exercise and Sport Sciences Reviews, 30*, 138–143.

Salsich, G. B., et al. (2000). Relationships between plantar flexor muscle stiffness, strength, and range of motion in subjects with diabetes-peripheral neuropathy compared to age–matched controls. *Journal of Orthopaedic and Sports Physical Therapy, 30*, 473–483.

Sandercock, T. G. (2000). Nonlinear summation of force in cat soleus muscle results primarily from stretch of the common–elastic elements. *Journal of Applied Physiology, 89*, 2206–2214.

Sanders, R. H. (1998). Lift or drag? Let's get skeptical about freestyle swimming. *Sportscience*, http://www.sportsci.org/jour/9901/med.html.

Sanders, R. H., & Allen, J. B. (1993). Changes in net joint torques during accommodation to change in surface compliance in a drop jumping task. *Human Movement Science, 12*, 299–326.

Sargent, D. A. (1921). The physical test of a man. *American Physical Education Review, 26*, 188–194.

Savelsbergh, G. J. P., & Whiting, H. T. A. (1988). The effect of skill level, external frame of reference and environmental changes on one–handed catching. *Ergonomics, 31*, 1655–1663.

Sayers, S. P., et al. (1999). Cross-validation of three jump power equations. *Medicine & Science in Sports & Exercise, 31*, 572–577.

Schieb, D. A. (1987, January). The biomechanics piezoelectric force plate. *Soma, 14*, 35–40.

Schlumberger, A., et al. (2006). Muscle imbalances–Fact or fiction? *Isokinetics and Exercise Science, 14*, 3–11.

Schmidt, R. A., & Wrisberg, C. A. (2007). *Motor learning and performance* (4th ed.). Champaign: Human Kinetics.

Schneider, K., et al. (1989). Changes in limb dynamics during the practice of rapid arm movements. *Journal of Biomechanics, 22*, 805–817.

Scott, W., et al. (2001). Human skeletal muscle fiber type classifications. *Physical Therapy, 81*, 1810–1816.

Segesser, B., & Pforringer, W. (Eds.). (1989). *The shoe in sport*. Chicago: Year Book Medical Publishers.

Selinger, J. C., et al. (2019). How humans initiate energy optimization and converge on their optimal gaits. *Journal of Experimental Biology, 222*(19), jeb198234. https://doi.org/10.1242/jeb.198234.

Semmler, J. G. (2002). Motor unit synchronization and neuromuscular performance. *Exercise and Sport Sciences Reviews, 30*, 8–14.

Shadwick, R. E., et al. (1998). Muscle dynamics in fish during steady swimming. *American Zoologist, 38*, 755–770.

Sheard, P. W. (2000). Tension delivery from short fibers in long muscles. *Exercise and Sport Sciences Reviews, 28*, 51–56.

Sheppard, L. M. (2006). Visual effects and video analysis lead to Olympics victories. *IEEE Computer Graphics and Applications, 26*(2), 6–11.

Shi, B., et al. (2019). Wearable ankle robots in post-stroke rehabilitation of gait: A systematic review. *Frontiers in Neurorobotics, 13*, 63. https://doi.org/10.3389/fnbot.2019.00063.

Shirier, I. (1999). Stretching before exercise does not reduce the risk of local muscle injury: A critical review of the clinical and basic science literature. *Clinical Journal of Sport Medicine, 9*, 221–227.

Shirier, I. (2004). Does stretching improve performance? A systematic and critical review of the literature. *Clinical Journal of Sport Medicine, 14*, 267–273.

Shirier, I. (2006). Recognizing and challenging dogma. *Clinical Journal of Sport Medicine, 16*, 93–94.

Shultz, S. J., et al. (2000). *Assessment of athletic injuries.* Champaign: Human Kinetics.

Siegler, S., & Moskowitz, G. D. (1984). Passive and active components of the internal moment developed about the ankle joint during human ambulation. *Journal of Biomechanics, 17*, 647–652.

Simic, L., et al. (2013). Does pre-exercise static stretching inhibit maximal muscular performance? A meta-analytic review. *Scandinavian Journal of Medicine & Science in Sports, 23*, 131–148.

Simonsen, E. B., et al. (1997). Mechanisms contributing to different joint moments observed during human walking. *Scandinavian Journal of Medicine & Science in Sports, 7*, 1–13.

Slifkin, A. B., & Newell, K. M. (2000). Variability and noise in continuous force production. *Journal of Motor Behavior, 32*, 141–150.

Smith, F. (1982). Dynamic variable resistance and the Universal system. *National Strength and Conditioning Association Journal, 4*(4), 14–19.

Smith, L. K., et al. (1996). *Brunnstrom's clinical kinesiology* (5th ed.). Philadelphia: F. A. Davis.

Soderberg, G. L., & Knutson, L. M. (2000). A guide for use and interpretation of kinesiologic eletcromyographic data. *Physical Therapy, 80*, 485–498.

Solomito, M. J. et al. (2020). Influence of stride length on upper extremity joint moments and ball velocity in collegiate baseball pitchers. Sports Biomechanics, doi:10.1080/14763141.2020.10809698.

Solomonow, M. (2004). Ligaments: A source of work-related musculoskeletal disorders. *Journal of Electromyography and Kinesiology, 14*, 49–60.

Sorensen, H., et al. (1996). Dynamics of the martial arts high front kick. *Journal of Sports Sciences, 14*, 483–495.

Sorensen, H., et al. (2000). Biomechanical reflections on teaching fast motor skills. In Y. Hong & D. P. Johns (Eds.), *Proceedings of XVIIIth international symposium on biomechanics in sports* (Vol. 1, pp. 84–88). Hong Kong: The Chinese University of Hong Kong.

Sprigings, E. J., & Neal, R. J. (2000). An insight into the importance of wrist torque in driving the golf ball: A simulation study. *Journal of Applied Biomechanics, 16*, 156–166.

Sprigings, E., et al. (1994). A three–dimensional kinematic method for determining the effectiveness of arm segment rotations in producing racquet-head speed. *Journal of Biomechanics, 27*, 245–254.

Stauber, W. T. (2004). Factors involved in strain induced injury in skeletal muscles and outcomes of prolonged exposure. *Journal of Electromyography and Kinesiology, 14*, 61–70.

Steindler, A. (1955). *Kinesiology of the human body under normal and pathological conditions.* Springfield: Charles C. Thomas.

Stergiou, N., et al. (2006). Optimal movement variability: A new theoretical perspective for neurologic physical therapy. *Journal of Neurologic Physical Therapy, 30*, 120–129.

Stodden, D. F., et al. (2005). Relationship of biomechanical factors to baseball pitching velocity: Within pitcher variation. *Journal of Applied Biomechanics, 21*, 44–56.

Stone, M., et al. (2002). Training principles: Evaluation of modes and methods of resistance training–A coaching perspective. *Sports Biomechanics, 1*, 79–103.

Subic, A. J., & Haake, S. J. (2000). *The engineering of sport: Research, development and innovation.* London: Blackwell Scientific.

Sugisaki, N., et al. (2005). Behavior of fasicle and tendinous tissue of medial gastrocnemius muscle during rebound exercise of ankle joint. *International Journal of Sport and Health Sciences, 3*, 100–109.

Szucs, K. A., & Borstad, J. D. (2013). Gender differences between muscle activation and onset timing of the four subdivisions of trapezius during humerothoracic elevation. *Human Movement Science, 32*, 1288–1298.

Talbot, J. A., & Morgan, D. L. (1996). Quantitative analysis of sarcomere non-uniformities in active muscle following a stretch. *Journal of Muscle Research and Cell Motility, 17*, 261–268.

Taylor, A., & Prochazka, A. (Eds.). (1981). *Muscle receptors and movement.* New York: Oxford University Press.

Taylor, P. G., et al. (2015). Determining the optimal trial size using sequential analysis. *Journal of Sports Sciences, 33*, 300–308.

Teichtahl, A. J., et al. (2006). Foot rotation–A potential target to modify the knee adduction moment. *Journal of Science and Medicine in Sport, 9*, 67–71.

Thelen, D. G., et al. (2006). Neuromusculoskeletal models provide insights into the mechanisms and rehabilitation of hamstring strains. *Exercise and Sport Sciences Reviews, 34*, 135–141.

Thomas, J. S., et al. (1998). The influence of gender on spine, hip, knee, and ankle motions during a reaching task. *Journal of Motor Behavior, 30*, 98–103.

Thomas, E., et al. (2018). The relation between stretching typology and stretching duration: The effects on range of motion. *International Journal of Sports Medicine, 39*, 243–354.

Thomee, R., et al. (1999). Patellofemoral pain syndrome: A review of current issues. *Sports Medicine, 28*, 245–262.

Tol, J. L., et al. (2002). The relationship of the kicking action in soccer and anterior ankle impingement syndrome: A biomechanical analysis. *American Journal of Sports Medicine, 30*, 45–50.

Tomioka, M., Owings, T. M., & Grabiner, M. D. (2001). Lower extremity strength and coordination are independent contributors to maximum vertical jump height. *Journal of Applied Biomechanics, 17*, 181–187.

Torg, J. S. (1992). Epidemiology, pathomechanics, and prevention of football-induced cervical spinal cord trauma. *Exercise and Sport Sciences Reviews, 20*, 321–338.

Toussaint, H. M., et al. (2002). "Pumped-up propulsion" during front crawl swimming. *Medicine & Science in Sports & Exercise, 34*, 314–319.

Tsaousidis, N., & Zatsiorsky, V. (1996). Two types of ball–effector interaction and their relative contribution to soccer kicking. *Human Movement Science, 15*, 861–876.

U.S. Department of Health and Human Services. (1996). *Physical activity and health: A report of the surgeon general.* Atlanta: U.S. Department of Health and Human Services, Centers for Disease

Control and Prevention, National Center for Chronic Disease Prevention and Health Promotion.

U.S. Department of Health and Human Services. (2018). *Physical activity guidelines for americans* (2nd ed.). Washington: US Dept of Health and Human Services.

van Dieen, J. H., et al. (1999). Stoop or squat: A review of biomechanical studies on lifting technique. *Clinical Biomechanics, 14,* 685–696.

van Donkelaar, P., & Lee, R. G. (1994). The role of vision and eye motion during reaching to intercept moving targets. *Human Movement Science, 13,* 765–783.

van Ingen Schenau, G. J., & Cavanagh, P. R. (1999). Power equations in endurance sports. *Journal of Biomechanics, 23,* 865–881.

van Ingen Schenau, G. J., et al. (1989). Biomechanics of speed skating. In C. Vaughan (Ed.), *Biomechanics of sport* (pp. 121–167). Boca Raton: CRC Press.

van Ingen Schenau, G. J., et al. (1995). The control of multi–joint movements relies on detailed internal representations. *Human Movement Science, 14,* 511–538.

van Ingen Schenau, G. J., et al. (1997). Does elastic energy enhance work and efficiency in the stretch–shortening cycle? *Journal of Applied Biomechanics, 13,* 389–415.

van Zandwijk, J. P., et al. (2000). Control of maximal and submaximal vertical jumps. *Medicine & Science in Sports & Exercise, 32,* 477–485.

Veloso, A., & Abrantes, J. (2000). Estimation of power output from leg extensor muscles in the acceleration phase of the sprint. In Y. Hong & D. P. Johns (Eds.), *Proceedings of XVIIIth international symposium on biomechanics in sports* (Vol. 1, pp. 97–101). Hong Kong: The Chinese University of Hong Kong.

Vezina, M. J., & Hubley-Kozey, C. L. (2000). Muscle activation in therapeutic exercises to improve trunk stability. *Archives of Physical Medicine and Rehabilitation, 81,* 1370–1379.

Vieira, T. M., & Botter, A. (2021). The accurate assessment of muscle excitation requires the detection of multiple surface electromyograms. *Exercise and Sport Sciences Reviews,* 49, 23–34.

Vigotsky, A. D., et al. (2019). Mechanical misconceptions: Have we lost the "mechanics" in "sports biomechanics"? *Journal of Biomechanics, 93,* 1–5.

Vint, P. F., & Hinrichs, R. N. (1996). Differences between one-foot and two-foot vertical jump performances. *Journal of Applied Biomechanics, 12,* 338–358.

Wakai, M., & Linthorne, N. P. (2005). Optimum take-off angle in the standing long jump. *Human Movement Science, 24,* 81–96.

Walshe, A. D., et al. (1996). The validity and reliability of a test of lower body musculotendinous stiffness. *European Journal of Applied Physiology, 73,* 332–339.

Ward, T., & Groppel, J. (1980). Sport implement selection: Can it be based upon anthropometric indicators? *Motor Skills: Theory into Practice, 4,* 103–110.

Watts, R. G., & Bahill, A. T. (2000). *Keep your eye on the ball: The science and folklore of baseball* (2nd ed.). New York: W.H. Freeman.

Wells, R. P. (1981). The projection of the ground reaction force as a predictor of internal joint moments. *Bulletin of Prosthetics Research, 18*(1), 15–19.

Wells, R. P. (1988). Mechanical energy costs of human movement: An approach to evaluating the transfer possibilities of two–joint muscles. *Journal of Biomechanics, 21,* 955–964.

Wells, K. F., & Dillon, K. E. (1952). The sit and reach: A test of back and leg flexibility. *Research Quarterly, 23,* 115–118.

Weppler, C. H., & Magnusson, S. P. (2010). Increasing muscle extensibility: A matter of increasing length or modifying sensation? *Physical Therapy, 90,* 438–449.

Whiting, W. C., & Zernicke, R. F. (2008). *Biomechanics of musculoskeletal injury* (2nd ed.). Champaign: Human Kinetics.

Whitting, J. W., et al. (2013). Passive dorsiflexion stiffness is poorly correlated with passive dorsiflexion range of motion. *Journal of Science and Medicine in Sport, 16,* 157–161.

Whittle, M. (1999). Generation and attenuation of transient impulsive forces beneath the foot: A review. *Gait and Posture, 10,* 264–275.

Whittle, M. (2001). *Gait analysis: An introduction* (3rd ed.). Oxford: Butterworth-Heinemann.

Wickham, J. B., & Brown, J. M. M. (1998). Muscles within muscles: The neuromotor control of intra–muscular segments. *European Journal of Applied Physiology, 78,* 219–225.

Wickham, J. B., et al. (2004). Muscles within muscles: Anatomical and functional segmentation of selected shoulder joint musculature. *Journal of Musculoskeletal Research, 8,* 57–73.

Wilkerson, J. D. (1997). Biomechanics. In J. D. Massengale & R. A. Swanson (Eds.), *The history of exercise and sport science* (pp. 321–366). Champaign: Human Kinetics.

Wilkinson, S. (1996). Visual analysis of the overarm throw and related sport skills: Training and transfer effects. *Journal of Teaching in Physical Education, 16,* 66–78.

Williams, P. E., & Goldspink, G. (1978). Changes in sarcomere length and physiological properties in immobilized muscle. *Journal of Anatomy, 127,* 459–468.

Williams, J. G., & McCririe, N. (1988). Control of arms and fingers during ball catching. *Journal of Human Movement Studies, 14,* 241–247.

Williams, K. R., & Sih, B. L. (2002). Changes in golf clubface orientation following impact with the ball. *Sports Engineering, 5,* 65–80.

Williams, E. U., & Tannehill, D. (1999). Effects of a multimedia performance principle training program on correct analysis and diagnosis of throwlike movements. *Physical Educator, 56,* 143–154.

Williams, K. R., et al. (1987). Biomechanical studies of elite female distance runners. *International Journal of Sports Medicine, 8,* 107–118.

Willwacher, S., et al. (2020). Surface stiffness and footwear affect the loading stimulus for lower extremity muscles when running. *Journal of Strength and Conditioning Research.* https://doi.org/10.1519/JSC.0000000000003410.

Wilson, G. J. (1994). Strength and power in sport. In T. R. Bloomfield et al. (Eds.), *Applied anatomy and biomechanics in sport* (pp. 110–208). Melbourne: Blackwell Scientific Publications.

Wilson, G. J., et al. (1991). The effect on performance of imposing a delay during a stretch-shortening cycle movement. *Medicine & Science in Sports & Exercise, 23,* 364–370.

Wilson, G. J., et al. (1993). The optimal training load for the development of dynamic athletic performance. *Medicine & Science in Sports & Exercise, 25,* 1279–1286.

Wilson, J. D., et al. (2006). Core strength and lower extremity alignment during single leg squat. *Medicine & Science in Sports & Exercise, 38,* 945–952.

Winter, D. A. (1984). Kinematic and kinetic patterns of human gait: Variability and compensating effects. *Human Movement Science, 3,* 51–76.

Winter, D. A. (1990). *Biomechanics and motor control of human movement* (2nd ed.). New York: Wiley Interscience.

Winter, D. A. (1995). Human balance and posture control during standing and walking. *Gait and Posture, 3,* 193–214.

Winter, E. W. (2005). Jumping: Power or impulse? *Medicine & Science in Sports & Exercise, 37,* 523.

Winter, D. A., & Eng, P. (1995). Kinetics: Our window into the goals and strategies of the central nervous system. *Behavioral Brain Research, 67,* 111–120.

Winter, D. A., et al. (1981). Errors in the use of isokinetic dynamometers. *European Journal of Applied Physiology, 46,* 397–408.

Winter, E. M., et al. (2016). Misuse of "Power" and other mechanical terms in sport and exercise science research. *Journal of Strength and Conditioning Research, 30*, 292–300.

Winters, J. M., & Woo, S. (Eds.). (1990). *Multiple muscle systems.* New York: Springer.

Woo, L.-Y., et al. (1999). Biomechanics of knee ligaments. *American Journal of Sports Medicine, 27*, 533–543.

Wu, G., & Cavanagh, P. R. (1995). ISB recommendations for standardization in the reporting of kinematic data. *Journal of Biomechanics, 28*, 1257–1261.

Yang, X., et al. (2019). Spine-inspired continuum soft exoskeleton for stoop lifting assistance. *IEEE Robotics and Automation Letters, 4*, 4547–4554.

Yeadon, M. R. (1991, January–March). Coaching twisting dives. *Sports Coach*, 6–9.

Yeadon, M. R. (1997). The biomechanics of human flight. *American Journal of Sports Medicine, 25*, 575–580.

Yeadon, M. R. (1998). Computer simulation in sports biomechanics. In H. J. Riehle & M. M. Vieten (Eds.), *Proceedings of the XVIth international society of biomechanics in sports symposium* (pp. 309–318). Konstanz: Universi-tatsverlag Konstanz.

Yeadon, M. R., & Challis, J. H. (1994). The future of performance–related sports biomechanics research. *Journal of Sports Sciences, 12*, 3–32.

Yeadon, M. R., & Hiley, M. J. (2017). Twist limits for late twisting double somersaults on trampoline. *Journal of Biomechanics, 58*, 174–178.

Yoshida, M., et al. (2004). Specificity of practice, visual information, and intersegmental dynamics in rapid-aiming limb movements. *Journal of Motor Behavior, 36*, 281–290.

Zajac, F. E. (1991). Muscle coordination of movement: A perspective. *Journal of Biomechanics, 26*(S1), 109–124.

Zajac, F. E. (2002). Understanding muscle coordination of the human leg with dynamical simulations. *Journal of Biomechanics, 35*, 1011–1018.

Zajac, F. E., & Gordon, M. E. (1989). Determining muscle's force and action in multi-articular movement. *Exercise and Sport Sciences Reviews, 17*, 187–230.

Zajac, F. E., et al. (2002). Biomechanics and muscle coordination of human walking, I: Introduction to concepts, power transfer, dynamics and simulations. *Gait and Posture, 16*, 215–232.

Zatsiorksy, V. M., & Kraemer, W. J. (2006). *Science and practice of strength training* (2nd ed.). Champaign: Human Kinetics.

Zatsiorsky, V. M. (1998). *Kinematics of human motion.* Champaign: Human Kinetics.

Zatsiorsky, V. (Ed.). (2000). *Biomechanics in sport: Performance enhancement and injury prevention.* London: Blackwell Science.

Zatsiorsky, V. M. (2002). *Kinetics of human motion.* Champaign: Human Kinetics.

Zernicke, R. F., & Roberts, E. M. (1976). Human lower extremity kinetic relationships during systematic variations in resultant limb velocity. In P. V. Komi (Ed.), *Biomechanics V–B* (pp. 41–50). Baltimore: University Park Press.

Zernicke, R. F., et al. (1977). Human patellar-tendon rupture: A kinetic analysis. *Journal of Bone and Joint Surgery, 59A*, 179–183.

Zhang, S., et al. (2000). Contributions of lower extremity joints to energy dissipation during landings. *Medicine & Science in Sports & Exercise, 32*, 812–819.

Zivkovic, M. Z. et al. (2017). A simple method for assessment of muscle force, velocity, and power producing capacities from functional movement tasks. *Journal of Sports Sciences, 35*(13), 1287–1293.

Index